Weak Convergence Methods for Semilinear Elliptic Equations

Weak Convergence Methods for Semilinear Elliptic Equations

Jan Chabrowski
University of Queensland, Australia

World Scientific
Singapore • New Jersey • London • Hong Kong

Published by

World Scientific Publishing Co. Pte. Ltd.

P O Box 128, Farrer Road, Singapore 912805

USA office: Suite 1B, 1060 Main Street, River Edge, NJ 07661

UK office: 57 Shelton Street, Covent Garden, London WC2H 9HE

British Library Cataloguing-in-Publication Data
A catalogue record for this book is available from the British Library.

WEAK CONVERGENCE METHODS FOR SEMILINEAR ELLIPTIC EQUATIONS

ISBN 981-02-4076-7

Printed in Singapore by Regal Press (S) Pte. Ltd.

Preface

The main objective of this work is the study of nonlinear boundary value problems for elliptic equations on unbounded domains. Solutions of these problems will be found as critical points of the corresponding variational functionals. Variational problems investigated in this book originate in many branches of applied sciences. A typical example is the nonlinear Schrödinger equation which appears in mathematical modelling of phenomena arising in nonlinear optics and plasma physics. In this book we mainly consider the equation

$$-\Delta u + \lambda u = f(x, u) \text{ in } \mathbb{R}_N$$

with a nonlinearity f having a subcritical growth at infinity, that is, $f(x, u) = O(|u|^p)$ as $|u| \to \infty$, $2 < p < 2^* = \frac{2N}{N-2}$, where 2^* is the critical Sobolev exponent. It is always assumed that $f(x, 0) = 0$, so that equation $(*)$ has a trivial solution. Solutions of equation $(*)$ will be found as critical points of a variational functional given by

$$G(u) = \frac{1}{2} \int_{\mathbb{R}_N} (|\nabla u|^2 + \lambda u^2)\, dx - \int_{\mathbb{R}_N} F(x, u)\, dx,$$

where $F(x, u) = \int_0^u f(x, s)\, ds$. One of the main problems of a constrained minimization for G is to show that minimizing sequences are relatively compact in a Sobolev space $H^1(\mathbb{R}_N)$. There are a number of factors creating some diffficulties in this approach: the lack of compactness, since the Sobolev embedding theorem is no longer valid on unbounded domains and the invariance of some variational problems with respect to a group of translations. In recent years some efficient methods have been developed to deal with these difficulties [76], [77]. P.L. Lions' concentration–compactness principles have been successfully used by various authors to study variational problems for equation $(*)$. In Chapter 1 we formulate the concentration–compactness principle at infinity [31] which is a natural starting point in the study of constrained minimization problems for $(*)$ with a homogeneous nonlinearity $f(x,u) = b(x)|u|^{p-2}u$, $2 < p < 2^*$. The interesting feature of the concentration–compactness principle at infinity is that it leads to a considerable simplification of proofs of many known results from the theory of nonlinear elliptic equations involving subcritical Sobolev exponents. If the coefficient $b(x)$ has some singularities then weakly convergent sequences can concentrate at finite points. This requires a version of concentration–compactness principle that provides description of concentration at infinity and at finite points in quantitative terms. This is applied to obtain the existence of a principal eigenvalue for an eigenvalue problem on unbounded domains involving singular potentials. The principal result of Chapter 2 states that under an assumption on b controlling its behaviour at infinity, equation $(*)$ has a bound (positive) state. We also discuss the existence and nonexistence of bound states on unbounded subsets of $\mathbb{R}_N$. We emphasize the role of symmetry of b in the existence theory for equation $(*)$. Chapters 3 and 4 are devoted to equation $(*)$ with more general nonlinearity of subcritical growth. The main tool in Chapter 3 is Strauss' compactness lemma [105] combined with a subadditivity property of a variational functional. In Chapter 4 we consider minimization problems for $(*)$ subject to artificial constraints. The range of level sets of a variational functional for which the Palais–Smale condition holds, is determined by the examination of the loss of mass of a sequence of gradients of minimizing sequences. In Chapter 5 we focus on the problem of existence of nonradial solutions. In the case where a nonlinearity f is independent of x, we construct nonradial solutions with the aid of the inverse power method. We make here a significant use of the concentration–compactness principle at infinity to find level sets of a variational functional where the Palais–Smale condition holds. Chapter 6 is devoted to the study of the effect of the graph topology of coefficients on the existence of multiple solutions. In the final chapters 7 and 8 we concentrate on the effect of the shape of the graphs of the coefficient of $|u|^{p-2}u$ on the number of solutions. The presence of nondegenerate critical points of this coefficient gives rise to multi-peak solutions concentrating at these points. In the case of a local minimum of this coefficient, we construct a multi-peak solution concentrating at this point ([27], [85]). In both cases we use a local reduction method due to Bahri [8] (see also [6], [96] and [98]). In Chapter 8 we discuss the effect of the shape of the graph of a coefficient of a homogeneous nonlinearity on the multiplicity of positive and nodal solutions. These solutions are obtained as localized minima of a corresponding variational functional [26].

Finally, I would like to express my gratitude to Dr K. Matthews (University of Queensland) for his help in improving the presentation of the material in this book.

Contents

Chapter 1

Concentration–compactness principle at infinity

This chapter supplies a basic tool needed for our study of weakly convergent sequences in Sobolev spaces on unbounded domains in $\mathbb{R}_N$. We define quantities which will be used to examine in quantitative terms loss of mass at infinity of weakly convergent sequences. These quantities allow us to formulate a concentration–compactness principle at infinity which will be frequently used to show a relative compactness of Palais–Smale sequences.

1.1 Preliminaries - Sobolev spaces

Let Ω be a domain in $\mathbb{R}_N$. For $1 \leq p < \infty$, we let $L^p(\Omega)$ denote the Banach space consisting of measurable functions on Ω that are p-integrable. This space is equipped with the norm

$$||u||_p = ||u||_{L^p(\Omega)} = \left(\int_\Omega |u(x)|^p\, dx\right)^{\frac{1}{p}}.$$

For $p = \infty$, $L^\infty(\Omega)$ denotes the Banach space of bounded measurable functions on Ω with the norm

$$||u||_\infty = ||u||_{L^\infty(\Omega)} = \sup_{x\in\Omega} |u(x)|.$$

Let V be a nonnegative and measurable function on Ω. By $L^p_V(\Omega)$ we denote a weighted Lebesgue space with norm

$$||u||^p_{L^p_V(\Omega)} = \int_\Omega |u(x)|^p\, V(x)\, dx.$$

By $H^{1,2}(\mathbb{R}_N)$ or $H_0^{1,2}(\Omega)$ for a domain $\Omega \subset \mathbb{R}_N$, we mean the usual *Sobolev space* defined as the completion of $C_0^\infty(\Omega)$ with respect to the norm

$$\|u\|_{1,2}^2 = \int_{\mathbb{R}_N} \left(|Du(x)|^2 + |u(x)|^2\right) dx.$$

In the case of $H_0^{1,2}(\Omega)$ the integration in the above formula is over Ω. The dual spaces are denoted by $H^{-1,2}(\mathbb{R}_N)$ and $H^{-1,2}(\Omega)$, respectively. We abbreviate our notation to $H^{1,2}(\mathbb{R}_N) = H^1(\mathbb{R}_N)$, $H_0^{1,2}(\Omega) = H_0^1(\Omega)$ and for dual spaces to $H^{-1,2}(\mathbb{R}_N) = H^{-1}(\mathbb{R}_N)$ and $H^{-1,2}(\Omega) = H^{-1}(\Omega)$. Due to the *Sobolev–Relich–Kondrachev embedding theorem* we have a continuous embedding: $H^{1,2}(\Omega) \subset L^q(\Omega)$, with $2 \leq q \leq 2^* = \frac{2N}{N-2}$ and 2^* being the so-called critical Sobolev exponent.

Finally, we denote by $D_0^{1,2}(\Omega)$ a Sobolev space obtained as the completion of $C_0^\infty(\Omega)$ with respect to the norm

$$\|u\|_{D^{1,2}}^2 = \int_\Omega \|\nabla u\|^2\, dx.$$

It is well-known that for bounded domains Ω, $H_0^{1,2}(\Omega) = D_0^{1,2}(\Omega)$. Moreover, for unbounded domains Ω, $D_0^{1,2}(\Omega)$ is continuously embedded in $L^{2^*}(\Omega)$.

We shall frequently use the *Gagliardo–Nirenberg inequality*: for every $u \in H^1(\mathbb{R}_N)$ we have

$$\|u\|_p \leq c\|\nabla u\|_2^\alpha \|u\|_r^{1-\alpha},$$

with $\frac{N}{p} = \alpha\frac{N-2}{2} + (1-\alpha)\frac{N}{r}$, $r \geq 1$, $\alpha \in [0,1]$, where c is a constant depending on p, α, r and N. We always denote weak convergence in a Banach space by " $\rightharpoonup$ " and strong convergence by " $\to$ ".

1.2 Weakly convergent sequences

In this book we seek solutions of nonlinear elliptic problems on unbounded domains. Solutions of these problems will be found as critical points of variational functionals. A starting point is to generate the Palais–Smale sequences of variational functionals whose limits are *critical points*. To illustrate this approach let $F : H^1(\mathbb{R}_N) \to \mathbb{R}$ be defined by

$$F(u) = \frac{1}{2}\int_{\mathbb{R}_N} (|Du|^2 + u^2)\, dx - \frac{1}{p}\int_{\mathbb{R}_N} |u|^p\, dx,$$

where $2 < p < 2^* = \frac{2N}{N-2}$ is a constant. Critical points of F are solutions of the equation

$$-\Delta u + u = |u|^{p-2}u \text{ in } \mathbb{R}_N. \tag{1.1}$$

We adopt here the following terminology for a given C^1-functional $\mathcal{F} : X \to \mathbb{R}$ on a Banach space X. A sequence $\{u_m\} \subset X$ such that $\mathcal{F}(u_m) \to c$ as $m \to \infty$, where c is a constant, and $\mathcal{F}'(u_m) \to 0$ in X^* will be called a *Palais–Smale sequence* for the functional $\mathcal{F}$. We say that a sequence $\{u_m\} \subset X$ satisfies the *Palais–Smale condition* ((PS)-condition for short) if it is a Palais–Smale sequence and is relatively compact.

The following result provides a picture of what can happen to a Palais–Smale sequence which is not relatively compact in $H^1(\mathbb{R}_N)$.

Proposition 1.1 *Let $\{u_m\} \subset H^1(\mathbb{R}_N)$ be a Palais–Smale sequence for the functional F. Then there exist an integer $k \geq 1$, k sequences $\{y_m^j\}$ in $\mathbb{R}_N$ $(j = 1, \ldots, k)$, $k + 1$ sequences of functions $\{u_m^j\}$ in $H^1(\mathbb{R}^N)$ $(j = 0, \ldots, k)$ and $k + 1$ weak non-zero solutions u^j $(j = 0, \ldots, k)$ to equation (1.1) such that for some subsequence of $\{u_m\}$, which we label again by $\{u_m\}$, we have*

$$u_m(x) = u_m^\circ(x) + [u_m^1(x - y_m^1) + u_m^2(x - y_m^1 - y_m^2) + \ldots + u_m^k(x - y_m^1 - \ldots - y_m^k)],$$

$u_m^j(x) = u^j(x) + o(1)$ *strongly in* $H^1(\mathbb{R}_N)$, $j = 0, \ldots, k$, *Moreover,*

$$\|u_m\|_{1,2}^2 = \|u^\circ\|_{1,2}^2 + \sum_{j=1}^{k} \|u^j\|_{1,2}^2 + o(1) \quad \textit{as } m \to \infty,$$

$$F(u_m) = F(u^\circ) + \sum_{j=1}^{k} F(u^j) + o(1) \quad \textit{as } m \to \infty.$$

The proof of Proposition 1.1 will be postponed to Chapter 3, where a more general result will be established.

As it turns out the functional F satisfies the Palais–Smale condition for c belonging to some intervals. To determine level sets of the variational functional where the Palais–Smale condition holds, we need the result which expresses in quantative terms *loss of mass at infinity* of a weakly convergent sequence in $H^1(\mathbb{R}_N)$.

Proposition 1.2 (Concentration–compactness principle at infinity) *Let $\{u_m\} \subset H_0^1(\Omega)$ be a sequence such that $u_m \rightharpoonup u$ in $H_0^1(\Omega)$, where $\Omega \subset \mathbb{R}_N$ is unbounded domain and define*

(a) $\alpha_\infty = \lim_{R\to\infty} \limsup_{m\to\infty} \int_{\Omega\cap(|x|>R)} |u_m(x)|^p\,dx$, *where* $2 < p < 2^*$,

(b) $\beta_\infty = \lim_{R\to\infty} \limsup_{m\to\infty} \int_{\Omega\cap(|x|>R)} (|Du_m(x)|^2 + u_m(x)^2)\,dx$. *These quantities are well defined and satisfy*

(c) $\limsup_{m\to\infty} \int_\Omega |u_m|^p\,dx = \int_\Omega |u|^p\,dx + \alpha_\infty$,

(d) $\limsup_{m\to\infty} \int_\Omega (|Du_m|^2 + u_m^2)dx \geq \int_\Omega (|Du|^2 + u^2)dx + \beta_\infty$ *and*

(e) $M\left(\alpha_\infty\right)^{\frac{2}{p}} \leq \beta_\infty$, *where*

$$M = M(\Omega) = \inf\{\int_\Omega (|Du_m|^2 + u^2)\,dx;\; u \in H_0^1(\Omega),\; \int_\Omega |u|^p\,dx = 1\}.$$

Proof. Let $\phi_R \in C^1(\mathbb{R}_N)$ be such that $\phi_R(x) = 0$ for $|x| < R$ and $\phi_R(x) = 1$ for $|x| > R+1$ and $0 \leq \phi_R(x) \leq 1$ on $\mathbb{R}_N$. It follows from the definition of M that

$$M\left(\int_\Omega |\phi_R u_m|^p\,dx\right)^{\frac{2}{p}} dx \leq \int_\Omega (|D(u_m\phi_R)|^2 + (u_m\phi_R)^2)\,dx.$$

Since

$$\lim_{R\to\infty} \limsup_{m\to\infty} \int_\Omega |u_m\phi_R|^p\,dx = \alpha_\infty,$$

$$\lim_{R\to\infty} \limsup_{m\to\infty} \int_\Omega (|Du_m|^2 + u_m^2)\phi_R^2\,dx = \beta_\infty,$$

$$\lim_{R\to\infty} \limsup_{m\to\infty} \int_\Omega |u_m D\phi_R|^2\,dx = 0$$

and

$$\lim_{R\to\infty} \limsup_{m\to\infty} \int_\Omega |Du_m u_m \phi_R D\phi_R|^2\,dx = 0$$

the inequality (e) follows. To show (c) we write for each $R > 0$

$$\limsup_{m\to\infty} \int_\Omega |u_m|^p\,dx = \int_{\Omega\cap(|x|<R)} |u|^p\,dx + \limsup_{m\to\infty} \int_{\Omega\cap(|x|>R)} |u_m|^p\,dx.$$

Letting $R \to \infty$, relation (c) follows. By a similar argument, using the weak lower semicontinuity of a norm with respect to the weak convergence, we deduce (d). □

We point out here that the cutoff function ϕ_R from the proof Proposition 1.2 will be used on many occasions throughout this work and its definition will not be repeated in the next sections.

In applications, it is convenient to use Proposition 1.2 together with the following lemma due to Lions ([76] Lemma I.1).

Lemma 1.1 *Let $\{u_n\}$ be a bounded sequence in $H^1(\mathbb{R}_N)$ such that*

$$\liminf_{n\to\infty} \sup_{y\in\mathbb{R}_N} \int_{B(y,R)} u_n^2\, dx = 0$$

for some $R > 0$. Then $u_n \to 0$ in $L^s(\mathbb{R}_N)$ for $2 < s < 2^*$

Proof. Without loss of generality we may assume that $R = 1$. It follows from the Hölder and Sobolev inequalities that

$$||u||_{L^s(B(y,1))} \leq A||u||_{L^2(B(y,1))}^{1-\alpha}||u||_{H^{1,2}(B(y,1))}^{\alpha},$$

where $A > 0$ is a constant independent of y and $\alpha = \frac{s-2}{2s}N$. If $s \geq \frac{4}{N} + 2$, then $\alpha s \geq 2$ and

$$\int_{B(y,1)} |u|^s\, dx \leq A^s ||u||_{L^2(B(y,1))}^{s(1-\alpha)} ||u||_{1,2}^{\alpha s-2} \int_{B(y,1)} \left(|\nabla u|^2 + u^2\right) dx.$$

Covering $\mathbb{R}_N$ by a family of balls $\{B(y_i, 1)\}$ such that each point is contained in at most k such balls and summing this inequality over this family of balls we get

$$||u||_s^s \leq kA^s \sup_{y\in\mathbb{R}_N} \left(\int_{B(y,1)} u^2\, dx\right)^{\frac{s(1-\alpha)}{2}} ||u||_{1,2}^{\alpha s}.$$

Substituting $u = u_n$ the result follows for $s \geq \bar{s} = 2 + \frac{4}{N}$. If $2 < s < \bar{s}$, we write $s = 2\theta + \bar{s}(1-\theta)$ for some $\theta \in (0,1)$. By the Hölder inequality we get

$$||u||_s^s \leq ||u||_2^{2\theta} ||u||_{\bar{s}}^{(1-\theta)\bar{s}}$$

and the result in this case follows from the case already established. □

To close this section we introduce a classification of bounded sequences $\{u_m\}$ in $H^1(\mathbb{R}_N)$ subject to the condition that p and ρ exist such that $2 < p < 2^*$ and

$$0 < \rho = \int_{\mathbb{R}_N} |u_m(x)|^p \, dx. \tag{1.2}$$

for all m. We may assume that $u_m \rightharpoonup u$ in $H^1(\mathbb{R}_N)$. According to Proposition 1.2(c) we have

$$\rho = \int_{\mathbb{R}_N} |u(x)|^p \, dx + \alpha_\infty.$$

We now distinguish three cases:

(1) **(vanishing)** $\alpha_\infty = \rho$, $u \equiv 0$,
(2) **(compactness)** $\alpha_\infty = 0$, $\int_{\mathbb{R}_N} |u|^p \, dx = \rho$,
(3) **(dichotomy)** $\alpha_\infty > 0$, $\int_{\mathbb{R}_N} |u|^p \, dx > 0$.

If (1) holds, then either (a) $\liminf_{m\to\infty} \sup_{y\in\mathbb{R}_N} \int_{B(y,R)} u_m^2 \, dx = 0$ for each $R > 0$ or (b) $\liminf_{m\to\infty} \sup_{y\in\mathbb{R}_N} \int_{B(y,R)} u_m^2 \, dx > 0$ for some $R > 0$. If (a) occurs, then by Lemma 1.1 $u_m \to 0$ in $L^\alpha(\mathbb{R}_N)$ for $2 < \alpha < \frac{2N}{N-2}$ which is impossible. Thus (b) prevails and there exists a sequence $\{y_m\} \subset \mathbb{R}_N$ such that $u_m(\cdot + y_m) \rightharpoonup u \not\equiv 0$ in $H^1(\mathbb{R}_N)$. In case (2) the sequence $\{|u_m|^p\}$ is tight (compactness). Finally, in case (3) we introduce the *Lévy concentration functions* Q_m (see [76]) defined by

$$Q_m(t) = \sup_{y\in\mathbb{R}_N} \int_{B(y,t)} |u_m(x)|^p \, dx.$$

Each function Q_m is increasing on $(0, \infty)$ and by Helly's selection theorem $Q(t) = \lim_{m\to\infty} Q_m(t)$ is an increasing function on $(0, \infty)$. Since $0 < \alpha_\infty < \rho$, it is easy to check that $0 < \lim_{t\to\infty} Q(t) < \rho$. In this case a sequence $\{u_n\}$ can be split into two parts concentrating on two disjoint sets whose distance tends to ∞ as $n \to \infty$. For more details we refer to P.L. Lions' concentration–compactness principle [76] pp. 116, 117.

This discussion leads to the following result.

Corollary 1.1 *Let $\{u_m\}$ be a sequence in $H^1(\mathbb{R}_N)$ satisfying (1.2) and such that $u_m \rightharpoonup 0$ in $H^1(\mathbb{R}_N)$. Then there exist a sequence $\{y_m\} \subset \mathbb{R}_N$ and a function $0 \not\equiv u \in H^1(\mathbb{R}^N)$ such that up to a subsequence $u_m(\cdot + y_m) \rightharpoonup u$ in $H^1(\mathbb{R}_N)$.*

We can always assume that $\{y_m\} \subset \mathbb{Z}_N$. Indeed, by Lemma 1.1 there exists $\{y_m\} \subset \mathbb{R}_N$ such that

$$\int_{B(0,R)} u_m(x+y_m)^2\,dx \geq \eta > 0$$

for some constant $\eta > 0$. For each m we choose $g_m \in \mathbb{Z}_N$ such that $|g_m - y_m| = \inf\{|g - y_m|;\, g \in \mathbb{Z}_N\}$. Thus $|g_m - y_m| \leq \frac{1}{2}\sqrt{N}$ and $v_m(x) = u_m(x+g_m)$ satisfies

$$\int_{B(0,R+\frac{1}{2}\sqrt{N})} v_m(s+g_m)^2\,dx \geq \eta.$$

1.3 The compactness lemma of Strauss

To examine minimizing sequences of variational problems we need a convergence result for the composition of a nonlinearity with a weakly convergent sequence in $H^1(\mathbb{R}_N)$. This will be established with the aid of Strauss' lemma.

Lemma 1.2 *Let $P : \mathbb{R}_N \times \mathbb{R} \to \mathbb{R}$ and $Q : \mathbb{R}_N \times \mathbb{R} \to \mathbb{R}$ be continuous functions such that $\sup_{x\in\mathbb{R}_N, |t|\leq a} |P(x,t)| < \infty$ and $\sup_{x\in\mathbb{R}_N, |t|\leq a} |Q(x,t| < \infty$ for each $a > 0$ and satisfying*

$$\lim_{|s|\to\infty} \frac{P(x,s)}{Q(x,s)} = 0 \quad \textit{uniformly in } x \in \mathbb{R}_N. \tag{1.3}$$

Suppose that $\{u_m\}$ and v are measurable functions on $\mathbb{R}_N$ such that

$$\sup_m \int_{\mathbb{R}_N} |Q(x,u_m(x))|\,dx < \infty, \tag{1.4}$$

and

$$\lim_{m\to\infty} P(x,u_m(x)) = v(x) \quad \textit{a.e. on } \mathbb{R}_N.$$

Then for every bounded Borel set $B \subset \mathbb{R}_N$, we have

$$\lim_{m\to\infty} \int_B |P(x,u_m(x)) - v(x)|\,dx = 0.$$

If additionally, we assume that

$$\lim_{s\to 0} \frac{P(x,s)}{Q(x,s)} = 0 \quad \textit{uniformly in } x \in \mathbb{R}_N \tag{1.5}$$

and

$$\lim_{|x|\to\infty} u_m(x) = 0 \tag{1.6}$$

uniformly in m, then $P(x, u_m) \to v$ in $L^1(\mathbb{R}_N)$.

Proof. (i) To prove the first part of the lemma, we need to show the uniform integrability of $\{P(x, u_m)\}$. It follows from (1.3) that for every $\epsilon > 0$ there exists $C_\epsilon > 0$ such that

$$|P(x, u_m(x))| \leq C_\epsilon + \epsilon |Q(x, u_m(x))| \quad \text{on } \mathbb{R}_N. \tag{1.7}$$

Denoting by $|A|$ the Lebesgue measure of a set $A \subset \mathbb{R}_N$, we have for each $K > 0$

$$\begin{aligned} |\{B \cap (|P(x, u_m(x))| \geq K)\}| &\leq \frac{1}{K} \int_{\{B\cap(|P(x,u_m(x))|\geq K)\}} |P(x, u_m(x))|\, dx \\ &\leq \frac{1}{K}\left(|B| C_\epsilon + \epsilon \sup_m \int_{\mathbb{R}_N} |Q(x, u_m(x))|\, dx\right). \end{aligned}$$

This, combined with (1.7) and (1.4), shows that

$$\lim_{K\to\infty} \int_{\{(|P(x,u_m(x))|\geq K)\cap B\}} |P(x, u_m(x))|\, dx = 0$$

and the uniform integrability of $\{P(x, u_m(x))\}$ follows.

(ii) Given $\epsilon > 0$, by (1.5) and (1.6) we can find $R_\circ > 0$ such that

$$|P(x, u_m(x))| \leq \epsilon |Q(x, u_m(x))|$$

for $|x| \geq R_\circ$ and all n. Applying Fatou's lemma we see that $v \in L^1(\mathbb{R}_N)$ and that

$$\int_{|x|\geq R_\circ} |v(x)|\, dx \leq \epsilon C.$$

By the part (i) of the proof, there exists $m_\circ$ such that

$$\int_{|x|\leq R_\circ} |P(x, u_m(x)) - v(x)|\, dx \leq \epsilon$$

for $m \geq m_\circ$. The last three inequalities and (1.4) imply that

$$\int_{\mathbb{R}_N} |P(x, u_m(x)) - v(x)|\, dx \leq 2\epsilon C + \epsilon$$

for $m \geq m_0$ and the result follows. $\square$

We now establish a convergence result that will be frequently used later.

Lemma 1.3 *Let $f \in C(\mathbb{R}_N \times \mathbb{R})$ and suppose that*

$$\lim_{|s|\to\infty} \frac{f(x,s)}{|s|^{\frac{N+2}{N-2}}} = 0 \tag{1.8}$$

uniformly in $x \in \mathbb{R}_n$ and

$$|f(x,t)| \leq C\big(|t| + |t|^{\frac{N+2}{N-2}}\big) \tag{1.9}$$

for all $x \in \mathbb{R}_N$ and $t \in \mathbb{R}$. If $u_m \rightharpoonup u_0$ in $H^1(\mathbb{R}_N)$ and $u_m \to u_0$ a.e. on $\mathbb{R}_N$, then

$$\lim_{m\to\infty} \left[\int_{\mathbb{R}_N} F(x,u_m)\,dx - \int_{\mathbb{R}_N} F(x,u_0)\,dx - \int_{\mathbb{R}_N} F(x,u_m-u_0)\,dx\right] = 0,$$

where $F(x,u) = \int_0^u f(x,t)\,dt$.

Proof. Let $R > 0$. Applying the mean value theorem we have

$$\begin{aligned}
\int_{\mathbb{R}_N} F(x,u_m)\,dx &= \int_{|x|\leq R} F(x,u_m)\,dx \\
&+ \int_{|x|\geq R} F(x,u_0+(u_m-u_0))\,dx \\
&= \int_{|x|\leq R} F(x,u_m)\,dx \\
&+ \int_{|x|\geq R} \big(F(x,u_m-u_0) + f(x,\theta u_0 + (u_m-u_0))u_0\big)\,dx,
\end{aligned}$$

where θ depends on x and R and satisfies $0 < \theta < 1$. We now write

$$\begin{aligned}
&\left|\int_{\mathbb{R}_N} F(x,u_m)\,dx - \int_{\mathbb{R}_N} F(x,u_0)\,dx - \int_{\mathbb{R}_N} F(x,u_m-u_0)\,dx\right| \\
&\leq \left|\int_{|x|\leq R} (F(x,u_m) - F(x,u_0))\,dx\right| + \left|\int_{|x|\geq R} F(x,u_0)\,dx\right| \\
&+ \left|\int_{|x|\leq R} F(x,u_m-u_0)\,dx\right| \\
&+ \left|\int_{|x|\geq R} f(x,\theta u_0+(u_m-u_0))u_0\,dx\right|.
\end{aligned}$$

It follows from Lemma 1.2 that for each fixed $R > 0$

$$\lim_{m\to\infty}\int_{|x|\leq R}\big(F(x,u_m) - F(x_m,u_\circ)\big)\,dx = 0 \tag{1.10}$$

and

$$\lim_{m\to\infty}\int_{|x|\leq R} F(x,u_m - u_0)\,dx = 0. \tag{1.11}$$

Applying (1.9) and the Hölder inequality we get that

$$\begin{aligned}
&\left|\int_{|x|\geq R} f(x,\theta u_\circ + (u_m - u_\circ))u_\circ\,dx\right| \\
&\qquad \leq C\int_{|x|\geq R}\left(|\theta u_\circ + (u_m - u_\circ)||u_\circ| + |\theta u_\circ + (u_m - u_\circ)|^{\frac{N+2}{N-2}}|u_\circ|\right)dx \\
&\qquad \leq C\left(\int_{|x|\geq R}|u_\circ|^2\,dx\right)^{\frac{1}{2}}\left(\int_{|x|\geq R}|\theta u_\circ + (u_m - u_\circ)|^2\,dx\right)^{\frac{1}{2}} \\
&\quad + \;\; C\left(\int_{|x|\geq R}|u_\circ|^{2^*}\,dx\right)^{\frac{1}{2^*}}\left(\int_{|x|\geq R}|\theta u_\circ + (u_m - u_\circ)|^{2^*}\,dx\right)^{\frac{N+2}{2N}}.
\end{aligned}$$

Since $\{u_m\}$ is bounded in $H^1(\mathbb{R}_N)$ we see that

$$\lim_{R\to\infty}\left|\int_{|x|\geq R} f(x,\theta u_\circ + (u_m - u_\circ))u_\circ\,dx\right| = 0. \tag{1.12}$$

The result follows from (1.10), (1.11) and (1.12). □

1.4 Concentration–compactness principle for singular eigenvalue problems

In Section 1.2 we have investigated only a possible concentration at infinity of weakly convergent sequences. Since the coefficient b was bounded and the exponent p subcritical, a concentration at finite points was excluded. It is known that a concentration at finite points can occur in the case of a coefficient b with singularities for a subcritical exponent $p < 2^*$ (see [111] and [103]).

In this section we examine a concentration phenomena for weakly convergent sequences for variational problems involving singular coefficients.

We follow the approach due to Smets [103]. As a model for this scenario we consider the following *eigenvalue problem with indefinite weight*

$$(\mathcal{P}_V) \qquad \begin{cases} -\Delta u &= \lambda V(x)u \text{ in } \Omega \\ u &\in D_0^{1,2}(\Omega), \ u \not\equiv 0, \end{cases}$$

where $\Omega \subset \mathbb{R}^N$ is an open set and $V \in L^1_{\text{loc}}(\Omega)$.

Solutions of problem $(\mathcal{P}_V)$ will be obtained as minimizers of the constrained minimization problem

$$S_V = \min\{\int_\Omega |\nabla u|^2\, dx;\ u \in D_0^{1,2}(\Omega),\ \int_\Omega V|u|^2\, dx = 1\}.$$

Throughout this section we make the assumption

$$(V_1) \qquad c\int_\Omega V^+u^2\, dx \le \int_\Omega |\nabla u|^2\, dx$$

for some constant $c > 0$ and all $v \in D_0^{1,2}(\Omega)$. This assumption guarantees that $S_V > 0$. If $\Omega = \mathbb{R}^N$, then this assumption is satisfied if $V(x) = \frac{1}{|x|^2}$ for $x \ne 0$. Indeed, this follows from the *Hardy inequality*

$$\int_{\mathbb{R}^N} \frac{u^2}{|x|^2}\, dx \le \left(\frac{N}{N-2}\right)^2 \int_{\mathbb{R}^N} |\nabla u|^2\, dx \tag{1.13}$$

for every $u \in D^{1,2}(\mathbb{R}^N)$. An elementary proof of the Hardy inequality can be found in [4]. This inequality shows that (V_1) is satisfied for a class of functions V with point sigularities and decay at infinity of the order $O(\frac{1}{|x|^2})$. We point out here that the validity of the condition (V_1) can be expressed in terms of a capacity and we refer to [80], p.112. We define the following quantities

$$S_{r,V} = \inf\{\int_\Omega |\nabla u|^2\, dx;\ u \in D_0^{1,2}(\Omega - B(0,r)),\ \int_\Omega V^+(x)u^2\, dx = 1\}$$

and

$$S^x_{r,V} = \inf\{\int_\Omega |\nabla u|^2\, dx;\ u \in H_0^1(\Omega \cap B(x,r)),\ \int_\Omega V^+(x)u^2\, dx = 1\}.$$

It is easy to see that $S_{r,V}$ and $S^x_{r,V}$ are increasing as $r \nearrow \infty$ and $r \searrow 0$, respectively. Therefore we can define

$$S_{\infty,V} = \sup_{r>0} S_{r,V} = \lim_{r\to\infty} S_{r,V}$$

and

$$S_V^x = \sup_{r>0} S_{r,V}^x = \lim_{r\to 0} S_{r,V}^x.$$

We also set $S_V^{\inf} = \inf_{x\in\bar{\Omega}} S_V^x$. To proceed further we define

$$\Sigma_V = \{x \in \Omega;\ S_V^x < \infty\}$$

and make the following assumption

(V_2) $\bar{\Sigma}_V$ is at most countable.

This assumption is justified because one can construct functions satisfying (V_1) such that Σ_V is not closed and not countable (Smets [103]). The set Σ_V is always a Borel set. Indeed, we can write

$$\Sigma_V = \bigcup_{n\in\mathbb{N}} \bigcap_{r>0} \{x;\ S_{r,V}^x \leq n\}$$

and it is easy to show that the sets $\bigcap_{r>0}\{x;\ S_{r,V}^x \leq n\}$ are closed.

We denote by $\mathcal{M}(\Omega)$ the Banach space of finite *Radon measure* on Ω equipped with norm

$$\|\mu\| = \sup_{\phi\in C_\circ(\Omega), \|\phi\|_\infty \leq 1} |\mu(\phi)|.$$

A sequence $\{\mu_n\} \subset \mathcal{M}(\Omega)$ is said to converge weakly to $\mu \in \mathcal{M}(\Omega)$ if for every $\phi \in C_\circ(\Omega)$ $\mu_n(\phi) \to \mu(\phi)$. By the *Banach–Alaoglu theorem* every bounded sequence $\{\mu\} \subset \mathcal{M}$ has a weakly convergent subsequence. By $\mathcal{M}^+(\Omega)$ we denote the cone of positive Radon measures on Ω and by δ_x the Dirac measure assigned to a point x.

We commence by a result which combines the concentration at infinity and at finite points. As in Section 1.2 we define the following two quantities controlling loss of mass of a weakly convergent sequence $u_n \rightharpoonup u$ in $D_\circ^{1,2}(\Omega)$:

$$\mu_\infty = \lim_{R\to\infty} \limsup_{n\to\infty} \int_{\Omega_R} |\nabla u_n|^2\, dx$$

and

$$\nu_\infty = \lim_{R\to\infty} \limsup_{n\to\infty} \int_{\Omega_R} V^+(x) u_n^2\, dx,$$

where $\Omega_R = \Omega - B(0,R)$.

Proposition 1.3 *Let $\{u_n\} \subset D_0^{1,2}(\Omega)$ be such that $u_n \rightharpoonup u$ in $D_0^{1,2}(\Omega)$, $u_n \to u$ a.e., $|\nabla u_n - \nabla u|^2 \rightharpoonup \mu$ in $\mathcal{M}^+(\Omega)$, $V^+(u_n-u)^2 \rightharpoonup \nu$ in $\mathcal{M}^+(\Omega)$ and $|\nabla u_n|^2 \rightharpoonup \tilde{\mu}$ in $\mathcal{M}^+(\Omega)$. Then*

(1) $\mu_\infty \geq S_{\infty,V}\nu_\infty$,
(2) $\nu = \sum_{i\in I}\nu_i\delta_{x_i}$ *for some* $x_i \in \Sigma_V$, $\nu_i > 0$,
(3) $\mu \geq \sum_{i\in I}\nu_i S_V^{x_i}\delta_{x_i}$,
(4) $\tilde{\mu} \geq |\nabla u|^2 + \sum_{i\in I}\nu_i S_V^{x_i}\delta_{x_i}$,
(5) $\limsup_{n\to\infty}\int_\Omega V^+u_n^2\,dx = \int_\Omega V^+u^2\,dx + \|\nu\| + \nu_\infty$,
(6) $\limsup_{n\to\infty}\int_\Omega |\nabla u_n|^2\,dx = \int_\Omega |\nabla u|^2\,dx + \|\mu\| + \mu_\infty$.

Moreover, if $S_{\infty,V} = S_V^{\text{inf}} = \infty$, then $\nu_\infty = \|\nu\| = 0$ and $\{u_n\}$ converges to u in the weighted Lebesgue space $L^2_{V^+}(\Omega)$.

Proof. To simpify our notation we assume that $V \geq 0$ on Ω. Since $u_n \rightharpoonup u$ in $D_0^{1,2}(\Omega)$, it is easy to check that

$$\lim_{R\to\infty}\limsup_{n\to\infty}\int_{\Omega_R} V(u_n-u)^2\,dx = \nu_\infty \tag{1.14}$$

and

$$\lim_{R\to\infty}\limsup_{n\to\infty}\int_{\Omega_R} |\nabla(u_n-u)|^2\,dx = \mu_\infty. \tag{1.15}$$

Let $\{\phi_R\}$, $R > 0$, be functions introduced in the proof of Proposition 1.1. As in this proposition we show that

$$\lim_{R\to\infty}\limsup_{n\to\infty}\int_{\Omega} V(u_n-u)^2\phi_R^2\,dx = \nu_\infty$$

and

$$\lim_{R\to\infty}\limsup_{n\to\infty}\int_{\Omega} |\nabla(u_n-u)|^2\phi_R^2\,dx = \mu_\infty.$$

Since $(u_n-u)\phi_R \in D_0^{1,2}(\Omega_R)$ we see that

$$S_{R,V}\int_\Omega V(u_n-u)^2\phi_R^2\,dx \leq \int_\Omega |\nabla((u_n-u)\phi_R)|^2\,dx.$$

Letting $R \to \infty$ and using (1.14) and (1.15) we get (1). To show that ν is supported by Σ_V we choose $\phi \in C_0^\infty(\mathbb{R}^N - \bar{\Sigma}_V)$. Since $\nu \geq 0$, we may assume that $\phi \geq 0$. We show that $\nu(\phi) = 0$. Let $M \geq 1$. To every $x \in \operatorname{supp}\phi$ there corresponds $r_x > 0$ such that

$$\inf\{\int_\Omega |\nabla u|^2\,dx;\ u \in H^1_0(\Omega \cap B(x,r_x)),\ \int_{B(x,r_x)} Vu^2\,dy = 1\} > M.$$

We select from the covering $\{B(x,r_x);\ x \in \operatorname{supp}\phi\}$ a finite subcovering $\{B(x_i, r_{x_i});\ i = 1,\ldots,m\}$. Let $\{\chi_i;\ i = 1,\ldots,m\}$ be a partition of unity subordinate to this subcovering and set $\phi_i = \chi_i\phi$. We then have

$$\begin{aligned}
\nu(\phi) &= \lim_{n\to\infty}\int_\Omega V(u_n-u)^2\phi\,dx = \sum_{i=1}^m \lim_{n\to\infty}\int_\Omega V(u_n-u)^2\phi_i\,dx \\
&\leq \frac{1}{M}\sum_{i=1}^m \limsup_{n\to\infty}\int_\Omega |\nabla(u_n-u)\phi_i^{\frac12}|^2\,dx \\
&= \frac{1}{M}\sum_{i=1}^m \limsup_{n\to\infty}\int_\Omega |\nabla(u_n-u)^2\phi\,dx \\
&= \frac{1}{M}\sum_{i=1}^m \mu(\phi_i) = \frac{1}{M}\mu(\phi) \leq \frac{1}{M}\|\mu\|\|\phi\|_\infty.
\end{aligned}$$

Letting $M \to \infty$, we get $\nu(\phi) = 0$. Since by assumption (V_2) $\bar\Sigma_V$ is countable, we have $\nu = \sum_{x\in\bar\Sigma_\nu}\nu_x\delta_x$. Let $x \in \Sigma_V$, and let $r_j \to 0$ as $j \to \infty$. We now choose $\psi_j \in C_0^\infty(B(x,r_j))$ such that $\psi_j \geq 0$, $\psi_j(x) = \max_{y\in B(x,r_j)}\psi_j(y) = 1$. We then have

$$\begin{aligned}
\mu(\{x\}) &= \lim_{j\to\infty}\mu(\psi_j^2) = \lim_{j\to\infty}\lim_{n\to\infty}\int_\Omega |\nabla(u_n-u)|^2\psi_j^2\,dy \\
&= \lim_{j\to\infty}\limsup_{n\to\infty}\int_\Omega |\nabla((u_n-u)\psi_j)|^2\,dy \\
&\geq S^x_{r_j,V}\lim_{j\to\infty}\limsup_{n\to\infty}\int_\Omega V(u_n-u)^2\psi_j^2\,dy \\
&= \lim_{j\to\infty} S^x_{r_j,V}\nu(\psi_j^2) = S^x_V\nu(\{x\}).
\end{aligned}$$

In a similar manner we show that if $x \in \bar\Sigma_V - \Sigma_V$, then $\mu(\{x\}) > j\nu(\{x\})$ for each $j \geq 0$ and so $\nu(\{x\}) = 0$. This shows that the Dirac masses belong to Σ_V. Since

$$\left|\int_\Omega |\nabla(u_n-u)|^2\psi_j^2\,dy - \int_\Omega |\nabla u_n|^2\psi_j^2\,dy\right|$$

$$\leq \; 2\Big(\int_\Omega |\nabla u_n|^2 \psi_j^2 \, dy\Big)^{\frac{1}{2}} \Big(\int_\Omega |\nabla u|^2 \psi_j^2 \, dy\Big)^{\frac{1}{2}} + \int_\Omega |\nabla u|^2 \psi_j^2 \, dy,$$

we get letting $j \to \infty$ that $\tilde{\mu}(\{x\}) = \mu(\{x\})$. From the lower semicontinuity of norm with respect to weak convergence we deduce that $\mu \geq |\nabla u|^2$ and (3), (4) follow. We now observe that for $\phi \in C_\circ^\infty(\Omega)$ we have

$$\begin{aligned} \int_\Omega V u_n^2 \phi \, dx &= \int_\Omega V(u_n - u)^2 \phi \, dx + \int_\Omega V u^2 \phi \, dx + o(1) \\ &= \nu(\phi) + \int_\Omega V u^2 \, dx + o(1), \end{aligned}$$

so that $V u_n^2 \rightharpoonup V u^2 + \nu$. Taking $\phi = 1 - \phi_R$ we deduce from this that

$$\begin{aligned} \limsup_{n\to\infty} \int_\Omega V u_n^2 \, dx &= \limsup_{n\to\infty} \Big[\int_\Omega V u_n^2 \phi_R \, dx + \int_\Omega V u_n^2 (1 - \phi_R) \, dx\Big] \\ &= \limsup_{n\to\infty} \int_\Omega V u_n^2 \phi_R \, dx + \int_\Omega V u^2 (1 - \phi_R) \, dx + \nu(1 - \phi_R) \\ &= \lim_{R\to\infty} \limsup_{n\to\infty} \int_\Omega V u_n^2 \phi_R \, dx + \int_\Omega V u^2 \, dx + \|\nu\| \\ &= \int_\Omega V u^2 \, dx + \|\nu\| + \nu_\infty. \end{aligned}$$

To show (6) let $\phi \in C_\circ^\infty(\Omega)$ with $\phi \geq 0$ on Ω. Since $\nabla u_n \rightharpoonup \nabla u$ in $L^2(\Omega)$ we easily get

$$\int_\Omega |\nabla u_n|^2 \phi \, dx = \int_\Omega |\nabla (u_n - u)|^2 \phi \, dx + \int_\Omega |\nabla u|^2 \, dx + o(1).$$

Applying this relation with $\phi = 1 - \phi_R$ we obtain

$$\begin{aligned} \limsup_{n\to\infty} \int_\Omega |\nabla u_n|^2 \, dx &= \limsup_{n\to\infty} \Big[\int_\Omega |\nabla u_n|^2 \phi_R \, dx + \int_\Omega |\nabla u_n|^2 (1 - \phi_R) \, dx\Big] \\ &= \limsup_{n\to\infty} \int_\Omega |\nabla u_n|^2 \phi_R \, dx + \int_\Omega |\nabla u|^2 (1 - \phi_R) \, dx + \mu(1 - \phi_R) \\ &= \lim_{R\to\infty} \limsup_{n\to\infty} \int_\Omega |\nabla u_n|^2 \phi_R \, dx + \int_\Omega |\nabla u|^2 \, dx + \|\mu\| \\ &= \int_\Omega |\nabla u|^2 \, dx + \|\mu\| + \mu_\infty. \end{aligned}$$

□

1.5 Application to eigenvalue problems

Proposition 1.4 can be used to formulate conditions on V guaranteeing the existence of the principal eigenvalue for problem $(\mathcal{P}_V)$.

Theorem 1.1 *Suppose that (V_1) and (V_2) hold. If $S_V < S_{\infty,V}$ and $S_V < S_V^{\inf}$, then there exits the principal eigenvalue for problem $(\mathcal{P}_V)$.*

Proof. Let $\{u_n\} \subset D_0^{1,2}(\Omega)$ be a minimizing sequence for S_V. It follows from (V_1) that $\int_\Omega V^+ u_n^2\,dx$ is bounded. We may assume that $u_n \rightharpoonup u$ in $D_0^{1,2}(\Omega)$, $u_n \to u$ a.e. on Ω, $|\nabla(u_n - u)|^2 \rightharpoonup \mu$, $V^+(u_n-u)^2 \rightharpoonup \nu$ and $|\nabla u_n|^2 \to \tilde{\mu}$ in $\mathcal{M}$. By Proposition 1.3 we have

$$\begin{aligned} S_V &= \limsup_{n\to\infty} \int_\Omega |\nabla u_n|^2\,dx = \int_\Omega |\nabla u|^2\,dx + S_V^{\inf}\|\nu\| + \mu_\infty \quad (1.16)\\ &\geq S_V \int_\Omega V u^2\,dx + S_V^{\inf}\|\nu\| + S_{\infty,V}\,\nu_\infty, \end{aligned}$$

and

$$\lim_{n\to\infty} \int_\Omega V^+ u_n^2\,dx = \int_\Omega V^+ u^2\,dx + \|\nu\| + \nu_\infty. \tag{1.17}$$

On the other hand, by Fatou's lemma we have

$$\lim_{n\to\infty} \int_\Omega V^+ u_n^2\,dx - 1 = \lim_{n\to\infty} \int_\Omega V^- u_n^2\,dx \geq \int_\Omega V^- u^2\,dx. \tag{1.18}$$

Combining (1.17) and (1.18) we obtain

$$1 \leq \int_\Omega V u^2\,dx + \|\nu\| + \nu_\infty.$$

If either $\|\nu\| \neq 0$ or $\nu_\infty \neq 0$, then (1.16) implies that

$$\begin{aligned} S_V &\geq S_V \int_\Omega V u^2\,dx + S_V^{\inf}\|\nu\| + S_{\infty,V}\,\nu_\infty \\ &> S_V\Big(\int_\Omega V u^2\,dx + \|\nu\| + \nu_\infty\Big) > S_V, \end{aligned}$$

which is impossible. Therefore $1 \leq \int_\Omega V u^2\,dx$ and hence $1 = \int_\Omega V u^2\,dx$, $\int_\Omega |\nabla u|^2\,dx = S_V$. □

We now formulate two conditions on V guaranteeing the validity of the inequalities

$$S_V < S_{\infty,V} \text{ and } S_V < S_V^{\inf}. \tag{1.19}$$

Let $v \in L^{\frac{N}{2}}(\Omega)$ and $V \geq 0$ and $V \not\equiv 0$ on Ω. If $u \in H_\circ^1(\Omega \cap B(0,r))$ with $\int_{\Omega\cap B(x,r)} u^2\,dx = 1$, then the Hölder and Sobolev inequalities imply that

$$\begin{aligned} 1 &= \int_{\Omega\cap B(x,r)} Vu^2\,dx \leq \left(\int_{\Omega\cap B(x,r)} V^{\frac{N}{2}}\,dy\right)^{\frac{2}{N}} \left(\int_{\Omega\cap B(x,r)} |u|^{2^*}\,dy\right)^{\frac{N-2}{N}} \\ &\leq \left(\int_{\Omega\cap B(x,r)} V^{\frac{N}{2}}\,dx\right)^{\frac{2}{N}} \left(\int_{\Omega\cap B(x,r)} |\nabla u|^2\,dy\right). \end{aligned}$$

Since $\int_{\Omega\cap B(x,r)} V^{\frac{N}{2}}\,dy \to 0$ as $r \to 0$, we see that $S_V^x = \infty$ and $S_V^{\inf} = \infty$. Similarly, we show that $S_{\infty,V} = \infty$.

The second condition implying the validity of (1.19) is taken from [105]. Let $V \in L^1_{\text{loc}}(\Omega)$, $V \geq 0$ and $V \not\equiv 0$ on Ω. Suppose that for every $y \in \bar{\Omega}$, $\lim_{x\to y, x\in\Omega} |x-y|^2 V(y) = 0$ and $\lim_{|x|\to\infty, x\in\Omega} |x|^2 V(x) = 0$. In this case the inequality (1.19) follows from the Hardy inequality. We only establish the validity of the first inequality. Let $\epsilon > 0$. We choose $R > 0$ so that $|x|^2 V(x) < \epsilon$ for $x \in \Omega_R$. It then follows from the Hardy inequality that

$$\int_{\Omega_R} Vu^2\,dx \leq \epsilon \int_{\Omega_R} \frac{u^2}{|x|^2}\,dx \leq \epsilon \left(\frac{2}{N-2}\right)^2 \int_\Omega |\nabla u|^2\,dx,$$

which implies that $S_{\infty,V} = \infty$.

Finally, we give a compactness criterion for the functional $u \to \int_\Omega Vu^2\,dx$ in $D_\circ^{1,2}(\Omega)$ in terms of $S_V^{\inf}$ and $S_{\infty,V}$.

Theorem 1.2 *Suppose that* (V_1) *and* (V_2) *hold. Then conditions* $S_V^{\inf} = S_{\infty,V} = \infty$ *are necessary and sufficient for weak continuity of mapping* $u \to \int_\Omega V^+ u^2\,dx$ *defined on* $D_\circ^{1,2}(\Omega)$.

Proof. Necessity. Suppose that $S_{\infty,V} < \infty$. Then for each $m \geq 1$ there exists $u_m \in D_\circ^{1,2}(\Omega \cap \mathbb{R}^N - B(0,m))$ such that $\int_\Omega V^+ u_m^2\,dx = 1$ and $\int_\Omega |\nabla u_m|^2\,dx \leq S_{\infty,V} + 1$. It is clear that $u_m \rightharpoonup 0$ in $D_\circ^{1,2}(\Omega)$. By the weak continuity $\int_\Omega V^+ u_m^2\,dx \to 0$, which is impossible. In a similar way we show that $S_V^{\inf} = \infty$.

Sufficiency. This follows from the last assertion of Proposition 1.3. □

Using Theorem 1.2 and the two observations in the paragraph preceding it, we can state the following existence result.

Corollary 1.2 *Suppose that* $V \in L^1_{loc}(\Omega)$, $V^+ = V^1 + V^2 \not\equiv 0$, *where* $V_1 \in L^{\frac{N}{2}}(\Omega)$ *and for every* $y \in \bar{\Omega}$, $\lim_{x\to y, x\in\Omega} |x-y|^2 V_2(x) = 0$ *and* $\lim_{|x|\to\infty, x\in\Omega} |x|^2 V_2(x) = 0$. *Then problem* $(\mathcal{P}_V)$ *has an eigenfunction* $\phi \geq 0$ *corresponding to the eigenvalue* $\lambda_1 = S_V = \int_\Omega |\nabla\phi|^2\,dx$.

We now illustrate the above existence result by some examples. First, we observe that if $V \in L^1_{\text{loc}}(\Omega) \cap C^1(\mathbb{R}^N)$ and u is an eigenfunction of $(\mathcal{P}_V)$ on $\mathbb{R}^N$, then

$$\frac{d}{ds}\left(\frac{\int_{\mathbb{R}^N} |\nabla u(sx)|^2\,dx}{\int_{\mathbb{R}^N} V(x)u(sx)^2\,dx}\right)\Big|_{s=1} = 0,$$

which implies that

$$\int_{\mathbb{R}^N} (2V(x) + x\nabla V(x))u(x)^2\,dx = 0.$$

The folllowing examples are taken from [108]. Let

$$V_1(x) = \frac{1}{1+|x|^2} \quad\text{and}\quad V_2(x) = \frac{1}{|x|^2(1+|x|^2)}$$

for $x \neq 0$. Then $2V_1(x) + x\nabla V_1(x) > 0$ and $2V_2(x) + x\nabla V_2(x) < 0$. By the above observation problems $(\mathcal{P}_{V_1})$ and $(\mathcal{P}_{V_2})$ have no eigenvalues.
As next examples consider

$$V_3(x) = \frac{1}{(1+|x|^2)[\log(2+|x|^2)]^{\frac{2}{N}}}$$

and

$$V_4(x) = \frac{1}{|x|^2(1+|x|^2)[\log(2+\frac{1}{|x|^2})]^{\frac{2}{N}}}.$$

Then V_3 and V_4 satisfy (V_1) and (V_2) although these functions are not in $L^{\frac{N}{2}}(\mathbb{R}^N)$.
If $V(x) = \frac{1}{|x|^2}$ for $x \in \mathbb{R}^N - \{0\}$, then $(\mathcal{P}_V)$ has no eigenvalue. This means that the best constant in the Hardy inequality is not achieved. Indeed, in this case $\Sigma_V = \{0\}$ and $\nu_o \geq \frac{(N-2)^2}{4}$, which means that the minimizing sequence for S_V concentrates at the origin.

1.6 Bibliographical notes

The concentration–compactnes principles were discovered by P.L. Lions [73], [74] for both subcritical and critical cases. Since then it has been used by many authors (see for example [7], [27], [50], [97], [98], [108], [53], [112], [113], [114]. The concentration–compactness principle at infinity was formulated by Chabrowski [30], [31]. It was also used be Troestler [109]. For the version of the concentration–compactness principle at infinity for critical case we refer to the papers [13], [19] and the monographs [32] and [118]. Lemma 1.1 which is sometimes called a vanishing lemma appears in the work of P.L. Lions [73]. The proof presented here is taken from the paper of Cotti–Zelati and Rabinowitz [42]. Various versions of Proposition 1.1 can be found in [15] and [104]. A more general version of this result is given in Chapter 3 (Theorem 3.5), where we followed the proof from the paper Zhu–Cao [119]. The results of Section 1.5 were proved by Smets [100]. In Theorems 1.1 and 1.2 we followed the approach from papers [100], [105].

Chapter 2
Constrained minimization

This chapter treats constrained minimization problems which lead to the existence of the so called ground states. We characterize unbounded sets in $\mathbb{R}_N$ which do not admit ground states. With the aid of concentration–compactness principle at infinity we find level sets of variational functionals for which the Palais–Smale condition holds. These level sets are used to find critical points of variational functionals which do not admit ground states. We also discuss the role of symmetry and periodicity of coefficients of variational functionals for the existence of critical points.

2.1 Preliminaries

Let $\Omega \subset \mathbb{R}_N$ be an unbounded domain and consider the minimization problem

$$(\mathcal{M}) \quad M_b = M(b,\Omega) = \inf\{\int_\Omega (|Du|^2+u^2)\,dx;\ u \in H_\circ^1(\Omega),\ \int_\Omega b(x)|u|^p\,dx = 1\},$$

where $2 < p < 2^*$ and b is a continuous, positive and bounded function. We associate with problem $(\mathcal{M})$ a *variational functional* $G : H_\circ^1(\Omega) \to \mathbb{R}$ given by

$$G(u) = \frac{1}{2}\int_\Omega (|Du|^2 + u^2)\,dx - \frac{1}{p}\int_\Omega b(x)|u|^p\,dx.$$

The *Fréchet derivative* of G is given by

$$\langle G'(u), \phi\rangle = \int_\Omega (DuD\phi + \phi)\,dx - \int_\Omega b(x)|u|^{p-2}u\phi\,dx$$

for each $\phi \in H_0^1(\Omega)$. Every *minimizing sequence* $\{u_m\}$ of problem $(\mathcal{M})$ after rescaling $v_m = (M_{\lambda,b})^{\frac{1}{p-2}} u_m$, is a Palais–Smale sequence for G, that is,

$$G(v_m) \to \Big(\frac{1}{2} - \frac{1}{p}\Big) M_b^{\frac{p}{p-2}} \text{ as } m \to \infty \tag{2.1}$$

and

$$G'(v_m) \to 0 \text{ in } H^{-1}(\Omega) \text{ as } m \to \infty. \tag{2.2}$$

A minimizer v of problem $(\mathcal{M})$, after rescaling $u = M_b^{\frac{1}{p-2}} v$, satisfies the equation

$$-\Delta u + u = b(x)|u|^{p-2}u \text{ in } \Omega \tag{2.3}$$

in a distributional sense, that is,

$$\int_\Omega (DuD\phi + u\phi)\, dx = \int_\Omega b(x)|u|^{p-2}u\phi\, dx$$

for each $\phi \in H_0^1(\Omega)$.

If $b \equiv 1$ on $\mathbb{R}_N$, this equation is a special case of the equation

$$-\Delta u = f(u) \text{ in } \mathbb{R}_N, \tag{2.4}$$

with a continuous and odd nonlinearity $f : \mathbb{R} \to \mathbb{R}$ satisfying

$$-\infty < \limsup_{s\to 0} \frac{f(s)}{s} \le \limsup_{s\to 0} \frac{f(s)}{s} = -m < 0,$$

$$-\infty \le \limsup_{s\to\infty} \frac{f(s)}{s^l} \le 0,$$

where $l = \frac{N+2}{N-2}$ and

$$\int_0^\zeta g(s)\, ds > 0 \text{ for some } \zeta > 0.$$

Equation (2.4) appears in the theory of stationary states of the nonlinear Schrödinger equation and has been extensively studied by P.L. Lions and Berestycki [18]. In particular, they showed that equation (2.4) has a positive and radially symmetric solution decreasing to 0 with $r = |x| \to \infty$. Their approach is based on the fact that the subspace of radially symmetric functions in $H^1(\mathbb{R}_N)$ is compactly embedded into $L^p(\mathbb{R}_N)$, with $2 < p < 2^*$

(see [75]). They also established the existence of infinitely many radially symmetric solutions varying in sign, using the *Lusternik–Schnirelman* theory of critical points. If $p \geq \frac{2N}{N-2}$, then it follows from the *Pohozhaev identity* (see [18]) that problem (2.3) has no solution.

The infimum $M_b(\Omega)$ depends on Ω. It is easy to see that $M_b(\Omega_1) \geq M_b(\Omega_2)$ whenever $\Omega_1 \subset \Omega_2$. Applying the maximum principle it is not difficult to show that if $M_b(\Omega_1) = M_b(\Omega_2)$ and $\Omega_1 \subsetneq \Omega_2$, then problem $(\mathcal{M})$ is not solvable on Ω_1. Such a situation, in case $b \equiv 1$, occurs for $\Omega_1 = \mathbb{R}_N - D$, where D is a bounded domain in $\mathbb{R}_N$. More generally, problem $(\mathcal{M})$, with $b \equiv 1$, is not solvable on *large-interior domains* (see [71]), whose definition as well as the nonexistence result for problem $(\mathcal{M})$ on such domains will be given in Section 2.2. We shall return later to the dependence of M_b on Ω. In what follows we always set $M = M_b(\Omega) = M_{b,\lambda}(\Omega)$ in case where $b \equiv 1$ on Ω.

2.2 Periodic domains

In this section we show that problem $(\mathcal{M})$ is solvable in case $b \equiv 1$ and $\Omega = \mathbb{R}_N$. This is a very well- known result (see [18]). The usual method applied in this situation consists in finding level sets of the variational functional G for which the Palais–Smale condition holds (see [14] or [71]). This can be done with the help of Proposition 1.1. Our approach is based on Proposition 1.2. We first show that every minimizing sequence of problem $(\mathcal{M})$ has a subsequence weakly convergent to a nonzero limit. Then we show that this subsequence is relatively compact. It will be clear from the proof that this existence result can be very easily extended to a class of periodic domains in $\mathbb{R}_N$, whose definition will be given later.

Theorem 2.1 *Let $b \equiv 1$ on $\mathbb{R}_N$. Then the minimization problem $(\mathcal{M})$ is solvable.*

Proof. Let $\{u_m\}$ be a minimizing sequence of problem $(\mathcal{M})$. Since $\{u_m\}$ is bounded in $H^1(\mathbb{R}_N)$ we may assume that $u_m \rightharpoonup u$ in $H^1(\mathbb{R}_N)$. We now distinguish two cases: (i) $u \equiv 0$ in $\mathbb{R}_N$ and (ii) $u \not\equiv 0$ in $\mathbb{R}_N$.
Case (i). By Corollary 1.1 there exists a sequence $\{y_m\}$ such that $v_m(\cdot) = u_m(\cdot + y_m) \rightharpoonup v \not\equiv 0$ in $H^1(\mathbb{R}_N)$. Let α_∞ and β_∞ be quantities related to

v_m by Proposition 1.2. It follows from (c) of this proposition that

$$\int_{\mathbb{R}_N} |v(x)|^p \, dx + \alpha_\infty = 1. \tag{2.5}$$

We show that $\alpha_\infty = 0$. Arguing indirectly we assume that $\alpha_\infty > 0$. Since $v \not\equiv 0$, it follows from (2.5) that

$$\alpha_\infty < 1. \tag{2.6}$$

Let ϕ_R be a function defined in the proof of Proposition 1.2. Since $G'(v_m M^{\frac{1}{p-2}}) \to 0$ in $H^{-1}(\mathbb{R}_N)$, we see that $\langle G'(v_m M^{\frac{1}{p-2}}), M^{\frac{1}{p-2}} v_m \phi_R\rangle \to 0$ as $m \to \infty$ uniformly in $R \geq 1$. Hence

$$\begin{aligned} o(1) &= \langle G'(v_m M^{\frac{1}{p-2}}), M^{\frac{1}{p-2}} v_m \phi_R\rangle \\ &= M^{\frac{2}{p-2}} \int_{\mathbb{R}_N} \left(|Dv_m|^2 \phi_R^2 + v_m^2 \phi_R^2\right) dx \\ &+ M_1^{\frac{2}{p-2}} \int_{\mathbb{R}_N} Dv_m D\phi_R v_m \, dx - M^{\frac{p}{p-2}} \int_{\mathbb{R}_N} |v_m|^p \phi_R \, dx. \end{aligned}$$

Letting $m \to \infty$ and then $R \to \infty$, we find that $M\alpha_\infty = \beta_\infty$. This, combined with inequality (i) of Proposition 1.2, yields $M\alpha_\infty^{\frac{2}{p}} \leq \alpha_\infty M$, that is, $\alpha_\infty \geq 1$ which contradicts (2.6). This means that $v_m \to v$ in $L^p(\mathbb{R}_N)$. To show that $v_m \to v$ in $H^1(\mathbb{R}_N)$, we write

$$\begin{aligned} o(1) &= \langle G'(M^{\frac{1}{p-2}} v_m) - G'(M^{\frac{1}{p-2}} v_n), M^{\frac{1}{p-2}}(v_m - v_n)\rangle \\ &= M^{\frac{2}{p-2}} \int_{\mathbb{R}_N} \left(|Dv_m - Dv_n|^2 + (v_m - v_n)^2\right) dx \\ &- M^{\frac{p}{p-2}} \int_{\mathbb{R}_N} \left(|v_m|^{p-2} v_m - |v_n|^{p-2} v_n\right)(v_m - v_n) \, dx. \end{aligned}$$

Since $v_m \to v$ in $L^p(\mathbb{R}_N)$, $\{v_m\}$ is a Cauchy sequence in $H^1(\mathbb{R}_N)$ and the result follows. Case (ii). If $u \not\equiv 0$, we apply Proposition 1.2 to $\{u_m\}$ and repeat the argument from Case (i). □

Remark 2.1 In case (i) the existence of a sequence $\{y_m\} \subset \mathbb{R}_N$ such that $w_m(\cdot) = u_m(\cdot + y_m) \rightharpoonup v \not\equiv 0$ in $H^1(\mathbb{R}_N)$ can also be shown without the use of Lemma 1.1 by the following argument.

The sequence $v_m = M^{\frac{1}{p-2}} u_m$ is a Palais–Smale sequence for the functional G. Let $0 < r < (\frac{1}{2} - \frac{1}{p})M^{\frac{p}{p-2}}$ and according to (2.1) we may assume that $G(v_m) \geq r$ for all m. Let $\{Q_m\}$, $m = (m_1, \ldots, m_N)$ be a family of all cubes of the form $\{x \in \mathbb{R}_N;\ m_i \leq x_i < m_i + 1, i = 1, \ldots, N\}$, where m_i are integers. We put

$$d_n = \max_{m \in \mathbb{Z}_N} ||v_n||_{L^p(Q_m)}.$$

We claim that there exists a $\delta > 0$ such that $d_n \geq \delta$ for all n. Indeed, by (2.1) and (2.2) we have

$$G(v_n) = (\frac{1}{2} - \frac{1}{p}) \int_{\mathbb{R}_N} |v_n|^p \, dx + o(1).$$

Let $l = \frac{2p}{p-2}$, we obtain by the Sobolev inequality that

$$\begin{aligned} lG(v_n) + o(1) &= \int_{\mathbb{R}_N} |v_n|^p \, dx = \sum_{m \in \mathbb{Z}_N} ||v_n||^p_{L^p(Q_m)} \\ &\leq \max_{m \in \mathbb{Z}_N} ||v_n||^{p-2}_{L^p(Q_m)} \sum_{m \in \mathbb{Z}_N} ||v_n||^2_{L^p(Q_m)} \\ &\leq Cd_n^{p-2} \sum_{m \in \mathbb{Z}_N} \int_{Q_m} (|Dv_n|^2 + v_n^2) \, dx = d_n^{p-2} C ||v_n||^2_{1,2} \\ &= Cd_n^{p-2} lG(v_n) + o(1) \end{aligned}$$

for some constant $C > 0$ independent of n. Since $G(v_n) \geq r > 0$ for all n, we see that there exists a constant $\delta > 0$ such that $d_n \geq \delta$ for all n. Therefore, for each n there exists a cube Q_{m_n} such that

$$||v_n||_{L^p(Q_{m_n})} \geq \frac{\delta}{2}.$$

This means that for each n there exists a point $y_n = y_{m_n}$ in $\mathbb{Z}_N$ whose coordinates are integers such that

$$||v_n||_{L^p(Q_{m_n})} = ||v_n(\cdot + y_n)||_{L^p(Q_\circ)} \geq \frac{\delta}{2}. \tag{2.7}$$

Since $\{v_n(\cdot + y_n)\}$ is also a Palais–Smale sequence for G, we see that $\{v_n(\cdot + y_n)\}$ must be bounded in $H^1(\mathbb{R}_N)$. Therefore, we can select a subsequence, relabelled again by $\{v_n(\cdot + y_n)\}$, such that

$$v_n(\cdot + y_n) \rightharpoonup v \text{ in } H^1(\mathbb{R}_N),$$

and

$$v_n(\cdot + y_n) \to v \text{ in } L^p_{\text{loc}}(\mathbb{R}_N).$$

By (2.7) we have $||v||_{L^p(Q_o)} \geq \frac{\delta}{2}$, that is, $v \not\equiv 0$.

Following [71] we introduce the definition of a *periodic domain* in $\mathbb{R}_N$. We say that $\Omega \subset \mathbb{R}_N$ is a periodic domain if there exists a partition $\{Q_m\}$ of Ω and a sequence of points $\{y_m\} \subset \mathbb{R}_N$ such that

(1) $\{y_m\}$ forms a subgroup of $\mathbb{R}_N$,
(2) Q_o is a bounded set and $Q_m = Q_o + y_m$.

Typical examples of periodic sets are: 1) $\Omega_1 = O \times \mathbb{R}_m$, where O is a bounded domain in $\mathbb{R}_p$ and $N = p+m$, 2) $\Omega_2 = \mathbb{R}_N$, 3) $\{(x,y) \in \mathbb{R}_2; \sin x < y < 1 + \sin x\}$.

Inspection of the proof of Remark 2.1 shows that

Corollary 2.1 *Let $b(x) \equiv 1$ on a periodic domain $\Omega \subset \mathbb{R}_N$. Then problem $(\mathcal{M})$ admits a solution.*

Obviously a solution to problem $(\mathcal{M})$ can be chosen to be positive since $G(u) = G(|u|)$. So we may assume that minimizers is nonnegative and it must be positive on its domain of definition by the maximum principle.

Applying the *Schwarz symmetrization* we can assume that a minimizer for $M_1(\mathbb{R}_N)$ is radially symmetric. We recall that given $f \in L^1(\mathbb{R}_N)$, the Schwarz symmetrization f^* of f, is a radial nonincreasing (in $r = |x|$) measurable function such that

$$m(f^* \geq \alpha) = m(|f| \geq \alpha)$$

for every $\alpha \geq 0$, where m is the Lebesgue measure on $\mathbb{R}_N$. It is obvious that

$$\int_{\mathbb{R}_N} F(f)\, dx = \int_{\mathbb{R}_N} F(f^*)\, dx$$

for every continuous function F such that $F(f) \in L^1(\mathbb{R}_N)$. Furthermore, for every $u \in H^1(\mathbb{R}_N)$ we have

$$\int_{\mathbb{R}_N} |\nabla u^*(x)|^2\, dx \leq \int_{\mathbb{R}_N} |\nabla u(x)|^2\, dx.$$

These two properties of the Schwarz symmetrization show that any minimizing sequence for $M_1(\mathbb{R}_N)$ can be replaced by a sequence of radially symmetric functions in $H^1(\mathbb{R}_N)$.

A minimizer w for $M_b(\Omega)$ has the smallest energy among all solutions of equation (2.3). This property can be expressed in the following way: let $v = M_b^{\frac{1}{p-2}} w$, then

$$G(v) = \inf\{G(u);\ u \in H_0^1(\Omega) - \{0\},\ \langle G'(u), u\rangle = 0\}.$$

Since v satisfies equation (2.3) we have $\langle G'(v), v\rangle = 0$ and hence

$$G(v) \geq \inf\{G(u); u \in H_0^1(\Omega) - \{0\}, \langle G'(u), u\rangle = 0\}.$$

On the other hand if $u \in H_0^1(\Omega)$ satisfies $\langle G'(u), u\rangle = 0$, then

$$M_b \leq \frac{\int_\Omega (|\nabla u|^2 + u^2)\, dx}{||u||_p^2}.$$

This inequality is equivalent to

$$\frac{\int_\Omega (|\nabla v|^2 + v^2)\, dx}{||v||_p^2} \leq \frac{\int_\Omega (|\nabla u|^2 + u^2)\, dx}{||u||_p^2},$$

from which, using the fact that $\langle G'(u), u\rangle = 0$ and $\langle G'(v), v\rangle = 0$, we easily deduce that $G(v) \leq G(u)$ and our claim easily follows.

It has been proved by Kwong [68] that a positive solution of the equation

$$-\Delta u + u = |u|^{p-2}u \text{ in } \mathbb{R}_N$$

is unique up to a translation. Consequently, combining the results of [58] and [18] we can state the following theorem.

Theorem 2.2 *Problem $(\mathcal{M})$ with $b \equiv 1$ has u unique up to a translation positive solution $\bar{u}$ which is radially symmetric such that $\frac{d}{dr}\bar{u}(r) < 0$ for $r > 0$, where $r = |x|$, and*

$$\lim_{|x|\to\infty} |x|^{\frac{N-1}{2}} e^{|x|} \bar{u}(|x|) > 0.$$

As we mentioned earlier, problem $(\mathcal{M})$ is not solvable for exterior domains $\mathbb{R}_N - D$, with D bounded. This means that a minimizing sequence is not relatively compact in $H_0^1(\mathbb{R}_N - D)$. The question can be asked what can be said about minimizing sequences on exterior domains? To answer this question we consider a more general class of domains in $\mathbb{R}_N$, namely large-interior domains (see [71] p.1285). We say that a domain $\Omega \subset \mathbb{R}_N$ is a *large-interior domain* if for every $R > 0$ there exists an $x \in \Omega$

such that $B(x, R) \subset \Omega$. For a large-interior domain $\Omega \subset \mathbb{R}_N$, we have $M(\mathbb{R}_N) = M(\Omega)$.

It is easy to see that $\Omega \subset \mathbb{R}_N$ is a large-interior domain if and only if there exists a sequence $\{y_n\} \subset \Omega$ such that $\lim_{n\to\infty} |y_n| = \infty$ and $\lim_{n\to\infty} \operatorname{dist}(y_n, \partial\Omega) = \infty$.

Proposition 2.1 *Let $\Omega \subset \mathbb{R}_N$ be a large-interior domain and let $b(x) \equiv 1$ on Ω. Then problem $(\mathcal{M})$ has no solution.*

Proof. Extending functions in $H_0^1(\Omega)$ by 0 outside Ω we have $H_0^1(\Omega) \subset H^1(\mathbb{R}_N)$. Hence $M(\Omega) \geq M(\mathbb{R}_N)$. Let $\bar{u}$ be a solution of problem $(\mathcal{M})$ on $\mathbb{R}_N$ and let $\{u_m\} \subset H^1(\mathbb{R}_N)$ be a sequence of functions with compact supports such that

$$\lim_{m\to\infty} \int_{\mathbb{R}_N} \left(|\nabla (u_m(x) - \bar{u}(x))|^2 + (u_m(x) - \bar{u}(x))^2 \right) dx = 0.$$

Let $K_n = \operatorname{supp} u_n$. Since each K_n is a compact set and Ω is a large-interior domain, for every n there exists $z_n \in \mathbb{R}_N$ such that $K_n - z_n \subset \Omega$ and $\lim_{n\to\infty} |z_n| = \infty$. Then the sequence $v_n(x) = u_n(x - z_n)$ satisfies

$$\begin{aligned} &\lim_{n\to\infty} \int_{\mathbb{R}_N} \left(|\nabla (v_n(x) - \bar{u}(x - z_n))|^2 + |v_n(x) - \bar{u}(x - z_n)|^2 \right) dx \\ &= \lim_{n\to\infty} \int_{\mathbb{R}_N} \left(|\nabla (u_n(x) - \bar{u}(x))|^2 + |u_n(x) - \bar{u}(x)|^2 \right) dx = 0. \end{aligned}$$

This implies that

$$\begin{aligned} &\lim_{n\to\infty} \int_{\Omega} (|\nabla v_n|^2 + v_n^2)\, dx \\ &\quad = \lim_{n\to\infty} \int_{\mathbb{R}_N} (|\nabla \bar{u}(x - z_n)|^2 + \bar{u}(x - z_n)^2)\, dx = M(\mathbb{R}_N). \end{aligned}$$

Since $\lim_{n\to\infty} \|u_n - \bar{u}\|_{1,2} = 0$, we have

$$\int_{\mathbb{R}_N} |u_n|^p\, dx = \int_{\mathbb{R}_N} |\bar{u}|^p\, dx + o(1) \text{ as } n \to \infty.$$

Obviously this implies that

$$\int_{\Omega} |v_n|^p\, dx = 1 + o(1) \text{ as } n \to \infty.$$

Hence

$$\frac{\int_{\Omega}\left(|\nabla v_n|^2 + v_n^2\right) dx}{\left(\int_{\Omega} |v_n|^p \, dx\right)^{\frac{2}{p}}} = M(\Omega) + o(1)$$

as $n \to \infty$ and this completes the proof. □

Examples of large-interior domains are: exterior domains, cones and half spaces. Large-interior domains were introduced in the paper [71]. However, the authors of this paper called them "ball up domains".

Lemma 2.1 *Let $\Omega \subset \mathbb{R}_N$ be unbounded domain and let $b(x) \equiv 1$ on Ω. If $\{u_m\} \subset H_0^1(\Omega)$ is a minimizing sequence for problem $(\mathcal{M})$ such that $u_m \to u \not\equiv 0$ in $H_0^1(\Omega)$, then problem $(\mathcal{M})$ has a solution.*

Proof. Indeed, we have by Proposition 1.2 that $1 = \int_{\mathbb{R}_N} |u|^p \, dx + \alpha_\infty$. Since $u \not\equiv 0$, we must have that $\alpha_\infty < 1$. On the other hand, since

$$\langle G'(u_m M^{\frac{1}{p-2}}), u_m M^{\frac{1}{p-2}} \phi_R \rangle \to 0,$$

as $m \to \infty$, uniformly in $R \geq 1$, where ϕ_R are functions from the proof of Proposition 1.2, we get that $M\alpha_\infty = \beta_\infty$. Therefore, by Proposition 1.2(e) we have $\alpha_\infty \geq 1$ and we have arrived at a contradiction. □

According to this lemma every weakly convergent sequence $\{u_m\} \subset H_0^1(\Omega)$ on a large-interior domain Ω must have a weak limit 0, that is, $\alpha_\infty = 1$. On the other hand if we extend every u_m by 0 outside Ω we obtain a sequence in $H^1(\mathbb{R}_N)$ which is a minimizing sequence for problem $(\mathcal{M})$ on $\mathbb{R}_N$. The proof of Theorem 2.1 shows that there exists a sequence of points $\{y_m\} \subset \mathbb{R}_N$, with integer coordinates, and a subsequence of $\{u_m\}$ relabelled again by $\{u_m\}$, such that $u_m(\cdot + y_m) \to u \not\equiv 0$ in $H^1(\mathbb{R}_N)$.

Therefore, we can state the following result.

Corollary 2.2 *Let $\Omega \subset \mathbb{R}_N$ be a large-interior domain and let $b(x) \equiv 1$ on Ω. Every convergent in $H_0^1(\Omega)$ minimizing sequence $\{u_m\}$ for problem $(\mathcal{M})$ has a weak limit 0. If we extend every u_m by 0 outside Ω, then there exists a sequence $\{y_m\} \subset \mathbb{R}_N$ and a subsequence of $\{u_m\}$, relabelled again by $\{u_m\}$, such that $u_m(\cdot + y_m) \to u \not\equiv 0$ in $H^1(\mathbb{R}_N)$ and u is a solution of problem $(\mathcal{M})$ on $\mathbb{R}_N$.*

The sequence $\{y_m\}$ in this corollary must be unbounded, since otherwise $u_m \rightharpoonup u \not\equiv 0$ and problem $(\mathcal{M})$ would be solvable for large-interior domains.

2.3 Extensions to arbitrary domains

In this section we derive some criteria for the solvability of problem $(\mathcal{M})$ on unbounded domains Ω. For simplicity, we always assume that $b \equiv 1$ on its domain of definition. The results, presented here, are known (see [71]), however we offer different and simpler proofs based on Proposition 1.2.

Proposition 2.2 *Let $\Omega = \Omega_1 \cup \Omega_2 \subset \mathbb{R}_N$ and suppose that Ω_2 is a bounded domain. If $M(\Omega) < M(\Omega_1)$, that is, $\Omega_1 \subsetneq \Omega$, then problem $(\mathcal{M})$ admits a solution.*

Proof. Let $\{u_m\} \subset H_0^1(\Omega)$ be a minimizing sequence for problem (M). By Lemma 2.1 it is sufficient to show that there exists a subsequence of $\{u_m\}$, denoted again by $\{u_m\}$, such that $u_m \rightharpoonup u \not\equiv 0$ in $H_0^1(\Omega)$. Arguing by contradiction, let us assume that $u_m \rightharpoonup 0$ in $H_0^1(\Omega)$. We then have

$$\alpha_\infty = 1. \tag{2.8}$$

Since the rescaled sequence $M^{\frac{1}{p-2}} u_m$ is a Palais–Smale sequence for G, we get, as in the proof of Lemma 2.1 that

$$\beta_\infty = M(\Omega). \tag{2.9}$$

On the other hand, since Ω_2 is bounded, we have that $\{u_m \phi_R\} \subset H_0^1(\Omega_1)$ for sufficiently large R. Therefore, we have

$$M(\Omega_1)\left(\int_{\Omega_1} |u_m\phi_R|^p\, dx\right)^{\frac{2}{p}} \le \int_{\Omega_1} \left(|D(u_m\phi_R)|^2 + (u_m\phi_R)^2\right) dx.$$

Letting $m \to \infty$ and then $R \to \infty$ we get that

$$M(\Omega_1)\alpha_\infty^{\frac{2}{p}} \le \beta_\infty,$$

which by (2.8) and (2.9) gives that $M(\Omega_1) \le M(\Omega)$ which is impossible. □

Proposition 2.3 *Suppose that $\Omega = \Omega_1 \cup \Omega_2 \subset \mathbb{R}_N$ and $\Omega_1 \cap \Omega_2$ is a bounded domain. If $M(\Omega_1) \le M(\Omega_2)$ and $M(\Omega_1)$ has a minimizer, then $M(\Omega)$ has a minimizer.*

Proof. This is trivial if $\Omega = \Omega_1$, so we may assume that $\Omega_1 \subsetneq \Omega$. Since $M(\Omega_1)$ admits a minimizer we must have $M(\Omega) < M(\Omega_1)$. Let $\{u_m\} \subset H_0^1(\Omega)$ be a minimizing sequence for $M(\Omega)$. According to Lemma 2.1 it is sufficient to show that $u_m \rightharpoonup u \not\equiv 0$ in $H_0^1(\Omega)$. Assuming that $u \equiv 0$ on Ω, we must have $\alpha_\infty = 1$. Since $\{u_m M(\Omega)^{\frac{1}{p-2}}\}$ is a Palais–Smale sequence for G, we deduce that $\beta_\infty = M(\Omega)$. Let $w_m = \phi_R u_m$, where ϕ_R is a function from the proof of Proposition 1.2. Since $\Omega_1 \cap \Omega_2$ is bounded, $\Omega_1 \cap (|x| > R)$ and $\Omega_2 \cap (|x| > R)$ are disjoint for large $R > 0$ and we put for $i = 1, 2$

$$w_{m,i}(x) = \begin{cases} w_m(x) & \text{on } \Omega_i \\ 0 & \text{elsewhere} \end{cases}$$

and we have $w_m = w_{m,1} + w_{m,2}$, with $w_{i,m} \in H_0^1(\Omega_i)$. Let

$$\alpha_{\infty,i} = \lim_{R\to\infty} \limsup_{m\to\infty} \int_{\Omega_i} |w_{m,i}|^p \, dx$$

and we obviously have that

$$1 = \alpha_\infty = \alpha_{\infty,1} + \alpha_{\infty,2}.$$

We now observe that

$$\begin{aligned} M(\Omega) = \beta_\infty &= \lim_{R\to\infty} \limsup_{m\to\infty} \int_{\Omega_1} \left(|Dw_{m,1}|^2 + w_{m,1}^2\right) dx \\ &\lim_{R\to\infty} \limsup_{m\to\infty} \int_{\Omega_2} \left(|Dw_{m,2}|^2 + w_{m,2}^2\right) dx \\ &\geq \alpha_{\infty,1}^{\frac{2}{p}} M(\Omega_1) + \alpha_{\infty,2}^{\frac{2}{p}} M(\Omega_2) \\ &\geq M(\Omega_1)\left(\alpha_{\infty,1}^{\frac{2}{p}} + \alpha_{\infty,2}^{\frac{2}{p}}\right) \geq M(\Omega_1), \end{aligned}$$

which is impossible. □

Corollary 2.3 *Under the assumptions of Proposition 2.3 if both $M(\Omega_1)$ and $M(\Omega_2)$ admit minimizers, then $M(\Omega)$ also admits a minimizer.*

Indeed, we have either $M(\Omega_1) \leq M(\Omega_2)$ or $M(\Omega_2) \leq M(\Omega_1)$ and in both cases we can apply Proposition 2.3. For examples illustrating the use of Propositions 2.2 and 2.3 we refer to the paper [71].

2.4 Constrained minimization with non-constant coefficient

To guarantee the solvability of problem $(\mathcal{M})$, with $b \not\equiv$ const, we must impose some restrictions on the behaviour of b at infinity. Throughout this section we assume that $b \not\equiv$ const . We start with an analogue of Lemma 2.1.

Lemma 2.2 *If $\{u_m\} \subset H^1(\mathbb{R}_N)$ is a minimizing sequence of problem $(\mathcal{M})$ such that $u_m \rightharpoonup u \not\equiv 0$ in $H^1(\mathbb{R}_N)$, then problem $(\mathcal{M})$ has a solution.*

Proof. It is enough to show that

$$\alpha_{b,\infty} = \lim_{R\to\infty} \limsup_{m\to\infty} \int_{(|x|>R)} b(x)|u_m(x)|^p \, dx = 0.$$

Assuming that $\alpha_{b,\infty} > 0$, we have

$$1 = \int_{\mathbb{R}_N} b(x)|u|^p \, dx + \alpha_{b,\infty}.$$

Since $\{M_b^{\frac{1}{p-2}} u_m\}$ is a Palais–Smale sequence for G, we easily derive that $\beta_\infty = \alpha_{b,\infty} M_b$. By the definition of problem $(\mathcal{M})$ we have

$$M_b \left(\int_{\mathbb{R}_N} b(x)|u(x)|^p \, dx \right)^{\frac{2}{p}} \leq \int_{\mathbb{R}_N} (|Du(x)|^2 + u(x)^2) \, dx. \tag{2.10}$$

Relations $G(M_b^{\frac{1}{p-2}} u_m) \to (\frac{1}{2} - \frac{1}{p}) M_b^{\frac{p}{p-2}}$ and $G'(M_b^{\frac{1}{p-2}} u_m) \to 0$ in $H^{-1}(\mathbb{R}_N)$ as $m \to \infty$ yield that

$$\left(\frac{1}{2} - \frac{1}{p}\right) M_b^{\frac{p}{p-2}} \geq \left(\frac{1}{2} - \frac{1}{p}\right) M_b^{\frac{2}{p-2}} \left(\int_{\mathbb{R}_N} (|Du|^2 + u^2) \, dx + \beta_\infty \right).$$

From this and the fact that $\beta_\infty = \alpha_{b,\infty} M_b$ we deduce that

$$\int_{\mathbb{R}_N} (|Du|^2 + u^2) \, dx \leq M_b \int_{\mathbb{R}_N} b(x)|u|^p \, dx.$$

This combined with (2.10) gives

$$M_b \left(\int_{\mathbb{R}_N} b(x)|u|^p \, dx \right)^{\frac{2}{p}} \leq M_b \int_{\mathbb{R}_N} b(x)|u|^p \, dx,$$

that is,

$$\int_{\mathbb{R}_N} b(x)|u|^p \, dx \geq 1$$

which is impossible. □

Lemma 2.2 implies that a dichotomy for minimizing sequences of problem $(\mathcal{M})$ never occurs. Indeed, using a quantity $\alpha_{b,\infty}$ from Lemma 2.2, we can distinguish the following three cases for a minimizing sequence

$\{u_m\} \subset H^1(\mathbb{R}_N)$ of problem (M) with $u_m \rightharpoonup u$ in $H^1(\mathbb{R}_N)$ and $1 = \int_{\mathbb{R}_N} b(x)|u|^p\,dx + \alpha_{b,\infty}$:

(1) $\alpha_{b,\infty} = 0$ and $\int_{\mathbb{R}_N} b(x)|u|^p\,dx = 1$. In this case problem $(\mathcal{M})$ is solvable (compactness).
(2) $u \equiv 0$, $\alpha_{b,\infty} = 1$. In this case there exists $\{y_m\} \subset \mathbb{R}_N$ such that up to a subsequence $u_m(\cdot + y_m) \rightharpoonup u \not\equiv 0$ in $H^1(\mathbb{R}_N)$.
(3) $\int_{\mathbb{R}_N} b(x)|u|^p\,dx > 0$, $\alpha_{b,\infty} > 0$. This case is not possible, because $u_m \rightharpoonup u \not\equiv 0$ in $H^1(\mathbb{R}_N)$ and by Lemma 2.2 problem (M) has a solution and necessarily $\alpha_{b,\infty} = 0$.

A starting point in establishing the solvability of problem $(\mathcal{M})$ is to show, as in Remark 2.1, that for every minimizing sequence of problem $(\mathcal{M})$ there exists a sequence $\{y_m\} \subset \mathbb{Z}_N$ such that $u_m(\cdot + y_m) \rightharpoonup u \not\equiv 0$ in $H^1(\mathbb{R}_N)$ as $m \to \infty$. To achieve this we modify d_n by

$$d_n = \max_{m \in \mathbb{Z}_N} M_b^{\frac{1}{p-2}} \left(\int_{Q_m} b(x)|u_n|^p\,dx \right)^{\frac{1}{p}}$$

and show that $d_n \geq \delta$ for all n and some constant $\delta > 0$.

Theorem 2.3 *Let $\{u_m\}$ be a minimizing sequence of problem $(\mathcal{M})$. Then there exist a subsequence of $\{u_m\}$, denoted again by $\{u_m\}$, and a sequence $\{y_m\} \subset \mathbb{Z}_N$ such that $u_m(\cdot + y_m) \rightharpoonup u \not\equiv 0$ in $H^1(\mathbb{R}_N)$ as $m \to \infty$.*

(i) If $\{y_m\}$ is bounded then u. is a solution to problem $(\mathcal{M})$.
(ii) If $\lim_{|x|\to\infty} b(x) = b(\infty)$ and $\{y_m\}$ is unbounded, then $u_m(\cdot + y_m) \to u$ in $H^1(\mathbb{R}_N)$ as $m \to \infty$ and u is a solution of problem $(\mathcal{M})$ with $b(x) \equiv b(\infty)$ on $\mathbb{R}_N$.

Proof. If $\{y_m\} \subset \mathbb{Z}_N$ is bounded, then obviously $u_m \rightharpoonup u \not\equiv 0$ in $H^1(\mathbb{R}_N)$ and the result follows from Lemma 2.2. We now consider the case where $\{y_m\}$ is unbounded. To show that $u(\cdot + y_m) \to u$ in $H^1(\mathbb{R}_N)$ it is enough to check that

$$\alpha_{b,\infty} = \lim_{R\to\infty} \limsup_{m\to\infty} \int_{|x|>R} b(x+y_m)|u_m(x+y_m)|^p\,dx = 0.$$

Arguing indirectly, let us assume that $\alpha_{b,\infty} > 0$. Thus setting $v_m(x) = u_m(x + y_m)$ we have for each $R > 0$

$$1 = \int_{\mathbb{R}_N} b(x)u_m(x)|^p\,dx = \int_{|x|<R} b(x+y_m)|v_m(x)|^p\,dx$$

$$+ \int_{|x|>R} b(x+y_m)|v_m(x)|^p \, dx$$

and then letting $m \to \infty$ and $R \to \infty$ we deduce that

$$1 = \int_{\mathbb{R}_N} b(\infty)|u(x)|^p \, dx + \alpha_{b,\infty}. \tag{2.11}$$

Since $u \not\equiv 0$, we have $1 > \alpha_{b,\infty} > 0$. Using the fact that $\{M_{b,\lambda}^{\frac{1}{p-2}} u_m\}$ is a Palais–Smale sequence for G we get that

$$\int_{\mathbb{R}_N} (|Dv_m|^2 + v_m^2) \, dx = M_b \int_{\mathbb{R}_N} b(x)|u_m(x)|^p \, dx + o(1).$$

Setting

$$\beta_\infty = \lim_{R\to\infty} \limsup_{m\to\infty} \int_{|x|>R} (|Dv_m|^2 + v_m^2) \, dx$$

we deduce from Proposition 1.2 that

$$\int_{\mathbb{R}_N} (|Du|^2 + u^2) \, dx + \beta_\infty \leq M_b. \tag{2.12}$$

Since we may assume that $|y_m| \to \infty$ as $m \to \infty$, we easily check that $M_b^{\frac{1}{p-2}} u$ is a weak solution of equation (2.3) on $\mathbb{R}_N$, with $b(x)$ replaced by $b(\infty)$, from which we deduce that

$$\int_{\mathbb{R}_N} (|Du|^2 + u^2) \, dx = M_b b(\infty) \int_{\mathbb{R}_N} |u|^p \, dx. \tag{2.13}$$

We associate with each y_m the variational problem

$$M(y_m) = \inf\{\int_{\mathbb{R}_N} (|Du|^2 + u^2) \, dx;\ u \in H^1(\mathbb{R}_N),\ \int_{\mathbb{R}_N} b(x+y_m)|u(x)|^p \, dx = 1\}.$$

We now show that $\lim_{m\to\infty} M(y_m) = M_b$. Since

$$\int_{\mathbb{R}_N} b(x+y_m)|v_m(x)|^p \, dx = \int_{\mathbb{R}_N} b(x)|u_m(x)|^p \, dx = 1,$$

we see that

$$M(y_m) \leq \int_{\mathbb{R}_N} (|Dv_m|^2 + v_m^2) \, dx = \int_{\mathbb{R}_N} (|Du_m|^2 + u_m^2) \, dx,$$

from which we deduce that $\limsup_{m\to\infty} M(y_m) \leq M_b$. Our claim will follow if we can show that $M_b \leq M(y_m)$ for each $y_m \in \mathbb{Z}_N$. Suppose that $M(y_m) < M_b$ for some m. Then there exists $z_m \in H^1(\mathbb{R}_N)$ such that

$$\int_{\mathbb{R}_N} b(x+y_m)|z_m(x)|^p\,dx = 1 \text{ and } \int_{\mathbb{R}_N} (|Dz_m|^2 + z_m^2)\,dx < M_{b,\lambda}.$$

This can be rewritten as

$$\int_{\mathbb{R}_N} b(x)|z_m(x-y_m)|^p\,dx = 1 \text{ and } \int_{\mathbb{R}_N} (|Dz_m|^2 + z_m^2)\,dx < M_b$$

and this contradicts the definition of M_b. Let ϕ_R be a function from the proof of Proposition 1.2. Then

$$\begin{aligned} M(y_m)\left(\int_{\mathbb{R}_N} b(x+y_m)|\phi_R(x)v_m(x)|^p\,dx\right)^{\frac{2}{p}} &\leq \int_{\mathbb{R}_N} \big(|D(v_m(x)\phi_R(x))|^2 \\ &+ (v_m(x)\phi_R(x))^2\big)\,dx. \end{aligned}$$

Letting $m \to \infty$ and then $R \to \infty$ we obtain

$$M_{b,\lambda}(\alpha_{b,\infty})^{\frac{2}{p}} \leq \beta_\infty.$$

This, in conjunction with (2.12) and (2.13), gives

$$\int_{\mathbb{R}_N} b(\infty)|u|^p\,dx + (\alpha_{b,\infty})^{\frac{2}{p}} \leq 1,$$

which combined with (2.11) gives $\alpha_{b,\infty} \geq 1$ which is impossible. Therefore $v_m \to u$ in $L^p(\mathbb{R}_N)$. The convergence of v_m to u in $H^1(\mathbb{R}_N)$ follows from the following estimate:

$$\begin{aligned} &\int_{\mathbb{R}_N} (|D(v_m - v_n)|^2 + (v_m - v_n)^2)\,dx \\ &= \int_{\mathbb{R}_N} \big(b(x+y_m)|v_m|^{p-2}v_m - b(x+y_n)|v_n|^{p-2}v_n\big)(v_m - v_n)\,dx \\ &\leq \sup_{\mathbb{R}_N} b\left[\left(\int_{\mathbb{R}_N} |v_m|^p\,dx\right)^{\frac{1}{p}} + \left(\int_{\mathbb{R}_N} |v_n|^p\,dx\right)^{\frac{1}{p}}\right] \|v_m - v_n\|_{L^p(\mathbb{R}_N)} \\ &+ o(1). \end{aligned}$$

□

Solutions obtained in Theorem 2.3 can be taken to be positive. This can be justified by an argument in the paragraph following Corollary 2.1. To show the existence of solutions to problem $(\mathcal{M})$ we must impose a condition on b that rules out part *(ii)* of Theorem 2.3. We denote by M_∞ the infimum of problem $(\mathcal{M})$ for $b(x) \equiv b(\infty)$ on $\mathbb{R}_N$, that is,

$$M_\infty = \inf\{\int_{\mathbb{R}_N} (|Du|^2 + u^2)\, dx;\ u \in H^1(\mathbb{R}_N), b(\infty) \int_{\mathbb{R}_N} |u|^p\, dx = 1\}.$$

Theorem 2.4 *Let $b(x) \geq b(\infty) = \lim_{|x|\to\infty} b(x) > 0$ on $\mathbb{R}_N$. Then problem $(\mathcal{M})$ has a positive solution.*

Proof. We may assume that $b(x) > b(\infty)$ on a set of positive measure in $\mathbb{R}_N$, since otherwise the result follows from Theorem 2.1. Let $\{u_m\}$ be a minimizing sequence for problem $(\mathcal{M})$. It is easy to see that $\{u_m\}$ is bounded in $H^1(\mathbb{R}_N)$ and we may assume that $u_m \rightharpoonup u$ in $H^1(\mathbb{R}_N)$, $u_m \to u$ a.e. on $\mathbb{R}_N$ and $u_m \to u$ in $L^p_{\text{loc}}(\mathbb{R}_N)$ as $m \to \infty$. First, we show that $u \not\equiv 0$. Arguing indirectly, we assume that $u \equiv 0$ on $\mathbb{R}_N$. Let α_∞ and β_∞ be quantities related to $\{u_m\}$ by (a) and (b) of Proposition 1.2. We write for each $R > 0$

$$1 = \int_{|x|<R} b(x)|u_m(x)|^p\, dx + \int_{|x|>R} b(x)|u_m(x)|^p\, dx.$$

Since $u_m \to 0$ in $L^p_{\text{loc}}(\mathbb{R}_N)$, we deduce from this identity, by letting $m \to \infty$ and then $R \to \infty$, that

$$1 = b(\infty)\alpha_\infty. \tag{2.14}$$

Since $\{M_b^{\frac{1}{p-2}} u_m\}$ is a Palais–Smale sequence for the functional G, we have $\langle G'(M_b^{\frac{1}{p-2}} u_m), M_b^{\frac{1}{p-2}} u_m\phi_R\rangle \to 0$ as $m \to \infty$ uniformly in $R \geq 1$. This implies, letting $m \to \infty$ and $R \to \infty$, that

$$M_b^{\frac{2}{p-2}} \beta_\infty = \alpha_\infty b(\infty) M_b^{\frac{p}{p-2}}.$$

By virtue (2.14) we get $\beta_\infty = M_b$. If follows from the definition of M_∞ that

$$M_\infty \left(\int_{\mathbb{R}_N} b(\infty)|u_m\phi_R|^p\, dx\right)^{\frac{2}{p}} \leq \int_{\mathbb{R}_N} (|D(u_m\phi_R)|^2 + (u_m\phi_R)^2)\, dx.$$

From this we deduce that

$$M_\infty (b(\infty)\alpha_\infty)^{\frac{2}{p}} \leq \beta_\infty = M_b,$$

which by (2.14) is equivalent to

$$M_\infty \leq M_b. \tag{2.15}$$

By Theorem 2.1, M_∞ is achieved by a positive function $w \in H^1(\mathbb{R}_N)$. We then have

$$M_b\Big(\int_{\mathbb{R}_N} b(x)|w|^p\,dx\Big)^{\frac{2}{p}} \leq \int_{\mathbb{R}_N} (|Dw|^2 + w^2)\,dx = M_\infty.$$

Since $w > 0$ on $\mathbb{R}_N$, $\int_{\mathbb{R}_N} b(\infty)|w(x)|^p\,dx = 1$ and $b(x) > b(\infty)$ on a set of positive measure on $\mathbb{R}_N$ we get that $M_b < M_\infty$ which contradicts (2.15). Obviously, $M_b^{\frac{1}{p-2}}u$ is a solution of (2.3) and by the maximum principle $u > 0$ on $\mathbb{R}_N$. We now show that $\int_{\mathbb{R}_N} |u_m|^p\,dx \to \int_{\mathbb{R}_N} |u|^p\,dx$ as $m \to \infty$, which implies that $u_m \to u$ in $L^p(\mathbb{R}_N)$. To prove this we must show that $\alpha_\infty = 0$. Suppose that $\alpha_\infty > 0$. As in the previous part of the proof we have that

$$1 = \int_{\mathbb{R}_N} b(x)|u(x)|^p\,dx + b(\infty)\alpha_\infty.$$

Since $u \not\equiv 0$ on $\mathbb{R}_N$, we have that $0 < b(\infty)\alpha_\infty < 1$. Using again the fact that $\langle G'(M_{b,\lambda}^{\frac{1}{p-2}}u_m), M_b^{\frac{1}{p-2}}u_m\phi_R\rangle \to 0$ as $m \to \infty$ uniformly in $R \geq 1$, we see that

$$\beta_\infty M_{b,\lambda}^{\frac{2}{p-2}} \doteq \alpha_\infty b(\infty) M_b^{\frac{p}{p-2}},$$

that is,

$$\beta_\infty = \alpha_\infty b(\infty) M_b. \tag{2.16}$$

On the other hand we have

$$M_b\Big(\int_{\mathbb{R}_N} b(x)|\phi_R u_m|^p\,dx\Big)^{\frac{2}{p}} \leq \int_{\mathbb{R}_N} (|D(u_m\phi_R)|^2 + (u_m\phi_R)^2)\,dx.$$

Letting $m \to \infty$ and then $R \to \infty$ we deduce

$$M_b\big(b(\infty)\alpha_\infty\big)^{\frac{2}{p}} \leq \beta_\infty.$$

This, combined with (2.16), gives $b(\infty)\alpha_\infty \geq 1$ and we have arrived at a contradiction. □

Theorem 2.4 can be reformulated in the following way.

Theorem 2.5 *Suppose that* $\lim_{|x|\to\infty} b(x) = b(\infty) > 0$ *and let* $M_b < M_\infty$*, then problem* $(\mathcal{M})$ *admits a positive solution.*

Obviously, assumption $b(x) \geq b(\infty)$ on $\mathbb{R}_N$ with strict inequality on the set of positive measure in $\mathbb{R}_N$, guarantees $M_b < M_\infty$. A weaker condition ensuring $M_b < M_\infty$ was given by Yi Li [70].

It is known that if $b(x) \leq b(\infty)$ on $\mathbb{R}_N$ with a strict inequality on a set of positive measure, then problem $(\mathcal{M})$ does not have a solution (see [101]). To justify this assertion suppose that problem $(\mathcal{M})$ has a solution $u \not\equiv 0$. Let $\{y_m\} \subset \mathbb{R}_N$ with $\lim_{m\to\infty} |y_m| = \infty$. Then by the Lebesgue dominated convergence theorem we have

$$\begin{aligned}\lim_{m\to\infty} \int_{\mathbb{R}_N} b(x)|u(x+y_m)|^p\,dx &= \lim_{m\to\infty} \int_{\mathbb{R}_N} b(x-y_m)|u(x)|^p\,dx \\ &= \int_{\mathbb{R}_N} b(\infty)|u(x)|^p\,dx > \int_{\mathbb{R}_N} b(x)|u(x)|^p\,dx = 1.\end{aligned}$$

This shows that there exists m such that

$$\int_{\mathbb{R}_N} b(x)|u(x+y_m)|^p\,dx > 1.$$

Hence, we can find $0 < t_m < 1$ such that

$$t_m^p \int_{\mathbb{R}_N} b(x)|u(x+y_m)|^p\,dx = 1,$$

and consequently

$$\begin{aligned}M_b &\leq t_m^2 \int_{\mathbb{R}_N} \left(|Du(x+y_m)|^2 + u(x+y_m)^2\right) dx \\ &= t_m^2 \int_{\mathbb{R}_N} \left(|Du(x)|^2 + u(x)^2\right) dx < M_b\end{aligned}$$

which is impossible. In this case assertion (i) of Theorem 2.3 cannot hold and assertion (ii) explains what happens to a minimizing sequence of problem $(\mathcal{M})$. In this case every weakly convergent minimizing sequence has a weak limit 0 and $\alpha_\infty = \frac{1}{b(\infty)}$.

Remark 2.2 It is worth mentioning that problem $(\mathcal{M})$ on $\mathbb{R}_N$ is also solvable if $b(x)$ is a periodic function on $\mathbb{R}_N$, that is, $b(x) = b(x+y)$ for all $x \in \mathbb{R}_N$ and $y \in \mathbb{Z}_N$. This can be justified as follows. For a given minimizing sequence $\{u_m\}$ we can always find a sequence $\{y_m\} \subset \mathbb{Z}_N$ such that $u_m(\cdot + y_m) \rightharpoonup u \not\equiv 0$ in $H^1(\mathbb{R}_N)$. Since $\{u_m(\cdot + y_m)\}$ is also a minimizing sequence for a problem (M), Lemma 2.2 implies the solvability of problem $(\mathcal{M})$. Obviously, this remark can be extended to periodic domains.

2.5 Solutions belonging to higher energy level

Condition $b(x) \geq b(\infty) = \lim_{|x|\to\infty} b(x) > 0$ on $\mathbb{R}_N$ guarantees the existence of a solution to problem $(\mathcal{M})$. Dao-Min Cao [22] observed that this condition can be relaxed by assuming that $b(x) \geq 2^{\frac{2-p}{2}} b(\infty)$ on $\mathbb{R}_N$. However, in this case a solution belongs to a higher energy level of the functional G. The key ingredient in [22] is the fact that the Palais–Smale condition holds for level c satisfying

$$(\frac{1}{2} - \frac{1}{p}) M_\infty^{\frac{p}{p-2}} < c < (\frac{1}{2} - \frac{1}{p}) (M_b^{\frac{p}{p-2}} + M_\infty^{\frac{p}{p-2}}), \tag{2.17}$$

which is a consequence of Proposition 1.1. In this section we reprove the result from [22] with the use of Proposition 1.2. Another generalization of the condition $b(x) \geq b(\infty)$ on $\mathbb{R}_N$, which gives the solvability of problem $(\mathcal{M})$, was given by Bahri and Yan Yan Li [7].

Proposition 2.4 *Suppose that* $\lim_{|x|\to\infty} b(x) = b(\infty)$. *Let* $\{u_m\} \subset H^1(\mathbb{R}_N)$ *be a sequence of nonnegative functions such that* $G(u_m) \to c$, *with* c *satisfying (2.17), and* $G'(u_m) \to 0$ *in* $H^{-1}(\mathbb{R}_N)$. *Then the sequence is relatively compact in* $H^1(\mathbb{R}_N)$.

Proof. Using the argument from Remark 2.1 we show that there exists a sequence $\{y_m\} \subset \mathbb{Z}_N$ such that $v_m(\cdot) = u_m(\cdot + y_m) \rightharpoonup u \not\equiv 0$ in $H^1(\mathbb{R}_N)$. First, we show that if $\{y_m\}$ is bounded then $\{u_m\}$ must be relatively compact in $H^1(\mathbb{R}_N)$. In the second step of the proof we show that $\{y_m\}$ must be bounded. If $\{y_m\}$ is bounded, we may assume that $u_m \rightharpoonup u$ in $H^1(\mathbb{R}_N)$. It is sufficient to show that $\alpha_\infty = 0$. Arguing indirectly, let us assume that $\alpha_\infty > 0$. We obviously have

$$\begin{aligned} G(u_m) &= (\frac{1}{2} - \frac{1}{p}) \int_{\mathbb{R}_N} (|Du_m|^2 + u_m^2)\, dx + o(1) \\ &= (\frac{1}{2} - \frac{1}{p}) \int_{\mathbb{R}_N} b(x)|u_m|^p\, dx + o(1). \end{aligned}$$

It follows from Proposition 1.2 that

$$(\frac{1}{2}-\frac{1}{p})\Big(\int_{\mathbb{R}_N}(|Du|^2+u^2)\,dx+\beta_\infty\Big) \tag{2.18}$$
$$\leq \quad c=(\frac{1}{2}-\frac{1}{p})\Big(\int_{\mathbb{R}_N} b(x)|u|^p\,dx+b(\infty)\alpha_\infty\Big).$$

Since $\langle G'(u_m), u_m\phi_R\rangle \to 0$ as $m \to \infty$ uniformly in $R \geq 1$ we get that $\beta_\infty=\alpha_\infty b(\infty)$. This implies that

$$\int_{\mathbb{R}_N}(|Du|^2+u^2)\,dx \leq \int_{\mathbb{R}_N} b(x)|u|^p\,dx. \tag{2.19}$$

It follows from the definition of M_∞ that

$$M_\infty\Big(\int_{\mathbb{R}_N} b(\infty)|\phi_R u_m|^p\,dx\Big)^{\frac{2}{p}} \leq \int_{\mathbb{R}_N}(|D(u_m\phi_R)|^2+(u_m\phi_R)^2)\,dx.$$

Letting $m\to\infty$ and then $R\to\infty$ we get

$$M_\infty\Big(b(\infty)\alpha_\infty\Big)^{\frac{2}{p}} \leq \beta_\infty=b(\infty)\alpha_\infty,$$

that is,

$$M_\infty^{\frac{p}{p-2}} \leq b(\infty)\alpha_\infty. \tag{2.20}$$

Also, we have

$$M_b\Big(\int_{\mathbb{R}_N} b(x)|u|^p\,dx\Big)^{\frac{2}{p}} \leq \int_{\mathbb{R}_N}(|Du|^2+u^2)\,dx,$$

which combined with (2.19) gives

$$M_b^{\frac{p}{p-2}} \leq \int_{\mathbb{R}_N} b(x)|u|^p\,dx. \tag{2.21}$$

Inequalities (2.18), (2.20) and (2.21) lead to

$$c \geq (\frac{1}{2}-\frac{1}{p})(M_b^{\frac{p}{p-2}}+M_\infty^{\frac{p}{p-2}})$$

and we have arrived at a contradiction. We now proceed to show that the sequence $\{y_m\}$ must be bounded. Suppose that $\{y_m\}$ is unbounded. We put

$$\alpha_{b,\infty} = \lim_{R\to\infty} \limsup_{m\to\infty} \int_{|x|>R} b(x+y_m)|v_m(x)|^p \, dx.$$

We commence by showing that $\alpha_{b,\infty} = 0$. Suppose that $\alpha_{b,\infty} > 0$. We have

$$\left(\frac{1}{2}-\frac{1}{p}\right)\int_{\mathbb{R}_N}(|Dv_m|^2+v_m^2)\,dx+o(1) = c = \left(\frac{1}{2}-\frac{1}{p}\right)\int_{\mathbb{R}_N} b(x)|u_m|^p\,dx+o(1)$$

and this implies that

$$\begin{aligned}\left(\frac{1}{2}-\frac{1}{p}\right)&\left[\int_{\mathbb{R}_N}(|Du|^2+u^2)\,dx+\beta_\infty\right] \\ \leq \quad c = \left(\frac{1}{2}-\frac{1}{p}\right)&\left[\int_{\mathbb{R}_N} b(\infty)|u|^p\,dx+\alpha_{b,\infty}\right],\end{aligned} \tag{2.22}$$

where

$$\beta_\infty = \lim_{R\to\infty}\limsup_{m\to\infty}\int_{|x|>R}(|Dv_m|^2+v_m^2)\,dx.$$

It is easy to check that u is a solution of the equation

$$-\Delta u + u = b(\infty)|u|^{p-2}u \quad \text{in } \mathbb{R}_N$$

and hence

$$\int_{\mathbb{R}_N}(|Du|^2+u^2)\,dx = b(\infty)\int_{\mathbb{R}_N}|u|^p\,dx. \tag{2.23}$$

This combined with (2.22) gives $\beta_\infty \leq \alpha_{b,\infty}$. It follows from the definition of M_b that

$$M_b\left(\int_{\mathbb{R}_N} b(x+y_m)|\phi_R v_m|^p\,dx\right)^{\frac{2}{p}} \leq \int_{\mathbb{R}_N}(|D(v_m\phi_R)|^2+(v_m\phi_R)^2)\,dx$$

which yields

$$M_b\alpha_{b,\infty}^{\frac{2}{p}} \leq \beta_\infty \leq \alpha_{b,\infty},$$

that is,

$$M_b \leq \left(\alpha_{b,\infty}\right)^{\frac{p-2}{p}}. \tag{2.24}$$

The obvious inequality

$$M_\infty \left(\int_{\mathbb{R}_N} b(\infty)|u|^p \, dx \right)^{\frac{2}{p}} \leq \int_{\mathbb{R}_N} (|Du|^2 + u^2) \, dx$$

combined with (2.23) gives

$$M_\infty^{\frac{p}{p-2}} < \int_{\mathbb{R}_N} b(\infty)|u|^p \, dx.$$

A contradiction is derived from (2.22), (2.24) and the last inequality. So $v_m \to u$ in $H^1(\mathbb{R}_N)$ and $u = \bar{u}$ up to a translation, where $\bar{u}$ is a ground state solution from Theorem 2.2. Hence

$$c = \int_{\mathbb{R}_N} |D\bar{u}|^2 + \bar{u}^2) \, dx = M_\infty^{\frac{p}{p-2}}$$

and we have arrived at a contradiction. □

Inspection of the proof shows that if $G(u_m) \to c \in \left(0, \left(\frac{1}{2} - \frac{1}{p}\right) M_\infty^{\frac{p}{p-2}}\right)$ and $G'(u_m) \to 0$ in $H^{-1}(\mathbb{R}_N)$, then $\{u_m\}$ is relatively compact in $H^1(\mathbb{R}_N)$.

To proceed further we introduce the notation

$$J(u) = \int_{\mathbb{R}_N} (|Du|^2 + u^2) \, dx$$

for $u \in H^1(\mathbb{R}_N)$ and

$$V = \{u \in H^1(\mathbb{R}_N);\ u \geq 0,\ \int_{\mathbb{R}_N} b(x)|u|^p \, dx = 1\}.$$

Obviously, $M_b = \inf_{u \in V} J(u)$. We denote by $C(\overline{B(0,R)}, V)$ the set of continuous functions f from $\overline{B(0,R)}$ into V and put

$$\mathcal{F} = \{f \in C(\overline{B(0,R)}, V);\ f\,|_{\partial B(0,R)} = h_y\},$$

where

$$h_y(x) = \frac{\bar{u}(x-y)}{\left(\int_{\mathbb{R}_N} b(x)|\bar{u}(x-y)|^p \, dx \right)^{\frac{1}{p}}} \quad \text{for } y \in \mathbb{R}_N,$$

with $\bar{u}$ being a positive solution to problem $(\mathcal{M})$ with $b(x) \equiv b(\infty)$ (see Theorem 2.2).

As a consequence of Proposition 2.4 we can state the following min-max principle (see [87]).

Proposition 2.5 *Suppose that $M_b = M_\infty$ and let*

$$c = \inf_{f\in\mathcal{F}} \max_{y\in\overline{B(0,R)}} J(f(y)).$$

If $\max_{y\in\partial B(0,R)} J(h_y) < c \in \left(M_\infty, 2^{\frac{p-2}{p}} M_\infty\right)$, *then* $J\,|_V$ *has a critical point* v *such that* $J(v) = c$.

We now define a functional $\beta : H^1(\mathbb{R}_N) \to \mathbb{R}_N$ by

$$\beta(u) = \int_{\mathbb{R}_N} u(x)^2 \frac{x}{|x|}\, dx$$

and put

$$\bar{V} = \{u \in V;\ \beta(u) = 0\} \ \text{ and } \ \bar{M}_b = \inf_{u\in\bar{V}} J(u).$$

Lemma 2.3 *Suppose that $M_b = M_\infty$ and that problem $(\mathcal{M})$ is not solvable. Then $M_b < \bar{M}_b$ and there exists $R > 0$ such that*

(i) $J(h_y) \in \left(M_\infty, \frac{M_\infty + \bar{M}_b}{2}\right)$ *if* $|y| \geq R$,
(ii) $\left(\beta \circ h_y, y\right) > 0$ *if* $|y| = R$.

Here $(\cdot,\cdot)$ denotes the scalar product in $\mathbb{R}_N$.

Proof. A result of this nature is very well-known in the current literature (see [22], [14], [61]). We only show the first part of our assertion providing the proof based on Proposition 2.4. We always have $M_b \leq \bar{M}_b$. Suppose that $M_b = \bar{M}_b$. Let $\{u_m\} \subset \bar{V}$ be a sequence such that $J(u_m) \to \bar{M}_b$ as $m \to \infty$. Since this sequence is bounded in $H^1(\mathbb{R}_N)$ we may assume that $u_m \rightharpoonup u$ in $H^1(\mathbb{R}_N)$. Since $(\mathcal{M})$ is not solvable, by virtue of Lemma 2.2, $u \equiv 0$. By Theorem 2.3 we may assume that there exists an unbounded sequence $\{y_m\} \subset \mathbb{Z}_N$ such that $u_m(\cdot + y_m) \to \bar{u}$ in $H^1(\mathbb{R}_N)$, where $\bar{u}$ is a positive solution of problem $(\mathcal{M})$ with $b(x) \equiv b(\infty)$ (see Theorem 2.2). We now observe that

$$\begin{aligned}
&\int_{\mathbb{R}_N} [|Du_m(x+y_m) - D\bar{u}(x)|^2 + (u_m(x+y_m) - \bar{u}(x))^2]\, dx \\
&\quad = \int_{\mathbb{R}_N} [|Du_m(x) - D\bar{u}(x-y_m)|^2 + (u_m(x) - \bar{u}(x-y_m))^2]\, dx \\
&\quad \to 0
\end{aligned}$$

as $m \to \infty$. Consequently, we can write $u_m(x) = \bar{u}(x - y_m) + w_m(x)$ with $|y_m| \to \infty$ and $w_m \to 0$ in $H^1(\mathbb{R}_N)$. Having this representation of u_m we can repeat the argument from [14] to show that $\beta(u_m) \neq 0$ for large m, which is impossible. For the proof of (i) and (ii) we refer to [14] or [61]. □

We are now in a position to state the main result of this section.

Theorem 2.6 *Suppose that* $\lim_{|x|\to\infty} b(x) = b(\infty)$ *and that* $b(x) \geq 2^{\frac{2-p}{2}} b(\infty)$ *on* $\mathbb{R}_N$. *Then equation (2.3) has a nontrivial solution.*

Proof. We only give a short outline of the proof and for more details we refer to the paper [22]. If $M_b < M_\infty$, the result follows from Theorem 2.5. Therefore, we can assume that $M_b = M_\infty$. It is enough to consider jointly with this the case, where (M) is not solvable. In this case we have by Lemma 2.3 that $M_b < \bar{M}_b$ and assertions (i) and (ii) of this lemma hold. Using the Brouwer degree, we show as in paper [14] that for each $f \in \mathcal{F}$ there exists $y \in B(0,R)$ such that $\beta(f(y)) = 0$, that is $f(y) \in \bar{V}$. Therefore,

$$c \geq \bar{M}_b > M_\infty.$$

According to our assumption on b we have

$$\begin{aligned}
\max_{y\subset\partial B(0,R)} J(h(y)) &= \max_{y\in\partial B(0,R)} \frac{\int_{\mathbb{R}_N}(|D\bar{u}|^2 + \bar{u}^2)\,dx}{\left(\int_{\mathbb{R}_N} b(x)|u(x-y)|^p\,dx\right)^{\frac{2}{p}}} \\
&< 2^{\frac{p-2}{p}} \frac{\int_{\mathbb{R}_N}(|D\bar{u}|^2 + \bar{u}^2)\,dx}{\left(\int_{\mathbb{R}_N} b(\infty)|\bar{u}|^p\,dx\right)^{\frac{2}{p}}} = 2^{\frac{p-2}{p}} M_\infty,
\end{aligned}$$

which means that

$$M_\infty < \bar{M}_b \leq c < 2^{\frac{p-2}{p}} M_\infty$$

and the result follows from Proposition 2.5. □

Proposition 2.4 and Theorem 2.6 can be extended to large-interior domains.

Proposition 2.6 *Suppose that* $\Omega \subset \mathbb{R}_N$ *is a large-interior domain and let* $\lim_{|x|\to\infty} b(x) = b(\infty) > 0$. *Suppose that* $\{u_m\} \subset H^1_\circ(\Omega)$ *is a sequence of nonnegative functions satisfying*

$$G(u_m) \to c \in \left(\left(\frac{1}{2} - \frac{1}{p}\right) M_\infty^{\frac{p}{p-2}}, \left(\frac{1}{2} - \frac{1}{p}\right)\left(M_\infty^{\frac{p}{p-2}} + M_b^{\frac{p}{p-2}}\right)\right)$$

and $G(u_m) \to 0$ *in* $H^{-1}(\Omega)$ *as* $m \to \infty$. *Then the sequence* $\{u_m\}$ *is relatively compact in* $H_0^1(\Omega)$.

Proof. The proof is similar to that of Proposition 2.4, so we only sketch some details. First we extend each function u_m by 0 outside Ω. For this modified sequence we still have $G'(u_m) \to 0$ in $H^{-1}(\mathbb{R}_N)$ and $G(u_m) \to c$. As in Remark 2.1 we show that there exists a sequence $\{y_m\} \subset \mathbb{Z}_N$ such that $v_m(\cdot) = u_m(\cdot + y_m) \rightharpoonup u \not\equiv 0$ in $H^1(\mathbb{R}_N)$. If $\{y_m\}$ is bounded we may assume that $u_m \rightharpoonup u \not\equiv 0$ in $H^1(\mathbb{R}_N)$ and the rest of the proof is the same as in the corresponding part of the proof of Proposition 2.4. Finally, assuming that $\{y_m\}$ is unbounded we show as in Proposition 2.4 that $v_m \to u$ in $H^1(\mathbb{R}_N)$ and consequently $u \equiv \bar{u}$ up to a translation and we get a contradiction. □

Finally, we consider the extension of Proposition 2.4 to higher energy levels of the functional G. By $G_\infty : H^1(\mathbb{R}_N) \to \mathbb{R}$ we denote a functional given by

$$G_\infty(u) = \frac{1}{2}\int_{\mathbb{R}_N} (|Du|^2 + u^2)\,dx - \frac{b(\infty)}{p}\int_{\mathbb{R}_N} |u|^p\,dx.$$

Proposition 2.7 *Suppose* $\lim_{|x|\to\infty} b(x) = b(\infty)$ *and let* $\{u_m\} \subset H^1(\mathbb{R}_N)$ *be a sequence of nonnegative functions such that* $G'(u_m) \to 0$ *in* $H^{-1}(\mathbb{R}_N)$ *and* $G(u_m) \to c$ *as* $m \to \infty$ *with*

$$c > \left(\frac{1}{2} - \frac{1}{p}\right)\left(M_\infty^{\frac{p}{p-2}} + M_b^{\frac{p}{p-2}}\right) \quad \text{and} \quad c \neq k\left(\frac{1}{2} - \frac{1}{p}\right)M_\infty^{\frac{p}{p-2}}$$

for each integer $k \geq 1$. *Then there exists subsequence of* $\{u_m\}$, *relabelled again by* $\{u_m\}$, *such that* $u_m \rightharpoonup u > 0$ *in* $H^1(\mathbb{R}_N)$ *and* u *is a solution of equation (2.3) with* $M_{b,\lambda}^{\frac{p}{p-2}} < \int_{\mathbb{R}_N} b(x)|u|^p\,dx$.

Proof. Let

$$k\left(\frac{1}{2} - \frac{1}{p}\right)M_\infty^{\frac{p}{p-2}} < c < (k+1)\left(\frac{1}{2} - \frac{1}{p}\right)M_\infty^{\frac{p}{p-2}}$$

for some integer $k \geq 1$. Since $\{u_m\}$ is bounded in $H^1(\mathbb{R}_N)$ we may assume that $u_m \rightharpoonup v$ in $H^1(\mathbb{R}_N)$. If $v \not\equiv 0$ we are done, so let $v \equiv 0$. By virtue of Remark 2.1 there exists a sequence $\{y_m\} \subset \mathbb{R}_N$ such that $u_m(\cdot + y_m) \rightharpoonup u > 0$ in $H^1(\mathbb{R}_N)$. We commence by showing that $\{y_m\}$ is a bounded sequence. Arguing indirectly we assume that $\{y_m\}$ is unbounded. Obviously, u is a ground state solution to problem $(\mathcal{M})$ with $b(x) \equiv b(\infty)$ on $\mathbb{R}_N$ and by Proposition 1.2 we have

$$c = \left(\frac{1}{2} - \frac{1}{p}\right)\left(b(\infty)\int_{\mathbb{R}_N} |u|^p\, dx + \alpha_{b,\infty}\right),$$

where

$$\alpha_{b,\infty} = \lim_{R\to\infty} \limsup_{m\to\infty} \int_{|x|>R} b(x+y_m)|u_m(x+y_m)|^p\, dx$$

and

$$M_\infty^{\frac{p}{p-2}} = \int_{\mathbb{R}_N} b(\infty)|u|^p\, dx.$$

We must have that $\alpha_{b,\infty} > 0$ since otherwise $c = \left(\frac{1}{2} - \frac{1}{p}\right)M_\infty^{\frac{p}{p-2}}$, which is impossible. We put

$$v_m^1(x) = v_m(x) - u(x) \text{ and } v_m(x) = u_m(x+y_m).$$

The weak convergence of v_m to u implies that

$$\int_{\mathbb{R}_N} (|Dv_m^1|^2 + (v_m^1)^2)\, dx = \int_{\mathbb{R}_N} (|Dv_m|^2 + v_m^2)\, dx - \int_{\mathbb{R}_N} (|Du|^2 + u^2)\, dx + o(1).$$

On the other hand, by Lemma 1.3 we have

$$\int_{\mathbb{R}_N} b(x)|v_m^1(x)|^p\, dx = \int_{\mathbb{R}_N} b(x)|v_m(x)|^p\, dx - \int_{\mathbb{R}_N} b(x)|u(x)|^p\, dx + o(1)$$

as $m \to \infty$ and hence

$$G(v_m^1) = G(v_m) - G(u) + o(1)$$

as $m \to \infty$. Since $v_m \to u$ in $L^p_{\text{loc}}(\mathbb{R}_N)$ and $\lim_{|x|\to\infty} b(x) = b(\infty)$, we deduce from this relation that

$$G(v_m^1) = G_\infty(v_m) - G_\infty(u) + o(1)$$

as $m \to \infty$. Since $u_m \to 0$ in $L^p_{\text{loc}}(\mathbb{R}_N)$, the last relation can be written as

$$G(v_m^1) = G(u_m) - G_\infty(u) + o(1) = c - \left(\frac{1}{2} - \frac{1}{p}\right)M_\infty^{\frac{p}{p-2}}$$

as $m \to \infty$. This equality implies that $v_m^1 \nrightarrow 0$ in $H^1(\mathbb{R}_N)$ and consequently $\alpha_\infty^1 = \lim_{R\to\infty} \limsup_{m\to\infty} \int_{|x|>R} |v_m^1(x)|^p\, dx > 0$. In a similar manner we show that

$$G'(v_m^1) = G'(u_m) - G'_\infty(u) + o(1) = o(1)$$

as $m \to \infty$. It then follows from Proposition 1.2

$$(\frac{1}{2} - \frac{1}{p})\alpha_{b,\infty} = c - (\frac{1}{2} - \frac{1}{p}) M_\infty^{\frac{p}{p-2}} = (\frac{1}{2} - \frac{1}{p}) M_\infty^{\frac{p}{p-2}} + (\frac{1}{2} - \frac{1}{p})\alpha_\infty^1 b(\infty).$$

We now apply again Remark 2.1 to obtain a sequence $\{y_m\} \subset \mathbb{R}_N$ such that

$$v_m^1(\cdot + y_m^1) \rightharpoonup u^1 > 0 \text{ in } H^1(\mathbb{R}_N).$$

The sequence $\{y_m^1\}$ must unbounded, since otherwise $v_m^1(\cdot) \rightharpoonup u^1 > 0$ in $H^1(\mathbb{R}_N)$, which is impossible. Obviously, we have $u \equiv u^1$ up to a translation and

$$M_\infty^{\frac{p}{p-p}} = \int_{\mathbb{R}_N} b(\infty)|u^1|^p\, dx.$$

Letting $v_m^2(\cdot) = v_m^1(\cdot + y_m^1) - u^1(\cdot)$, we show as before that

$$G(v_m^2) = G_\infty(v_m^1) - G_\infty(u^1) + o(1) = G(v_m^1) - G_\infty(u^1) + o(1)$$

and

$$G'(v_m^2) = G'(v_m^1) - G'_\infty(u^1) + o(1) = o(1)$$

as m$\to \infty$. This implies that $v_m^2 \nrightarrow 0$ in $H^1(\mathbb{R}_N)$ and

$$\alpha_\infty^2 = \lim_{R\to\infty} \limsup_{m\to\infty} \int_{|x|>R} |v_m^2(x)|^p\, dx > 0$$

provided $c - 2(\frac{1}{2} - \frac{1}{p}) M_\infty^{\frac{p}{p-2}} > 0$. Hence by Proposition 1.2 we have

$$b(\infty)\alpha_\infty^1 = M_\infty^{\frac{p}{p-2}} + \alpha_\infty^2 b(\infty).$$

Continuing this process we obtain sequences $\{y_m^j\} \subset \mathbb{Z}_N$ and $\{v_m^j\} \subset H^1(\mathbb{R}_N)$, such that $v_m^j(\cdot + y_m^j) \rightharpoonup u^j$ in $H^1(\mathbb{R}_N)$ and $u^j = u$ up to a translation and $b(\infty)\alpha_\infty^{j-1} = M_\infty^{\frac{p}{p-2}} + b(\infty)\alpha_\infty^j$, where

$$v_m^j(\cdot) = v_m^{j-1}(\cdot + y_m^{j-1}) - u^{j-1}(\cdot) \text{ and } \alpha_\infty^j = \lim_{R\to\infty} \limsup_{m\to\infty} \int_{|x|>R} |v_m^j|^p \, dx.$$

The construction of these sequences must terminate at $j = k$, otherwise $v_m^k(\cdot + y_m^k) \rightharpoonup u^k > 0$ with $u^k = u$ up to a translation and with $\{y_m^k\}$ being unbounded. Letting $v_m^{k+1}(\cdot) = v_m^k(\cdot + y_m^k) - u^k(\cdot)$, we get

$$G(v_m^{k+1}) = G(v_m^k) - G_\infty^k(u^k) < 0$$

which is impossible. This means that c is an integer multiple of $\left(\frac{1}{2} - \frac{1}{p}\right) M_\infty^{\frac{p}{p-2}}$ which contradicts our assumption on c. Therefore, $\{y_m\}$ is bounded and $u_m \rightharpoonup u > 0$ in $H^1(\mathbb{R}_N)$ and this completes the proof. □

2.6 Large-interior domains

According to Proposition 2.1 a solution of equation (2.3) on a large-interior domain cannot be obtained as a minimizer of problem $(\mathcal{M})$. The aim of this section is to show that a solution of equation (2.3) can be found in a higher energy level of the functional G. This will be established by a suitable modification of the method from the Section 2.5. However this approach requires a further restriction on a domain Ω.

Throughout this section we assume that Ω is a large-interior domain satisfying the following condition:

(A) $\Omega = \mathbb{R}_N - \omega$ and there exist a domain $\tilde{\omega}$ of finite Lebesgue measure with $\omega \subset \tilde{\omega}$ and a function $\phi \in H_\circ^1(\tilde{\omega})$ such that $\phi(x) = 1$ on ω.

Later we shall give an example of a domain Ω having property (A). As in Section 2.5 we consider the functional J on a set V. Let ϕ be a function postulated by assumption (A) and we set

$$\psi(x) = \begin{cases} 1 - \phi(x) & \text{if } x \in \tilde{\omega}, \\ 1 & \text{if } x \in \mathbb{R}_N - \tilde{\omega}. \end{cases}$$

For $\epsilon > 0$ we define a mapping $h_\epsilon : B(0, R) \to V$ by

$$h_\epsilon(y)(x) = \frac{\psi\left(\frac{x}{\epsilon}\right)\bar{u}(x - y)}{\|\psi\left(\frac{\cdot}{\epsilon}\right)\bar{u}(\cdot - y)\|_p}$$

for $y \in B(0, R)$, where $\bar{u}$ is a minimizer of $M = M(\mathbb{R}_N)$. Let

$$J(u) = \int_\Omega (|\nabla u|^2 + u^2)\,dx \text{ for } u \in H_0^1(\Omega)$$

and set

$$V = \{u \in H_0^1(\Omega);\ u \geq 0,\ \int_\Omega |u|^p\,dx = 1\}.$$

We commence with some technical lemmas.

Lemma 2.4 *Let $y_R \in \partial B(0, R)$, then*

$$\lim_{R\to\infty} J(h_\epsilon(y_R)) = M.$$

Proof. If $y_R \in \partial B(0, R)$, then $\bar{u}(\epsilon x - y_R)^p \to 0$ as $R \to \infty$ for all $x \in \tilde{\omega}$ and

$$|\psi(x) - 1|^p \bar{u}(\epsilon x - y_R)^p \leq C\bar{u}(0)\big(|\psi(x)| + 1\big)^p \in L^1(\tilde{\omega}).$$

Hence by the Lebesgue dominated convergence theorem

$$\begin{aligned}
\int_{\mathbb{R}_N} |\psi(\frac{x}{\epsilon})\bar{u}(x - y_R) &- \bar{u}(x - y_R)|^p\,dx = \int_{\epsilon\tilde{\omega}} |\psi(\frac{x}{\epsilon}) - 1|^p \bar{u}(x - y_R)^p\,dx \\
&= \epsilon^N \int_{\tilde{\omega}} |\psi(x) - 1|^p \bar{u}(\epsilon x - y_R)^p\,dx = o(1)
\end{aligned}$$

as $R \to \infty$. This obviously yields

$$\int_\Omega |\psi(\frac{x}{\epsilon})\bar{u}(x - y_R)|^p\,dx = \int_{\mathbb{R}_N} \bar{u}(x - y_R)^p\,dx + o(1) = 1 + o(1)$$

as $R \to \infty$. We now estimate

$$\begin{aligned}
&\int_{\mathbb{R}_N} |\nabla\big(\psi(\frac{x}{\epsilon})\bar{u}(x - y_R) - \bar{u}(x - y_R)\big)|^2\,dx \\
&+ \int_{\mathbb{R}_N} |\psi(\frac{x}{\epsilon})\bar{u}(x - y_R) - \bar{u}(x - y_R)|^2\,dx \\
&\leq C\int_{\epsilon\tilde{\omega} - \epsilon\omega} |\nabla\psi(\frac{x}{\epsilon})|^2 \bar{u}(x - y_R)^2\,dx \\
&+ C\int_{\epsilon\tilde{\omega} - \epsilon\omega} |\psi(\frac{x}{\epsilon}) - 1|^2\big(|\nabla\bar{u}(x - y_R)|^2 \\
&+ \bar{u}(x - y_R)^2\big)\,dx = I_1 + I_2.
\end{aligned}$$

For I_1 we have

$$\begin{aligned} I_1 &= \frac{1}{\epsilon^2}\int_{\epsilon\tilde{\omega}-\epsilon\omega} |\nabla\psi(\frac{x}{\epsilon})|^2\tilde{u}(x-y_R)^2\,dx \\ &= \epsilon^{N-2}\int_{\tilde{\omega}-\omega} |\nabla\psi(x)|^2\bar{u}(\epsilon x-y_R)^2\,dx \\ &= o(1) \end{aligned}$$

as $R\to\infty$. In a similar way we show $I_2=o(1)$ as $R\to\infty$. Hence

$$\begin{aligned} &\int_\Omega |\nabla(\psi(\frac{x}{\epsilon})\bar{u}(x-y_R))|^2\,dx + \int_{\mathbb{R}_N} |\psi(x)\bar{u}(x-y_R)|^2\,dx \qquad (2.25) \\ &= \int_{\mathbb{R}_N} \left(|\nabla\bar{u}(x-y_R)|^2 + |\bar{u}(x-y_R)|^2\right)dx + o(1) = M + o(1). \end{aligned}$$

and the result follows from (2.25) and Theorem 2.1. □

Lemma 2.5 *We have*

$$\lim_{\epsilon\to 0} J(h_\epsilon(y)) = M$$

uniformly in $y\in\overline{B(0,R)}$.

Proof. First we show that

$$\lim_{\epsilon\to 0}\int_\Omega |\psi(\frac{x}{\epsilon})\bar{u}(x-y_R)|^p\,dx = 1$$

uniformly in y. Indeed, this follows from the following estimate

$$\begin{aligned} \int_{\mathbb{R}_N} |\psi(\frac{x}{\epsilon})\bar{u}(x-y)-\bar{u}(x-y)|^p\,dx &= \int_{\epsilon\tilde{\omega}} |\psi(\frac{x}{\epsilon})-1|^p\bar{u}(x-y)^p\,dx \\ \le \bar{u}(0)^p\int_{\epsilon\tilde{\omega}} |\psi(\frac{x}{\epsilon})-1|^p\,dx &= \bar{u}(0)^p\epsilon^N\int_{\tilde{\omega}} |\psi(x)-1|^p\,dx = o(1) \end{aligned}$$

uniformly in $y\in\mathbb{R}_N$. Similarly we have

$$\begin{aligned} &\int_{\mathbb{R}_N} |\nabla(\psi(\frac{x}{\epsilon})\bar{u}(x-y)-\bar{u}(x-y))|^2\,dx \\ &+\int_{\mathbb{R}_N} |\psi(\frac{x}{\epsilon})\bar{u}(x-y)-\bar{u}(x-y)|^2\,dx \\ &=\int_{\epsilon\tilde{\omega}-\epsilon\omega} |\nabla\psi(\frac{x}{\epsilon})|^2\bar{u}(x-y)^2\,dx \end{aligned}$$

$$+\int_{\epsilon\tilde{\omega}-\epsilon\omega} |\psi(\frac{x}{\epsilon}) - 1|^2 (|\nabla \bar{u}(x-y)|^2 + \bar{u}(x-y)^2)\, dx$$
$$= \epsilon^{N-2} \int_{\tilde{\omega}-\omega} |\nabla\psi(x)|^2 \bar{u}(\epsilon x - y)^2\, dx + o(1) = o(1)$$

as $\epsilon \to 0$ uniformly in y and this yields

$$\int_\Omega |\nabla (\psi(\frac{x}{\epsilon})\bar{u}(x-y))|^2\, dx + \int_\Omega |\psi(\frac{x}{\epsilon})\bar{u}(x-y)|^2\, dx$$
$$= \int_\Omega (|\nabla\bar{u}(x-y)|^2 + \bar{u}(x-y)^2)\, dx + o(1) = M + o(1)$$

as $\epsilon \to 0$ uniformly in y and the assertion easily follows. □

As in Section 2.5 we introduce a class of functions $\mathcal{F}$ defined by

$$\mathcal{F} = \{f \in C(\overline{B(0,R)}, V);\ f\ |_{\partial B(0,R)} = h_\epsilon\}$$

and set

$$c = \inf_{f\in\mathcal{F}} \max_{y\in B(0,R)} J(f(y)).$$

By virtue of Lemma 2.5 there exists $\epsilon_o > 0$ such that

$$c < 2^{\frac{p-2}{p}} M \text{ for } \epsilon < \epsilon_o. \tag{2.26}$$

In what follows we fix $\epsilon < \epsilon_o$. Let $\beta : V \to \mathbb{R}_N$ be defined by

$$\beta(u) = \int_{\mathbb{R}_N} |u(x)|^p \frac{x}{|x|}\, dx.$$

It is clear that the functional β is continuous on V.

Lemma 2.6 *For every $f \in \mathcal{F}$ the equation $\beta(f(x)) = 0$ admits at least one solution $z \in \overline{B(0,R)}$ for R sufficiently large.*

Proof. We give here an outline of the proof. Let $H : \overline{B(0,R)} \to \mathbb{R}_N$ be a mapping defined by

$$H(y) = \beta(f(y)).$$

The mapping H is homotopic to the identity mapping I, that is,

$$G(t,\cdot) = tH(\cdot) + (1-t)I, \quad 0 \le t \le 1.$$

One can show that $G(t,x) \neq 0$ for $0 \leq t \leq 1$ and $x \in \partial B(0,R)$ provided R is sufficiently large. By the homotopic invariance of the Brouwer degree, we have

$$\deg(H, \overline{B(0,R)}, 0) = \deg(I, \overline{B(0,R)}, 0) = 1.$$

Therefore there exists $z \in B(0,R)$ such that $\beta(f(z)) = 0$ if $R > 0$ is sufficiently large. □

Lemma 2.7 *We have* $\inf_{u\in V} J(u) > M$.

Proof. Arguing by contradiction, there exists a sequence $\{u_m\} \subset V$ such that $\lim_{m\to\infty} J(u_m) = M$. Since $M(\Omega)$ is not achieved and $M = M(\Omega)$, the sequence $\{u_m\}$ is not relatively compact in $H_0^1(\Omega)$. Letting $v_m = M^{\frac{1}{p-2}} u_m$ we see that

$$G(v_m) = \Big(\frac{1}{2} - \frac{1}{p}\Big) M^{\frac{p}{p-2}} + o(1) \ \text{ and } \ G'(v_m) = o(1)$$

in $H^{-1}(\Omega)$ as $m \to \infty$. By Corollary 2.2 there exist a sequence $\{x_m\} \subset \mathbb{R}_N$ such that

$$v_m(x) = M^{\frac{1}{p-2}} \bar{u}(x - x_m) + w_m(x),$$

where $w_m \to 0$ in $H^1(\mathbb{R}_N)$ and $|x_m| \to \infty$ and $\bar{u}$ is a minimizer for $M(\mathbb{R}_N)$. By Lemma 1.3 we have

$$|\beta\big(M^{\frac{1}{p-2}} \bar{u}(\cdot - x_m) + w_m\big) - M^{\frac{p}{p-2}} \int_{\mathbb{R}_N} \bar{u}(x - x_m)^p \frac{x}{|x|}\, dx| = o(1)$$

as $m \to \infty$. On the other hand

$$\beta\big(M^{\frac{1}{p-2}} \bar{u}(\cdot - x_m) + w_m\big) = M^{\frac{p}{p-2}} \int_{\mathbb{R}_N} \bar{u}(x - x_m)^p \frac{x}{|x|}\, dx + o(1) = \frac{x_m}{|x_m|} + o(1),$$

which is impossible, because $\beta(u_m) = 0$. □

Corollary 2.4 *We have*

$$\max_{y\in\overline{B(0,R)}} J(f(y)) \geq \inf_{u\in V} J(u) > M(\mathbb{R}_N)$$

for every $f \in \mathcal{F}$ *and* R *sufficiently large.*

We are now in a position to formulate the existence result for equation (2.3).

Theorem 2.7 *If Ω is a large-interior domain satisfying (A) then equation (2.3) amits a nontrivial solution.*

Proof. By (2.27) and Corollary 2.4 the $(PS)_c$ condition holds for level c. To apply the min-max principle (see Proposition 2.5 we must show that

$$\max_{y\in \overline{B(0,R)}} J(f(y)) > \max_{y\in\partial B(0,R)} J(f(y)) = \max_{y\in\partial B(0,R)} J(h_\epsilon(y)).$$

This follows from Lemma 2.4 and Corollary 2.4 if we choose R sufficiently large. □

We now give an example of a large interior domain Ω satisfying condition (A). For this we need notion of the capacity of a given set.

Let $D \subset \mathbb{R}_N$ be a domain contained in the strip $E = \{x = (x_1, \ldots, x_N) \in \mathbb{R}_N; a < x_N < b\}$. If G is a bounded smooth domain such that $\bar{G} \subset D$, then the *capacity* of G with respect to D ($\mathrm{Cap}_D G$) is defined by

$$\mathrm{Cap}_D G = \inf\{\int_D |\nabla u|^2\, dx;\ u \in H_\circ^1(D) \text{ and } u(x) \equiv 1 \text{ a.e. on } G\}. \quad (2.27)$$

In this definition we do not assume that D is bounded which is a usual requirement in the definition of the capacity (see [67], p.44,45). Since D is contained in the strip, it is known that the Poincaré inequality holds in D, that is,

$$\|u\|_{H_\circ^1} = \left(\int_D |\nabla u|^2\, dx\right)^{\frac{1}{2}}$$

is a norm in $H_\circ^1(D)$. Using this fact it is not difficult to show that the infimum in the above definition is attained by a function $u \in H_\circ^1(D)$.

We now formulate necessary and sufficient condition for the existence of a minimizer in (2.28) in the case where G is unbounded subset of D such that $\bar{G} \subset D$.

Lemma 2.8 *Let G and D be two smooth domains contained in the strip E with $\bar{G} \subset D$. If there exists a constant $K > 0$ such*

$$\mathrm{Cap}_D\big(G \cap B(0,n)\big) \le K \text{ for all } n \ge 1, \quad (2.28)$$

then there exists a function $\phi \in H_\circ^1(D)$ such that $\phi \equiv 1$ a.e. on G.

Proof. If (2.29) holds then there exists a sequence $\{\phi_n\} \subset H_0^1(D)$ such that

$$\int_D |\nabla \phi_n|^2\, dx = \mathrm{Cap}_D\big(G \cap B(0,n)\big) \leq K$$

and $\phi_n(x) = 1$ a.e. on $G \cap B(0,n)$. Taking a subsequence if necessary we may assume that there exists a $\phi \in H_0^1(D)$ such that $\phi_n \rightharpoonup \phi$ in $H_0^1(D)$. Since $\phi_n = 1$ a.e. on $G \cap B(0,n)$ we see that $\phi = 1$ a.e. on G. □

Obviously, if there exists $\phi \in H_0^1(D)$ with $\phi \equiv 1$ a.e. on G, then $\mathrm{Cap}_D\big(G \cap B(0,n)\big) \leq \int_D |\nabla\phi|^2\, dx$ for every $n \geq 1$, which means that (2.28) is a necessary and sufficient condition for the existence of a minimizer in (2.27). This remark will be used to give an example of domains G and D satisfying (2.28). Let C be domain defined by

$$C = \{x = (x_1, x') \in \mathbb{R}_N;\ \rho = |x'| < \frac{1}{1 + |x_1|^\alpha} \text{ for } \alpha > \frac{1}{N-3} \text{ and } N \geq 4\}.$$

We check that C satisfies (2.29) with $G = C$ and $D = 2C$. To achieve this we consider a function ϕ given by

$$\phi(x) = \begin{cases} 1 & \text{if } x \in C, \\ 0 & \text{if } \in x \notin 2C, \\ 2^\alpha \frac{\left(1+\frac{|x_1|^\alpha}{2^\alpha}\right)(1+|x_1|^\alpha)|x'|}{(1-2^{\alpha+1})|x_1|^\alpha - 2^\alpha} - 2^{\alpha+1}\frac{1+|x_1|^\alpha}{(1-2^{\alpha+1})|x_1|^\alpha - 2^\alpha} & \text{if } x \in 2C - C. \end{cases}$$

By straightforward computations we check that

$$\left(\frac{\partial\phi}{\partial x_1}\right)^2 \leqq \gamma |x_1|^{2\alpha-2}|x'|^2 + \frac{1}{|x_1|^{2\alpha}} \quad \text{for } x_1 > \tau$$

and

$$\left(\frac{\partial\phi}{\partial x_i}\right)^2 \leqq \gamma \frac{(1+|x_1|^\alpha)^2 x_i^2}{|x'|^2} \quad \text{if } i \geqq 2 \text{ and } x_1 > \tau,$$

where $\gamma > 0$ and $\tau > 0$ are constants. Then

$$\begin{aligned} \int_{2C} |\nabla\phi|^2\, dx &\leqq \gamma \int_{2C-C} \left[|x_1|^{2\alpha-2}|x'|^2 + \frac{1}{|x_1|^{2\alpha}} + (1+|x_1|^\alpha)^2\right] dx \leqq \gamma \\ &+ \gamma \int_{-\infty}^{\infty} \int_{(1+|x_1|^\alpha)^{-1}}^{2\left(1+\frac{|x_1|^\alpha}{2^\alpha}\right)^{-1}} \left(|x_1|^{2\alpha-2}\rho^2 + (1+|x_1|^\alpha)^2\right)\rho^{N-2}\, d\rho\, dx_1 \\ &\leq \gamma + \gamma \int_{-\infty}^{\infty} \frac{|x_1|^{2\alpha-2}}{(1+|x_1|^\alpha)^{N+1}}\, dx_1 + \gamma \int_{-\infty}^{\infty} \frac{dx_1}{(1+|x_1|^\alpha)^{N-3}} < \infty, \end{aligned}$$

if $\alpha > \frac{1}{N-3}$. Since $\phi \in H_0^1(D)$ and $\phi = 1$ a.e. on C, condition (2.27) is satisfied.

2.7 Perturbation of large interior domains

In this section we introduce a class of perturbated large interior domains for which the problem $(\mathcal{M})$ is solvable. However the solvability of this constrained minimization problem will be achieved by restricting this problem to a subspace of functions satisfying some symmetry conditions.

Let $N \geq 3$. We denote by $x = (x', x_N)$ a generic point of $\mathbb{R}_N$, with $x' \in \mathbb{R}_{N-1}$ and $x_N \in \mathbb{R}$. Let $\mathcal{A} = \{x; (x', x_N) \in \mathbb{R}_N, x' = 0\}$ be the x_N-axis.

A set Ω' is said to be symmetric with respect to $\mathcal{A}$ (or *axially symmetric)* if it has the following property: $(x', x_N) \in \Omega'$ if and only if $(|x'|, 0, \ldots, 0, x_N) \in \Omega'$.

By C we always denote a ball $B_{N-1}(r) = B_{N-1}(0, r)$ in $\mathbb{R}_{N-1}$ or an annulus $B_{N-1}(r) - B_{N-1}(s)$, $r > s$. We set $C_{b,\infty} = C \times (b, \infty)$ and $C_{-\infty,b} = C \times (-\infty, b)$ for $b \in \mathbb{R}$.

Throughout this section we always assume that Ω' is an axially symmetric large interior domain in $\mathbb{R}_N$ satisfying the following condition:

(A) There exist constants a and a' such that

$$\bar{C}_{-\infty,a} \subset \mathbb{R}_N - \Omega', \quad \bar{C}_{a',\infty} \subset \Omega'.$$

Let D be another axially symmetric domain satisfying

$$\bar{C}_{b,\infty} \subset \bar{D} \subset \Omega'$$

for some $b \in \mathbb{R}$. For $q \in \mathbb{R}$ we consider a domain $\Omega_q = \Omega' - (D + qe_N)$, where $e_N = (0, \ldots, 0, 1)$.

Given an axially symmetric domain E we denote by $H_s^1(E)$ a subspace of $H_0^1(E)$ consisting of axially symmetric functions.

We now consider the following minimization problem

$$(\mathcal{I}_q) \quad I_q = \inf\{\int_{\Omega_q} (|Du|^2 + u^2)\, dx;\ u \in H_s^1(\Omega_q),\ \int_{\Omega_q} |u|^p\, dx = 1\},$$

where $2 < p < 2^*$. We aim to show that problem $(\mathcal{I}_q)$ has a solution for a suitably chosen q. A starting point is to compare I_q with the energy level

$M^s(H)$, where $H = \mathbb{R}_N - \bar{C}_{-\infty,\infty}$, defined by

$$M^s(H) = \inf\{\int_H (|Du|^2 + u^2)\,dx;\ u \in H_s^1(H),\ \int_H |u|^p\,dx = 1\}.$$

We commence by showing that this problem has a solution.

Proposition 2.8 *There exists a function $v \in H_s^1(H)$ such that $\int_H |v|^p\,dx = 1$ and*

$$M^s(H) = \int_H (|Dv|^2 + v^2)\,dx.$$

Proof. The proof is similar to that of Theorem 2.1 (see the approach described in Remark 2.1. We introduce the collection of sets $A_k = \big(\mathbb{R}_{N-1} - B_{N-1}(r)\big) \times [-k,k)$, $k \in \mathbb{Z}$ (or $A_k = \big(\mathbb{R}_{N-1} - (B_{N-1}(r) - B_{N-1}(s))\big) \times [-k,k)$). Then $H = \bigcup_{k\in\mathbb{Z}} A_k$. Let $\{v_m\}$ be a minimizing sequence for $M^s(H)$. Since $\{v_m\}$ is bounded in $H_s^1(H)$ we may assume that $v_m \rightharpoonup v$ in $H_s^1(H)$. By Theorem 2 in [55] we can also assume that $v_m \to v$ in $L^p(A_1)$. Using the Sobolev inequality, we show as in Remark 2.1 that $\|v_m\|_{L^p(A_k)} \geq \delta$ for some constant $\delta > 0$ independent of k and m. This implies that $v_m \rightharpoonup v \not\equiv 0$ in $H^1(H)$. Repeating the argument of Theorem 2.1 we prove that v is a minimizer for $M^s(H)$. □

Let M be the energy level of problem $(\mathcal{M})$ with $\Omega = \mathbb{R}_N$ and $b \equiv 1$ (see Section 2.1). As an immediate consequence of Proposition 2.8 we obtain that

$$M < M^s(H). \tag{2.29}$$

This inequality is crucial in establishing the existence of a minimizer for I_q.

Theorem 2.8 *There exists a constant $q_\circ$ such that I_q, $q \geq q_\circ$, has a minimizer.*

Proof. First we show that there exists a constant $q_\circ$ such that

$$I_q < M^s(H) \tag{2.30}$$

for all $q \geq q_\circ$. Since Ω' is a large interior domain we can find a sequence $\{\phi_n\} \subset H_s^1(\Omega')$ with compact supports, $\|\phi_n\|_{L^p} = 1$ and $\int_{\mathbb{R}_N}(|D\phi_n|^2 + \phi_n^2)\,dx \to M$. By virtue of (2.29) there exists an integer $\bar{n} > 0$ such that

$$\int_{\mathbb{R}_N} (|D\phi_{\bar{n}}|^2 + \phi_{\bar{n}}^2)\,dx < M^s(H).$$

Let $K_{\bar{n}} = \text{supp } \phi_{\bar{n}}$. We then choose q_o such that $(\bar{D}+q_o e_N)\cap K_{\bar{n}} = \emptyset$. With this choice of q_o we see that $\phi_{\bar{n}} \in H_s^1(\Omega_q)$ for $q \geq q_o$. This shows that (2.30) holds. We now fix $q \geq q_o$. Let $\{u_m\}$ be a minimizing sequence for problem $(\mathcal{I}_q$. Since $\{u_m\}$ is bounded in $H_s^1(\Omega_q)$, we can assume that $u_m \rightharpoonup u$ in $H_s^1(\Omega_q)$. Let α_∞ and β_∞ be quantities associated with $\{u_m\}$ defined in Proposition 1.2. We now show that $u \not\equiv 0$. Suppose to the contrary that $u \equiv 0$. Then $\alpha_\infty = 1$. As in the proof of Theorem 2.1 we show that $\beta_\infty = I_q$. Let $\{\phi_R\}$ be a family of functions defined in Proposition 1.2. Then $u_m\phi_R \in H_s^1(H)$ for R sufficiently large. On the other hand by the definition of $M^s(H)$ we see that

$$M^s(H)\left(\int_{\mathbb{R}_N} |\phi_R u_m|^p \, dx\right)^{\frac{2}{p}} \leq \int_{\mathbb{R}_N} \left(|D(\phi_R u_m)|^2 + (\phi_R u_m)^2\right) dx.$$

Letting $m \to \infty$ and then $R \to \infty$ we deduce from this inequality that $M^s(H) \leq I_q$ which is impossible. □

2.8 Role of symmetry

If $b(x)$ is a positive, bounded, continuous and radially symmetric function, then due to the compact embedding of $H_r^1(\mathbb{R}_N)$ into $L^p(\mathbb{R}_N)$, $2 < p < 2^*$, problem $(\mathcal{M})$ is solvable and consequently equation (2.3) has a nontrivial solution. This suggests that if $b(x)$ has some form of symmetry then equation (2.3) has solution in the case where the condition $b(x) \geq b(\infty) = \lim_{|x|\to\infty} b(x)$ on $\mathbb{R}_N$ is not satisfied. In this section we show how this condition can be weakened if $b(x)$ has some kind of symmetry.

Let $O(N)$ be a group of all *orthogonal transformations* on $\mathbb{R}_N$ and let $G \subset O(N)$ be a subgroup. For $x \neq 0$ we denote the cardinality of $G_x = \{gx,\, g \in G\}$ by $|G_x|$ and set $|G| = \inf_{x\in\mathbb{R}_N-\{0\}} |G_x|$. We say that a function $f : \mathbb{R}_N \to \mathbb{R}$ is *G-symmetric* if $f(gx) = f(x)$ for all $g \in G$ and $x \in \mathbb{R}_N$ and in the context of Sobolev spaces this equality is understood a.e. on $\mathbb{R}_N$. For example, if f is an even function, that is, $f(x) = f(-x)$ for all $x \in \mathbb{R}_N$, then taking $G = \{-\text{id}, \text{id}\}$ we see that $|G| = 2$. We denote by $H_G^1(\mathbb{R}_N)$ the subspace of $H^1(\mathbb{R}_N)$ consisting of all G-symmetric functions and its dual space by $H_G^{-1}(\mathbb{R}_N)$. To find a nontrivial solution we consider a functional $F : H_G^1(\mathbb{R}_N) \to \mathbb{R}$ given by

$$G(u) = \frac{1}{2}\int_{\mathbb{R}_N} (|\nabla u|^2 + u^2)\, dx - \frac{1}{p}\int_{\mathbb{R}_N} b(x)|u|^p \, dx.$$

Lemma 2.9 below shows that critical points of G are solutions of equation (2.3).

Lemma 2.9 *Let $b(x)$ be a G-symmetric bounded and continuous function on $\mathbb{R}_N$. Let $\{u_m\} \subset H^1_G(\mathbb{R}_N)$ be such that $G'(u_m) \to 0$ in $H^{-1}_G(\mathbb{R}_N)$, then $G'(u_m) \to 0$ in $H^{-1}(\mathbb{R}_N)$. Moreover, $G'(u) = 0$ in $H^1_G(\mathbb{R}_N)$ implies $G'(u) = 0$ in $H^1(\mathbb{R}_N)$.*

Proof. Since $H^1_G(\mathbb{R}_N)$ is a closed subspace of a Hilbert space $H^1(\mathbb{R}_N)$, we can write

$$H^1(\mathbb{R}_N) = H^1_G(\mathbb{R}_N) \oplus H^1_G(\mathbb{R}_N)^\perp.$$

Then for each $v \in H^1(\mathbb{R}_N)$ we have a decomposition $v = v_G + v_G^\perp$ with $v_G \in H^1_G(\mathbb{R}_N)$ and $v_G^\perp \in H^1_G(\mathbb{R}_N)^\perp$. Since $H^1_G(\mathbb{R}_N)$ is also a Hilbert space, we can identify $G'(u_m)$, through the duality, with an element in $H^1_G(\mathbb{R}_N)$ and hence $\langle G'(u_m), v_G^\perp\rangle = 0$. Consequently

$$\langle G'(u_m), v\rangle = \langle G'(u_m), v_G\rangle.$$

On the other hand $\langle G'(u_m), v_G\rangle \to 0$ uniformly on $B(0,1) \subset H^1_G(\mathbb{R}_N)$ and the result easily follows. □

We use Proposition 1.1 to determine level sets for which the $(PS)_c$-condition holds. Let

$$J^\infty = \inf\{G_\infty(u);\ u \in H^1(\mathbb{R}_N) - \{0\}, \langle G_\infty{}'(u), u\rangle = 0\},$$

where

$$G_\infty(u) = \frac{1}{2}\int_{\mathbb{R}_N} (|\nabla u|^2 + u^2)\, dx - \frac{1}{p}\int_{\mathbb{R}_N} b(\infty)|u|^p\, dx.$$

Proposition 2.9 *Let $\{u_m\} \subset H^1_G(\mathbb{R}_N)$ be sequence from Proposition 1.1 and suppose that $k \geq 1$, then $k \geq |G|$. This means that the $(PS)_c$-condition holds for $c \in (0, |k|J^\infty)$.*

Proof. Let $\{u^j\}$, $j = 0, \ldots, k$, and $\{y^j_m\}$, $j = 1, \ldots, k$ be sequences satisfying the assertion of Proposition 1.1. To every $\epsilon > 0$ there corresponds $R > 0$ such that

$$\|u^j\|_{1,2} - \|u^j\|_{H^1(B(0,R))} < \epsilon.$$

Therefore for $j = 1, \ldots, k$ we have

$$\|u^j(\cdot - y^j_m)\|_{1,2} - \|u^j(\cdot - y^j_m)\|_{H^1(B(0,R))} < \epsilon. \tag{2.31}$$

In what follows we shall use abbreviations

$$\|u\| = \|u\|_{1,2} \text{ and } \|u\|_\Omega = \|u\|_{H^1(\Omega)}.$$

Since $|y_m^j| \to \infty$ as $m \to \infty$, we may suppose, by choosing a subsequence if necessary, that $\{y_m^j\}$ has an asymptotic direction l_j, where l_j is a unit vector in $\mathbb{R}_N$. Without loss of generality, we may assume that $l_i \neq l_j$, $i \neq j$, $i, j = 1, \ldots, k$. For m sufficiently large we can construct a cone $P_j \subset \mathbb{R}_N$ with vertex at the origin, axis l_j and vertex angle ν such that $B(y_m^j, R) \subset P_j$, $j = 1, \ldots, k$, where ν will be determined later. Thus for m sufficienltly large we deduce from (2.31) that

$$\|u^j\| - \|u^j(\cdot - y_m^j)\|_{P_j} \leq \|u^j\| - \|u^j(\cdot - y_m^j)\|_{B(y_m^j,R)} < \epsilon. \tag{2.32}$$

Arguing indirectly we assume that $|G| > k$. Then there exist at least $g_1, \ldots, g_{k+1}$ such that $g_i(l_1) \neq g_j(l_j)$, $i \neq j$, $i, j = 1, \ldots, k+1$. We set $P_j^* = g_j(P_1)$, $l_j^* = g_j(l_j)$. For ν sufficiently large it is easy to check that $P_j^* \cap P_i^* = \emptyset$, $i \neq j$, $i, j = 1, \ldots, k+1$ and for every P_j, $j = 1, \ldots, k$, there exists at most one $P_{j^*}^* \in \{P_i^*; i = 1, \ldots, k+1\}$ such that $P_{j^*}^* \cap P_j \neq \emptyset$. Since $k+1 > k$, there exists at least one $P_{j'}^* \in \{P_i^*, i = 1, \ldots, k+1\}$ such that $P_{j'}^* \cap P_j = \emptyset$, $j = 1, \ldots, k$, that is,

$$P_{j'}^* \subset \mathbb{R}_N - \bigcup_{i=1}^k P_i. \tag{2.33}$$

It follows from Proposition 1.1, (2.31), (2.32), (2.33) and the fact that u_m are G-symmetric that

$$\begin{aligned}
o(1) &= \|u_m\|^2 - \|u^\circ\| - \sum_{j=1}^k \|u^j(\cdot - y_m^j)\| \\
&\geq \|u_m\|^2 - \left(\|u^\circ\|_{B(0,R)} + \epsilon\right) - \left(\sum_{j=1}^k \|u^j(\cdot - y_m^j)\|_{P_j} + k\epsilon\right) \\
&= \|u_m\|_{\mathbb{R}_N - B(0,R) \cap \bigcup_{i=1}^k P_i} - (k+1)\epsilon + o(1) \\
&\geqq \|u_m\|_{P_j^* - B(0,R)} - (k+1)\epsilon + o(1) \\
&= \|u_m\|_{P_1^* - B(0,R)} - (k+1)\epsilon + o(1) \\
&= \|u^1(\cdot - y_m^1)\|_{P_1}^2 - (k+1)\epsilon + o(1).
\end{aligned}$$

This means that

$$(k+1)\epsilon+o(1) \geqq \|u^1(\cdot-y_m^1)\|_{P_1}^2 \geq \|u^1(\cdot-y_m^1)\|_{B(y_m^1,R)}^2 = \|u^1\|_{B(0,R)}^2 \equiv a > 0$$

and we have arrived at a contradiction, since $\epsilon > 0$ is arbitrary small number. □

Theorem 2.9 *Suppose that $b(\infty) = \lim_{|x|\to\infty} b(x)$ is a G-symmetric function and that $b(x) \geqq b(\infty)|G|^{-\frac{p-2}{2}}$ on $\mathbb{R}_N$. Then equation (2.3) admits a nontrivial solution.*

Proof. We may assume that $|G| > 1$. Problem (M_∞) has a positive solution u. This means that $\bar{u} \in H_G^1(\mathbb{R}_N)$ and moreover

$$G_\infty(\bar{u}) = \sup_{t\geqq 0} G_\infty(t\bar{u})$$

and $G(t_\circ\bar{u}) < 0$ for $t_\circ$ sufficiently large. We set

$$\Gamma_G = \{\gamma \in C([0,1], H_G^1(\mathbb{R}_N));\ \gamma(0) = 0,\ \gamma(1) = t_\circ\bar{u}\},$$

then

$$0 < c = \inf_{\gamma\in\Gamma_G} \max_{t\in[0,1]} G(\gamma(t)). \tag{2.34}$$

Let $t^* > 0$ be a unique number satisfying

$$G(t^*\bar{u}) = \sup_{t\geqq 0} G(t\bar{u})$$

and

$$(t^*)^2 \int_{\mathbb{R}_N} (|\nabla\bar{u}|^2 + \bar{u}^2)\,dx = (t^*)^p \int_{\mathbb{R}_N} b(x)|\bar{u}|^p\,dx.$$

Since $\langle G_\infty(\bar{u}), \bar{u}\rangle = 0$, we have

$$\begin{aligned} G(t^*\bar{u}) &= \frac{(t^*)^2}{2} \int_{\mathbb{R}_N} (|\nabla\bar{u}|^2 + \bar{u}^2)\,dx \\ &- \frac{(t^*)^p}{p} \int_{\mathbb{R}_N} b(x)|\bar{u}|^p\,dx \\ &= (t^*)^p\Big(\frac{1}{2} - \frac{1}{p}\Big) \int_{\mathbb{R}_N} b(x)|\bar{u}|^p\,dx \\ &= (t^*)^2\Big(\frac{1}{2} - \frac{1}{p}\Big) \int_{\mathbb{R}_N} (|\nabla\bar{u}|^2 + \bar{u}^2)\,dx \end{aligned} \tag{2.35}$$

$$= \quad (t^*)^2\Big(\frac{1}{2} - \frac{1}{p}\Big)\int_{\mathbb{R}_N} b(\infty)|\bar{u}|^p\,dx.$$

Since $b(x) \geqq b(\infty)|G|^{-\frac{p-2}{2}}$, we get

$$\Big(\frac{1}{2} - \frac{1}{p}\Big)\int_{\mathbb{R}_N} b(\infty)|\bar{u}|^p\,dx = (t^*)^{p-2}\Big(\frac{1}{2} - \frac{1}{p}\Big)\int_{\mathbb{R}_N} b(x)|\bar{u}|^p\,dx$$
$$> \quad (t^*)^{p-2}\Big(\frac{1}{2} - \frac{1}{p}\Big)|G|^{-\frac{p-2}{2}}\int_{\mathbb{R}_N} b(\infty)|\bar{u}|^p\,dx,$$

which implies $t^* < \sqrt{|G|}$. Therefore by virtue of (2.35) we get

$$\begin{aligned} G(t^*\bar{u}) \quad &= \quad (t^*)^2\Big(\frac{1}{2} - \frac{1}{p}\Big)\int_{\mathbb{R}_N} b(\infty)|\bar{u}|^p\,dx \\ &< |G|\Big(\frac{1}{2} - \frac{1}{p}\Big)\int_{\mathbb{R}_N} b(\infty)|\bar{u}|^p\,dx = |G|J^\infty. \end{aligned}$$

Combining this with (2.34) we obtain $c < |G|J^\infty$. Consequently the $(PS)_c$-condition holds. By the mountain pass theorem the functional G has a nontrivial critical point $u \in H^1_G(\mathbb{R}_N)$ which by Lemma 2.9 is a solution of equation (2.3). □

2.9 Extensions to linear elliptic equations

The results of Section 2.3 can be easily extended to the following problem

$$M_K \quad = \quad \inf\Big\{\int_{\mathbb{R}_N}\Big(\sum_{i,j}^{N} a_{ij}(x)D_iuD_ju + u^2\Big)\,dx;\ u \in H^1(\mathbb{R}_N), \tag{2.36}$$
$$\int_{\mathbb{R}_N} K(x)|u|^p\,dx = 1\Big\},$$

where $2 < p < 2^*$.

Throughout this section we assume that the coefficients a_{ij} are continuous on $\mathbb{R}_N$ and $\lim_{|x|\to\infty} a_{ij}(x) = \bar{a}_{ij}$, $i, j = 1, \ldots N$ and $\{a_{ij}(x)\} \leq \{\bar{a}_{ij}\}$. This inequality is understood in the sense of quadratic forms, that is, $\sum_{i,j=1}^N \big(a_{ij}(x) - \bar{a}_{ij}\big)\xi_i\xi_j \leq 0$ for all $x, \xi \in \mathbb{R}_N$. Further we assume that

$$(\mathcal{E}) \qquad a_1|\xi|^2 \leq \sum_{i,j=1}^{N} a_{ij}(x)\xi_i\xi_j \leq a_2|\xi|^2, \quad a_{ij} = a_{ji},$$

for all $x, \xi \in \mathbb{R}_N$, where $a_1 > 0$ and $a_2 > 0$ are constants. The coefficient K is assumed to be continuous positive and bounded on $\mathbb{R}_N$ with $\lim_{|x|\to\infty} K(x) = K(\infty) > 0$. Let $\mathcal{H} : H^1(\mathbb{R}_N) \to \mathbb{R})$ be a variational functional for problem (2.37) given by

$$\mathcal{H}(u) = \frac{1}{2}\int_{\mathbb{R}_N}\left(\sum_{i,j=1}^{N} a_{ij}(x)D_iuD_ju + u^2\right)dx - \frac{1}{p}\int_{\mathbb{R}_N} K(x)|u|^p\,dx.$$

If $\{u_m\}$ is a minimizing sequence for problem (2.36), then a sequence $\{M_{K,\lambda}^{\frac{1}{p-2}} u_m\}$ is a Palais–Smale sequence for the functional $\mathcal{H}$. A solution u to problem (2.35) after rescaling $M_{K,\lambda}^{\frac{1}{p-2}} u$ satisfies the equation

$$-\sum_{i,j=1}^{N} D_i\big(a_{ij}(x)D_jv\big) + v = K(x)|v|^{p-2}v \text{ on } \mathbb{R}_N.$$

For simplicity, we have chosen the coefficient of u to be a constant 1. Obviously, we can replace 1 by a nonnegative and continuous function $c(x)$ on $\mathbb{R}_N$ such that the expression

$$\left(\int_{\mathbb{R}_N}\sum_{i,j}^{N}\big(a_{ij}(x)D_iuD_ju + c(x)u^2\big)\,dx\right)^{\frac{1}{2}}$$

defines an equivalent norm on $H^1(\mathbb{R}_N)$.

It is convenient for our purposes to modify the definition of the quantity β_∞ from Proposition 1.2. Let $\{u_m\} \subset H^1(\mathbb{R}_N)$ be such that $u_m \rightharpoonup u$ in $H^1(\mathbb{R}_N)$ and put

$$\beta_\infty = \lim_{R\to\infty}\limsup_{m\to\infty}\int_{|x|>R}\left(\sum_{i,j=1}^{N} a_{ij}(x)D_iu_mD_ju_m + u_m^2\right)dx.$$

Since $\lim_{|x|\to\infty} a_{ij}(x) = \bar{a}_{ij}$, $i,j = 1,\ldots,N$, it is easy to see that

$$\beta_\infty = \lim_{R\to\infty}\limsup_{m\to\infty}\int_{|x|>R}\left(\sum_{i,j=1}^{N} \bar{a}_{ij}D_iu_mD_ju_m + u_m^2\right)dx.$$

We need the following technical lemma.

Lemma 2.10 *Under our assumptions on coefficients $a_{ij}(x)$, $i,j = 1,\dots,N$, we have*

$$\beta_\infty = \lim_{R\to\infty}\limsup_{m\to\infty}\int_{\mathbb{R}_N}\left(\sum_{i,j=1}^N a_{ij}(x)D_i(\phi_R u_m)D_j(\phi_R u_m) + (\phi_R u_m)^2\right)dx$$

$$= \lim_{R\to\infty}\limsup_{m\to\infty}\int_{\mathbb{R}_N}\left(\sum_{i,j=1}^N \bar{a}_{ij}D_i(\phi_R u_m)D_j(\phi_R u_m) + (\phi_R u_m)^2\right)dx,$$

where ϕ_R is a function defined in the proof of Proposition 1.2.

Proof. It is sufficient to notice that

$$\sum_{i,j=1}^N a_{ij}(x)D_i(\phi_R u_m)D_j(\phi_R u_m) = \sum_{i,j=1}^N a_{ij}(x)D_i\phi_R D_j\phi_R u_m^2$$

$$+ \; 2\sum_{i,j=1}^N a_{ij}(x)D_i\phi_R u_m\phi_R D_j u_m + \sum_{i,j=1}^N a_{ij}(x)D_i u_m D_j u_m\phi_R^2,$$

$$\lim_{R\to\infty}\limsup_{m\to\infty}\int_{\mathbb{R}_N}\sum_{i,j=1}^N |a_{ij}(x)D_i\phi_R u_m\phi_R D_j u_m|\,dx = 0$$

and that

$$\lim_{R\to\infty}\limsup_{m\to\infty}\int_{\mathbb{R}_N}\sum_{i,j=1}^N a_{ij}(x)u_m^2 D_i\phi_R D_j\phi_R\,dx = 0.$$

□

As an immediate consequence of this lemma we obtain the inequality

$$M_K\left(K(\infty)\alpha_\infty\right)^{\frac{2}{p}} \le \beta_\infty$$

for every minimizing sequence of problem (2.36). We define the variational problem at infinity

$$M_\infty = \inf\{\int_{\mathbb{R}_N}\left(\sum_{i,j=1}^N \bar{a}_{ij}D_i u D_j u + u^2\right)dx;$$

$$u \in H^1(\mathbb{R}_N),\ \int_{\mathbb{R}_N} K(\infty)|u|^p\,dx = 1\}.$$

It follows from an obvious modification of Theorem 2.1 that this problem has a positive solution.

Theorem 2.10 *Suppose that $K(x) \geq K(\infty)$ on $\mathbb{R}_N$. Then problem (2.37) admits a positive solution.*

It suffices to consider the case $K(x) > K(\infty)$ on a set of positive measure in $\mathbb{R}_N$. The proof of this theorem, with obvious modifications, is similar to that of Theorem 2.4 and is omitted.

2.10 Periodic coefficients

In the final paragraph of Section 2.4 we pointed out that problem (M) has a solution if $b(x)$ is a periodic function on $\mathbb{R}_N$. This obviously can be extended to problem (2.36).

Proposition 2.10 *Suppose that the coefficients $a_{ij}(x)$, $i, j = 1, \ldots, N$, are continuous periodic and satisfy condition (E). Furthermore we assume that $K(x)$ is continuous positive and periodic on $\mathbb{R}_N$. Then problem (2.36) admits a solution.*

In the next two theorems, we remove the periodicity assumption either on K or on a_{ij}, replacing it by a suitable assumption about the behaviour of these coefficients at ∞.

Theorem 2.11 *Let K be a continuous positive function on $\mathbb{R}_N$. Suppose that $a_{ij}(x)$ are continuous periodic and satisfy (E) and that there exists $\lim_{|x|\to\infty} K(x) = K(\infty) > 0$ such that $K(x) \geq K(\infty)$ on $\mathbb{R}_N$. Then problem (2.36) admits a solution.*

Proof. We may assume that $K(x) \geq K(\infty)$ on $\mathbb{R}_N$, with strict inequality on a set of positive measure. Let $\{u_n\}$ be a minimizing sequence for problem (2.37). Since $\{u_n\}$ is bounded in $H^1(\mathbb{R}_N)$, we may assume that $u_n \rightharpoonup u$ in $H^1(\mathbb{R}_N)$. If $u \equiv 0$, then $\alpha_\infty K(\infty) = 1$. Since $\langle \mathcal{H}(u_n M_K^{\frac{1}{p-2}}), u_n M_K^{\frac{1}{p-2}} \phi_R \rangle \to 0$ as $n \to \infty$ uniformly in $R \geq 1$, we see that $\beta_\infty = M_K$. According to Proposition 2.10 there exists a positive minimizer w of the variational problem

$$M_P = \inf\{\int_{\mathbb{R}_N} (\sum_{i,j=1}^N a_{ij} D_i u D_j u + u^2)\, dx; \qquad u \in H^1(\mathbb{R}_N),$$

$$K(\infty) \int_{\mathbb{R}_N} |u|^p\, dx = 1\}.$$

It then follows that

$$M_P\left(\int_{\mathbb{R}_N} K(\infty)|u_n\phi_R|^p\,dx\right)^{\frac{1}{p}}$$
$$\leq \int_{\mathbb{R}_N} \big(\sum_{i,j=1}^{N} a_{ij}D_i(u_n\phi_R)D_j(u_n\phi_R) + (u_n\phi_R)^2\big)\,dx.$$

Then Lemma 2.10 gives

$$M_P \leq \beta_\infty = M_K.$$

On the other hand, by the definition of M_K, we get

$$M_K < M_K\left(\int_{\mathbb{R}_N} K(x)|w|^p\,dx\right)^{\frac{2}{p}} \leq \int_{\mathbb{R}_N} \big(\sum_{i,j=1}^{N} a_{ij}D_iwD_jw + w^2\big)\,dx = M_P$$

which is impossible. As in the proof of Theorem 2.4, we show that $u_n \to u$ in $L^p(\mathbb{R}_N)$ and this completes the proof. □

By a straightforward modification of the above argument we can also assert the following existence result.

Theorem 2.12 *Let $K(x)$ be a positive continuous and periodic function on $\mathbb{R}_N$. Suppose that $\{a_{ij}\}$ are continuous bounded and satisfy (E) and that*

$$\lim_{|x|\to\infty} a_{ij}(x) = \bar{a}_{ij}, \quad i,j = 1,\ldots,N$$

such that $\{a_{ij}(x)\} \leq \{\bar{a}_{ij}\}$ on $\mathbb{R}_N$. Then problem (2.36) admits a solution.

2.11 Bibliographical notes

The results on constrained minimization Theorems 2.1, 2.4 and 2.5 were established by P.L. Lions [76] with the aid of his version of the concentration–compactness principle. The approach offered here, based on [31], seems to be much simpler. The notion of periodic and large interior domains stem from the paper of Lien–Tzeng–Wang [74]. However, we present here a different and simplified approach based upon Proposition 1.2. The existence results for large interior domains are taken from the papers of Grossi [63] and Dall'Aglio–Grossi [48]. For related results on the exterior Dirichlet problem we refer to the papers of Benci–Cerami, [15], [16], Cerami–Passaseo

[17], [28], [29], Grossi–Passasseo [63] and Yang [111]. The proof of Propositions 2.6, which has been noted by Benci–Cerami and P.L. Lions, differs from those in [15] and [73] in that we use Proposition 1.2 instead the global compactness theorem 3.5 (see also Proposition 1.1). The results on the role of a symmetry in Section 2.8 are due to Deng–Zhou–Zhu [48] and Zhu–Cao [119]. Theorems 2.11 and 2.12 on constrained minimization under mixed assumptions on the periodicity and behaviour of coefficients at infinity were noted by Chabrowski [31] and the earlier form of these results is due to P.L. Lions [74] and P.L. Lions–Esteban [53].

Chapter 3

Nonlinear eigenvalue problem

Until now we have concentrated on constrained minimization problems involving homogeneous nonlinearities. In this chapter we consider nonlinear eigenvalue problems with more general nonlinearities. These eigenvalue problems are solved by constrained minimization depending on a positive parameter λ. The main point in this approach is the derivation of a subadditive condition for a variational functional, which combined with Lemma 1.3, leads to the existence results.

3.1 Eigenvalue problem

In this section we consider a minimization problem motivated by the question of the existence of standing waves for a nonlinear Schrödinger equation [102]. A solution to this problem is an eigenfunction of the following *eigenvalue problem*

$$\begin{cases} -\Delta u + \omega u & = f(x, u) \text{ in } \mathbb{R}_N \\ u \in H^1(\mathbb{R}_N), & u \not\equiv 0, \quad \omega \in \mathbb{R}. \end{cases} \tag{3.1}$$

Throughout this section we assume:

(A) $f \in C(\mathbb{R}_N \times, \mathbb{R})$, $\lim_{t\to 0} \frac{f(x,t)}{|t|} = 0$ and $\lim_{|t|\to\infty} \frac{f(x,t)}{|t|^{1+\frac{4}{N}}} = 0$ uniformly in $x \in \mathbb{R}_N$,

(B) $\lim_{|x|\to\infty} f(x,t) = \bar{f}(t)$ uniformly for t in bounded subsets of $\mathbb{R}$.

Let

$$f(x,t) = \int_0^t f(x,s)\,ds \text{ and } \bar{F}(t) = \int_0^t \bar{f}(s)\,ds$$

and we define variational functionals $I : H^1(\mathbb{R}_N) \to \mathbb{R}$ and $I^\infty :$ $H^1(\mathbb{R}_N) \to \mathbb{R}_N$ by

$$I(u) = \frac{1}{2}\int_{\mathbb{R}_N} |\nabla u|^2\,dx - \int_{\mathbb{R}_N} F(x,u)\,dx$$

and

$$I^\infty(u) = \frac{1}{2}\int_{\mathbb{R}_N} |\nabla u|^2\,dx - \int_{\mathbb{R}_N} \bar{F}(u)\,dx.$$

Solutions to problem (3.1) will be found as minimizers of the variational problem

$$(\mathcal{I}_\lambda) \qquad I_\lambda = \inf\{I(u);\ u \in H^1(\mathbb{R}_N), \int_{\mathbb{R}_N} |u|^2\,dx = \lambda\}, \quad \lambda > 0,$$

To find a solution of problem $(\mathcal{I}_\lambda)$ we introduce the (limit) variational problem

$$(\mathcal{I}_\lambda^\infty) \qquad I_\lambda^\infty = \inf\{I^\infty(u);\ u \in H^1(\mathbb{R}_N), \int_{\mathbb{R}_N} |u|^2\,dx = \lambda\}.$$

A crucial step in the proof of the existence of minimizers for I_λ and I_λ^∞ is a *subadditivity condition* which will be established with the aid of Lemma 1.3.

We commence by studying properties of functions $I_\lambda, I_\lambda^\infty : (0,\infty) \to \mathbb{R}$.

Lemma 3.1 *Functions I_λ and I_λ^∞ are continuous on $(0,\infty)$ and minimizing sequences for problems $(\mathcal{I}_\lambda)$ and $(\mathcal{I}_\lambda^\infty)$ are bounded in $H^1(\mathbb{R}_N)$.*

Proof. First we observe that for each $\epsilon > 0$ there exists $C_\epsilon > 0$ such that

$$\begin{aligned} |\bar{F}(u)|, |F(x,u)| \ &\le\ \epsilon \int_{\mathbb{R}_N} u^2\,dx + \epsilon \int_{\mathbb{R}_N} |u|^{2+\frac{4}{N}}\,dx \\ &+\ C_\epsilon \int_{\mathbb{R}_N} |u|^\alpha\,dx, \end{aligned}$$

where $2 < \alpha < 2 + \frac{4}{N}$ and $C_\epsilon > 0$. This is an easy consequence of (A). By the Hölder and Sobolev inequalities we have

$$\int_{\mathbb{R}_N} |u|^{2+\frac{4}{N}}\,dx \ =\ \int_{\mathbb{R}_N} |u|^{2\frac{2^*-2-\frac{4}{N}}{2^*-2}+2^*\frac{\frac{4}{N}}{2^*-2}}\,dx$$

$$
\begin{aligned}
&\leq \left(\int_{\mathbb{R}_N} u^2\,dx\right)^{\frac{2^*-2-\frac{4}{N}}{2^*-2}} \left(\int_{\mathbb{R}_N} |u|^{2^*}\,dx\right)^{\frac{4}{N(2^*-2)}} \\
&\leq S^{-1}\left(\int_{\mathbb{R}_N} u^2\,dx\right)^{\frac{2}{N}} \int_{\mathbb{R}_N} |\nabla u|^2\,dx.
\end{aligned}
$$

Similarly we have

$$
\begin{aligned}
\int_{\mathbb{R}_N} |u|^\alpha\,dx &= \int_{\mathbb{R}_N} |u|^{2\frac{2^*-\alpha}{2^*-2}+2^*\frac{\alpha-2}{2^*-2}}\,dx \\
&\leq \left(\int_{\mathbb{R}_N} u^2\,dx\right)^{\frac{2^*-\alpha}{2^*-2}} \left(\int_{\mathbb{R}_N} |u|^{2^*}\,dx\right)^{\frac{\alpha-2}{2^*-2}} \\
&\leq S^{-\frac{2^*(\alpha-2)}{2(2^*-2)}} \left(\int_{\mathbb{R}_N} u^2\,dx\right)^{\frac{2^*-\alpha}{2^*-2}} \left(\int_{\mathbb{R}_N} |\nabla u|^2\,dx\right)^{\frac{2^*(\alpha-2)}{2(2^*-2)}}.
\end{aligned}
$$

Consequently by the Young inequality we have

$$
\int_{\mathbb{R}_N} |u|^\alpha\,dx \leq \eta \int_{\mathbb{R}_N} |\nabla u|^2\,dx + K(\eta)\left(\int_{\mathbb{R}_N} u^2\,dx\right)^{\frac{2(2^*-\alpha)}{4\cdot 2^*-4-2^*\alpha}}
$$

for $\eta > 0$, where $K(\eta) > 0$ is a constant. Letting $\lambda = \int_{\mathbb{R}_N} u^2\,dx$, we derive the following estimates for $I(u)$ and $I^\infty(u)$

$$
\begin{aligned}
I(u), I^\infty(u) &\geq \left(\frac{1}{2} - \epsilon S^{-1}\lambda^{\frac{2}{N}} - C_\epsilon \eta\right) \int_{\mathbb{R}_N} |\nabla u|^2\,dx \\
&- \epsilon\lambda - K(\eta) C_\epsilon \lambda^{\frac{2(2^*-\alpha)}{4(2^*-1)-2^*\alpha}}.
\end{aligned}
$$

Choosing $\epsilon > 0$ and $\eta > 0$ so that

$$
\frac{1}{2} - \epsilon S^{-1}\lambda^{\frac{2}{N}} - C_\epsilon \eta > 0,
$$

we see that I_λ and I_λ^∞ are finite and moreover minimizing sequences for problems $(\mathcal{I}_\lambda)$ and $(\mathcal{I}_\lambda^\infty)$ are bounded. Since $\|\sigma u\|_2 = \sigma\|u\|_2$ for $\sigma > 0$, it is easy to check that I_λ and ${}^\infty_\lambda$ are continuous on $(0, \infty)$. $\square$

To proceed further, we observe that $I_\mu^\infty \leq 0$ for all $\mu > 0$. Indeed, let $u \in C_0^\infty(\mathbb{R}_N)$ and $\|u\|_2^2 = \mu$, then for each $\sigma > 0$ we have

$$
I_\mu^\infty \leq \frac{1}{2\sigma^2} \int_{\mathbb{R}_N} |\nabla u|^2\,dx - \sigma^N \int_{\mathbb{R}_N} \bar{F}(\sigma^{-\frac{N}{2}} u)\,dx \to 0
$$

as $\sigma \to \infty$.

Lemma 3.2 *Suppose that $I_\lambda^\infty < 0$ for some $\lambda > 0$, then $\frac{I_\mu^\infty}{\mu}$ is nonincreasing on $(0,\infty)$ and $\lim_{\mu\to 0+}\frac{I_\mu^\infty}{\mu} = 0$. Moreover there exists $\lambda^* \le \lambda$ such that*

$$\frac{I_\mu^\infty}{\mu} > \frac{I_\lambda^\infty}{\lambda} \quad \textit{for } \mu \in (0,\lambda^*). \tag{3.2}$$

Proof. First we observe that if $\sigma > 0$ and $\alpha > 0$ with $\|u\|_2^2 = \alpha$, then $\|u(\frac{\cdot}{\sigma^{\frac{1}{N}}})\|_2^2 = \sigma\alpha$. Consequently, for all $\alpha_1 > 0$ and $\alpha_2 > 0$ we have

$$I_{\alpha_1}^\infty = \inf\{\frac{1}{2}\big(\frac{\alpha_1}{\alpha_2}\big)^{\frac{N-2}{N}}\int_{\mathbb{R}_N}|\nabla u|^2\,dx - \frac{\alpha_1}{\alpha_2}\int_{\mathbb{R}_N}\bar{F}(u)\,dx;\ \|u\|_2^2 = \alpha_2\}. \tag{3.3}$$

If $0 < \alpha_1 < \alpha_2$, then for each $\epsilon > 0$ there exists $u \in H^1(\mathbb{R}_N)$ with $\|u\|_2^2 = \alpha_2$ such that

$$\begin{aligned} I_{\alpha_1}^\infty \ + \ \epsilon &> \frac{1}{2}\big(\frac{\alpha_1}{\alpha_2}\big)^{\frac{N-2}{N}}\int_{\mathbb{R}_N}|\nabla u|^2\,dx - \frac{\alpha_1}{\alpha_2}\int_{\mathbb{R}_N}\bar{F}(u)\,dx \\ &\ge \frac{\alpha_1}{\alpha_2}\big(\frac{1}{2}\int_{\mathbb{R}_N}|\nabla u|^2\,dx - \int_{\mathbb{R}_N}\bar{F}(u)\,dx\big) \ge \frac{\alpha_1}{\alpha_2}I_{\alpha_2}^\infty. \end{aligned} \tag{3.4}$$

This inequality yields $\frac{I_{\alpha_1}^\infty}{\alpha_1} \ge \frac{I_{\alpha_2}^\infty}{\alpha_2}$. Since $I_\mu^\infty \le 0$ for all $\mu > 0$, we see that

$$\lim_{\mu\to 0}\frac{I_\mu^\infty}{\mu} = a \le 0.$$

We claim that $a = 0$. Indeed, it follows from (3.4) and from the estimate obtained in the proof of Lemma 3.1 that for every $0 < \mu < \lambda$ there exists an $u_\mu \in H^1(\mathbb{R}_N)$, with $\|u_\mu\|_2^2 = \lambda$ such that

$$\begin{aligned} I_\mu^\infty + \mu^2 &> \frac{1}{2}\big(\frac{\mu}{\lambda}\big)^{\frac{N-2}{N}}\int_{\mathbb{R}_N}|\nabla u_\mu|^2\,dx - \frac{\mu}{\lambda}\int_{\mathbb{R}_N}\bar{F}(u_\mu)\,dx \\ &\ge \frac{\mu}{\lambda}\big(\frac{1}{2}\int_{\mathbb{R}_N}|\nabla u_\mu|^2\,dx - \int_{\mathbb{R}_N}\bar{F}(u_\mu)\,dx\big) \\ &\ge \frac{\mu}{\lambda}\big(C_1(\lambda)\int_{\mathbb{R}_N}|\nabla u_\mu|^2\,dx - C_2(\lambda)\big), \end{aligned}$$

where $C_1(\lambda) > 0$ and $C_2(\lambda) > 0$ are constants. Hence

$$\mu^2 \ge \frac{\mu}{\lambda}\big(C_1(\lambda)\int_{\mathbb{R}_N}|\nabla u_\mu|^2\,dx - C_2(\lambda)\big),$$

that is,

$$\int_{\mathbb{R}_N} |\nabla u_\mu|^2\, dx \le C_3(\lambda)$$

for some constant $C_3(\lambda) > 0$ independent of μ. If there exists $\epsilon_o > 0$ and a sequence $\mu_n \to 0$ such that

$$\int_{\mathbb{R}_N} |\nabla u_{\mu_n}|^2\, dx \ge \epsilon_o,$$

then, using the fact that $\int_{\mathbb{R}_N} \bar{F}(u_{\mu_n})\, dx \le C$ for some constant $C > 0$, we get

$$\frac{I^\infty_{\mu_n}}{\mu_n} + \mu_n \ge \frac{1}{2}\lambda^{-\frac{N-2}{N}} \mu_n^{-\frac{2}{N}} \epsilon_o - \frac{C}{\lambda} \to \infty$$

as $\mu_n \to 0$ and this contradicts the fact that $\lim_{\mu\to 0+} \frac{I^\infty_\mu}{\mu} = a \le 0$. Therefore $\lim_{\mu\to 0} \int_{\mathbb{R}_N} |\nabla u_\mu|^2\, dx = 0$ and this implies that $\lim_{\mu\to 0} \int_{\mathbb{R}_N} \bar{F}(u_\mu)\, dx = 0$ and consequently

$$\frac{I^\infty_\mu}{\mu} + \mu \ge -\frac{1}{\lambda} \int_{\mathbb{R}_N} \bar{F}(u_\mu)\, dx \to 0.$$

This shows that $\lim_{\mu\to 0+} \frac{I^\infty_\mu}{\mu} = 0$. Finally, we observe that $\lim_{\mu\to 0} \frac{I^\infty_\mu}{\mu} = 0 > \frac{I^\infty_\lambda}{\lambda}$ which implies (3.2). □

We can now formulate the existence result for problem (I^∞_λ).

Theorem 3.1 *Suppose that $I^\infty_\lambda < 0$ for some $\lambda > 0$. Then there exists $0 < \alpha_o \le \lambda$ such that problem $(\mathcal{I}^\infty_{\alpha_o})$ has a solution. Moreover each minimizing sequence for $(\mathcal{I}^\infty_{\alpha_o})$ up to a translation is relatively compact in $H^1(\mathbb{R}_N)$.*

Proof. According to Lemma 3.2 the set

$$\{\alpha_1;\ \frac{I^\infty_\alpha}{\alpha} > \frac{I^\infty_\lambda}{\lambda} \text{ for each } \alpha \in (0, \alpha_1)\}$$

is nonempty. We define

$$\alpha_o = \sup\{\alpha_1;\ \frac{I^\infty_\alpha}{\alpha} > \frac{I^\infty_\lambda}{\lambda} \text{ for each } \alpha \in (0, \alpha_1)\}.$$

It follows from the continuity of I^∞_λ that

$$0 < \alpha_o \le \lambda,$$

$$I^{\infty}_{\alpha_\circ} = \frac{\alpha_\circ}{\lambda} I^{\infty}_{\lambda} < 0,$$

and

$$I^{\infty}_{\alpha} > \frac{\alpha}{\lambda} I^{\infty}_{\lambda} \text{ for all } 0 < \alpha < \alpha_\circ.$$

This yields

$$I^{\infty}_{\alpha_\circ} = \frac{\alpha_\circ}{\lambda} I^{\infty}_{\lambda} = \frac{\alpha_\circ - \alpha}{\lambda} I^{\infty}_{\lambda} + \frac{\alpha}{\lambda} I^{\infty}_{\lambda} < I^{\infty}_{\alpha_\circ - \alpha} + I^{\infty}_{\alpha} \tag{3.5}$$

for each $\alpha \in (0, \alpha_\circ)$.
Let $\{u_m\} \subset H^1(\mathbb{R}_N)$ be a minimizing sequence for $I^{\infty}_{\alpha_\circ}$. Since $\{u_m\}$ is bounded we may assume that $u_m \rightharpoonup u$ in $H^1(\mathbb{R}_N)$, $u_m \to u$ a. e. on $\mathbb{R}_N$. First we consider the case $u \equiv 0$. In this case by Lemma 1.1 either (a) $u_m \to 0$ for $2 < \alpha < 2^*$ or (b) there exists a sequence $\{y_m\} \subset \mathbb{R}_N$ such that $u_m(\cdot + y_m) \rightharpoonup v \not\equiv 0$ in $H^1(\mathbb{R}_N)$.
In the first case $\lim_{m\to\infty} \int_{\mathbb{R}_N} \bar{F}(u_m)\, dx = 0$ and consequently

$$I^{\infty}_{\alpha_\circ} = \lim_{m\to\infty} I^{\infty}(u_m) = \lim_{m\to\infty} \Big(\frac{1}{2}\int_{\mathbb{R}_N} |\nabla u_m|^2\, dx - \int_{\mathbb{R}_N} \bar{F}(u_m)\, dx\Big) \geq 0$$

which is impossible. Hence (b) holds and letting $v_m(x) = u_m(x + y_m)$ we have

$$\int_{\mathbb{R}_N} u_m^2\, dx = \int_{\mathbb{R}_N} v_m^2\, dx = \int_{\mathbb{R}_N} v^2\, dx + \int_{\mathbb{R}_N} (v_m - v)^2\, dx + o(1). \tag{3.6}$$

We now show that $\int_{\mathbb{R}_N} v^2\, dx = \alpha_\circ$. In the contrary case $0 < \int_{\mathbb{R}_N} v^2\, dx < \alpha_\circ$. By (3.6) we have $\lim_{m\to\infty} \int_{\mathbb{R}_N} (v_m - v)^2\, dx \to \alpha_\circ - \lambda$, $\lambda = \int_{\mathbb{R}_N} v^2\, dx$. On the other hand, by Lemma 1.3 we have

$$\int_{\mathbb{R}_N} \bar{F}(v_m)\, dx = \int_{\mathbb{R}_N} \bar{F}(v)\, dx + \int_{\mathbb{R}_N} \bar{F}(v_m - v)\, dx + o(1)$$

and this implies that

$$I^{\infty}_{\alpha_\circ} = I^{\infty}(v) + I^{\infty}(u_m - v) + o(1) \geq I^{\infty}_{\lambda} + I^{\infty}_{\alpha_\circ - \lambda_\circ} + o(1).$$

Letting $m \to \infty$ we get $I^{\infty}_{\alpha_\circ} \geq I^{\infty}_{\lambda} + I^{\infty}_{\alpha_\circ - \lambda}$ which contradicts (3.5). Therefore $\int_{\mathbb{R}_N} u^2\, dx = \alpha_\circ$. It then follows from (3.6) that $u_m \to u$ in $L^2(\mathbb{R}_N)$. By the Gagliardo–Nirenberg inequality $v_m \to u$ in $L^p(\mathbb{R}_N)$, $2 \leq p < \infty$. Obviously this implies that $I^\infty_{\alpha_\circ} = I^\infty(u)$ and $\int_{\mathbb{R}_N} u^2\,dx = \alpha_\circ$. To complete the proof we show that $v_m \to u$ in $H^1(\mathbb{R}_N)$. Indeed, we have

$$\begin{aligned} I^\infty_{\alpha_\circ} &= \frac{1}{2}\int_{\mathbb{R}_N} |\nabla v_m|^2\,dx + \int_{\mathbb{R}_N} \bar{F}(v_m)\,dx + o(1) \qquad (3.7)\\ &= \frac{1}{2}\int_{\mathbb{R}_N} |\nabla u|^2\,dx + \frac{1}{2}\int_{\mathbb{R}_N} |\nabla(v_m - u)|^2\,dx - \int_{\mathbb{R}_N} \bar{F}(u)\,dx \\ &+ \int_{\mathbb{R}_N} (\bar{F}(u) - \bar{F}(v_m))\,dx + o(1). \end{aligned}$$

Since

$$\lim_{m\to\infty} \int_{\mathbb{R}_N} (\bar{F}(u) - \bar{F}(v_m))\,dx = 0,$$

we deduce from (3.7) that $\nabla v_m \to \nabla u$ in $L^2(\mathbb{R}_N)$ and hence $v_m \to u$ in $H^1(\mathbb{R}_N)$.
If $u \not\equiv 0$, we repeat the previous argument to show that $I^\infty_{\alpha_\circ}$ is attained. □

We are now in a position to establish the existence result for problem (I_λ).

Theorem 3.2 *Suppose that $F(x,t) \geq \bar{F}(t)$ on $\mathbb{R}_N \times \mathbb{R}$ and that $I_\lambda < 0$ for some $\lambda > 0$. Then the infimum $I_{\lambda_\circ}$ is attained for some $0 < \lambda_\circ \leq \lambda$.*

Proof. Since $F(x,t) \geq \bar{F}(t)$ on $\mathbb{R}_N \times \mathbb{R}$ we have $I_\mu \leq I^\infty_\mu$ for $\mu \geq 0$. We distinguish two cases: (i) $I_\lambda = I^\infty_\lambda < 0$ and (ii) $I_\lambda < I^\infty_\lambda$.
Case (i). By Theorem 3.1 there exists $\lambda_\circ \in (0, \lambda]$ such that $I^\infty_{\lambda_\circ} = I^\infty(\bar{u})$, $\int_{\mathbb{R}_N} \bar{u}^2\,dx = \lambda_\circ$ for some $\bar{u} \in H^1(\mathbb{R}_N)$. Thus

$$\begin{aligned} I_{\lambda_\circ} &\leq I(\bar{u}) = \frac{1}{2}\int_{\mathbb{R}_N} |\nabla \bar{u}|^2\,dx - \int_{\mathbb{R}_N} F(x,\bar{u})\,dx \\ &\leq \frac{1}{2}\int_{\mathbb{R}_N} |\nabla \bar{u}|^2\,dx - \int_{\mathbb{R}_N} \bar{F}(\bar{u})\,dx = I^\infty(\bar{u}) = I^\infty_{\lambda_\circ}. \end{aligned}$$

If $I_{\lambda_\circ} = I^\infty_{\lambda_\circ}$, then I also attains its infimum $I_{\lambda_\circ}$ at $\bar{u}$. Therefore it remains to consider the case $I_{\lambda_\circ} < I^\infty_{\lambda_\circ}$. Consequently we need to prove the following claim:
If $I_\lambda < I^\infty_\lambda$ for some $\lambda > 0$, then there exists $\alpha_\circ \in (0, \lambda]$ such that problem $(\mathcal{I}_{\alpha_\circ})$ has a solution. This obviously completes the proof of the case (i) and also provides the proof of the case (ii).

By virtue of Lemma 3.1 $I_\beta + I^\infty_{\lambda-\beta}$ is continuous for $\beta \in [0,\lambda]$ and also $I_0 = I^\infty_0 = 0$. If $I_\lambda < I^\infty_\lambda$ for some $\lambda > 0$, then there exists $\nu > 0$ such that

$$I_\lambda < I_\beta + I^\infty_{\lambda-\beta}$$

for all $\beta \in [0,\nu)$. Let

$$\alpha_0 = \sup\{\nu;\ I_\lambda < I_\beta + I^\infty_{\lambda-\beta},\ \text{ for } 0 \le \beta < \nu\}.$$

Then we have

$$I_\lambda = I_{\alpha_0} + I^\infty_{\lambda-\alpha_0}$$

and

$$I_\lambda < I_\alpha + I^\infty_{\lambda-\alpha} \tag{3.8}$$

for $0 \le \alpha < \alpha_0$. This implies that

$$I_{\alpha_0} + I^\infty_{\lambda-\alpha_0} = I_\lambda < I^\infty_\lambda \le 0$$

and hence

$$I_{\alpha_0} < I^\infty_\lambda - I^\infty_{\lambda-\alpha_0} \le I^\infty_{\alpha_0} \le 0. \tag{3.9}$$

We show that I_{α_0} is attained by a $\bar{u} \in H^1(\mathbb{R}_N)$ and every minimizing sequence for I_{α_0} is relatively compact in $H^1(\mathbb{R}_N))$. Let $\{u_m\}$ be a minimizing sequence for I_{α_0}. Since $\{u_m\}$ is bounded, we may assume that $u_m \rightharpoonup \bar{u}$ in $H^1(\mathbb{R}_N)$, $u_m \to \bar{u}$ a.e. on $\mathbb{R}_N$. Arguing indirectly we assume that $\bar{u} \equiv 0$ on $\mathbb{R}_N$. By Lemma 1.2 we see that

$$\lim_{m\to\infty} \int_{B(0,R)} |F(x,u_m)|\,dx = \lim_{m\to\infty} \int_{B(0,R)} |\bar{F}(u_m)|\,dx = 0$$

for each $R > 0$. We now write

$$\begin{aligned} I(u_m) &= \frac{1}{2}\int_{\mathbb{R}_N} |\nabla u_m|^2\,dx - \int_{\mathbb{R}_N} F(x,u_m)\,dx \\ &= I^\infty(u_m) + \int_{\mathbb{R}_N} \left(\bar{F}(u_m) - F(x,u_m)\right)dx. \end{aligned} \tag{3.10}$$

We show that

$$\lim_{m\to\infty} \int_{\mathbb{R}_N} \left(\bar{F}(u_m) - F(x,u_m)\right)dx = 0. \tag{3.11}$$

Towards this end we write

$$\int_{\mathbb{R}_N} |F(x,u_m) - \bar{F}(u_m)|\,dx \le \int_{B(0,R)} |F(x,u_m)|\,dx + \int_{B(0,R)} |\bar{F}(u_m)|\,dx \qquad (3.12)$$

$$+ \quad \left(\int_{|x|\ge R, |u_m|\le\delta} + \int_{|x|\ge R, \delta\le|u_m|\le\frac{1}{\delta}} + \int_{|x|\ge R, |u_m|>\frac{1}{\delta}} \right) |F(x,u_m) - \bar{F}(u_m)|\,dx.$$

We now define the following quantities

$$\epsilon_1(\delta) = \sup_{0<|t|<\delta, x\in\mathbb{R}_N} \frac{|F(x,t)-\bar{F}(t)|}{|t|^2},$$

$$\epsilon(R) = \sup_{\delta\le|t|\le\frac{1}{\delta}, |x|\ge R} |F(x,t)-\bar{F}(t)|$$

and

$$\epsilon_2(\delta) = \sup_{|t|\ge\frac{1}{\delta}} \frac{|F(x,t)-\bar{F}(t)|}{|t|^{\frac{2N}{N-2}}}.$$

It follows from assumption (A) that $\lim_{\delta\to 0}\epsilon_1(\delta) = \lim_{\delta\to 0}\epsilon_2(\delta) = 0$ and by (B) $\lim_{R\to\infty}\epsilon(R) = 0$ for each fixed $\delta > 0$. Inserting these quantities into (3.12) we derive the following estimate

$$\int_{\mathbb{R}_N} |F(x,u_m) - \bar{F}(u_m)|\,dx \le \epsilon_1(\delta) \int_{\mathbb{R}_N} u_m^2\,dx + \frac{\epsilon(R)}{\delta^2} \int_{\mathbb{R}_N} u_m^2\,dx$$

$$+ \quad \epsilon_2(\delta) \int_{\mathbb{R}_N} |u_m|^{\frac{2N}{N-2}}\,dx + \int_{B(0,R)} |F(x,u_m)|\,dx + \int_{B(0,R)} |\bar{F}(u_m)|\,dx.$$

First letting $m\to\infty$, $R\to\infty$ and then $\delta\to 0$, the relation (3.11) readily follows. Combining (3.10) and (3.11) we obtain

$$I(u_m) = I^\infty(u_m) + o(1),$$

which implies $I(u_m) \ge I^\infty_{\alpha_\circ} + o(1)$ and consequently $I_{\alpha_\circ} \ge I^\infty_{\alpha_\circ}$ and this contradicts (3.9). Therefore $0 < \int_{\mathbb{R}_N} \bar{u}^2\,dx \le \alpha_\circ$. Suppose that $\lambda = \int_{\mathbb{R}_N} \bar{u}^2\,dx < \alpha_\circ$. Writing

$$\int_{\mathbb{R}_N} |\nabla u_m|^2\,dx = \int_{\mathbb{R}_N} |\nabla \bar{u}|^2\,dx + \int_{\mathbb{R}_N} |\nabla(u_m - \bar{u})|^2\,dx + o(1),$$

$$\int_{\mathbb{R}_N} u_m^2\,dx = \int_{\mathbb{R}_N} \bar{u}^2\,dx + \int_{\mathbb{R}_N} (u_m - \bar{u})^2\,dx + o(1) \qquad (3.13)$$

and

$$\int_{\mathbb{R}_N} F(x, u_m)\, dx = \int_{\mathbb{R}_N} F(x, \bar{u})\, dx + \int_{\mathbb{R}_N} F(x, u_m - \bar{u})\, dx + o(1)$$

we deduce that

$$I_{\alpha_\circ} = I(\bar{u}) + I(u_m - \bar{u}) + o(1).$$

By a similar method used to obtain (3.11) we also have

$$I(u_m - \bar{u}) = I^\infty(u_m - \bar{u}) + o(1).$$

Hence the last two relations yield

$$I_{\alpha_\circ} = I(\bar{u}) + I^\infty(u_m - \bar{u}) + o(1) \geq I_\lambda + I^\infty_{\alpha_\circ - \lambda}$$

and this contradicts (3.9). Consequently $\int_{\mathbb{R}_N} \bar{u}^2\, dx = \alpha_\circ$ and (3.13) yields $u_m \to \bar{u}$ in $L^2(\mathbb{R}_N)$. By the Gagliardo–Nirenberg inequality we have $u_m \to u$ in $L^p(\mathbb{R}_N)$, $2 \leq p < 2^*$. This obviously shows that $I_{\alpha_\circ} = I(\bar{u})$ and $\int_{\mathbb{R}_N} \bar{u}^2\, dx = \alpha_\circ$, that is, $\bar{u}$ is a solution of problem $(\mathcal{I}_{\alpha_\circ})$. Finally, writing

$$\begin{aligned} I_{\alpha_\circ} &= \frac{1}{2}\int_{\mathbb{R}_N} |\nabla \bar{u}|^2\, dx + \frac{1}{2}\int_{\mathbb{R}_N} |\nabla(u_n - \bar{u})|^2\, dx \\ &- \int_{\mathbb{R}_N} \bar{F}(\bar{u})\, dx - \int_{\mathbb{R}_N} \left(F(x, u_m) - \bar{F}(u_m)\right) dx \\ &+ \int_{\mathbb{R}_N} \left(\bar{F}(\bar{u}) - \bar{F}(u_m)\right) dx + o(1) \end{aligned}$$

and using (3.11) we deduce from this that $\nabla u_m \to \nabla \bar{u}$ in $L^2(\mathbb{R}_N)$ and hence $u_m \to \bar{u}$ in $H^1(\mathbb{R}_N)$. □

Corollary 3.1 *Suppose that $F(x,t) \geq \bar{F}(t)$ on $\mathbb{R}_N \times \mathbb{R}$ and that $\bar{F}(\zeta) > 0$ for some $\zeta \in \mathbb{R}$. Then problem $(\mathcal{I}_\lambda)$ is solvable for some $\lambda > 0$.*

Proof. The condition $\bar{F}(\zeta) > 0$ for some $\zeta > 0$ implies that $\int_{\mathbb{R}_N} \bar{F}(u(x))\, dx > 0$ for some $u \in H^1(\mathbb{R}_N)$. Letting $v(x) = u(\frac{x}{\sigma})$, $\sigma > 0$, we have

$$I^\infty(v) = \sigma^N \left(\frac{1}{2\sigma^2} \int_{\mathbb{R}_N} |\nabla u(x)|^2\, dx - \int_{\mathbb{R}_N} \bar{F}(u(x))\, dx \right) < 0$$

for $\sigma > 0$ sufficiently large. Hence there exists $\lambda > 0$ such that $I_\lambda \leq I^\infty_\lambda < 0$ and the result follows from Theorem 3.2. □

3.2 Nonlinear scalar field equations

We study here a minimization problem subject to the constraint involving a gradient. It turns out that a solution of this problem is an eigenfunction of the following eigenvalue problem:

find $u \in H^1(\mathbb{R}_N)$, $u \not\equiv 0$, and $\mu \in \mathbb{R}$ satisfying

$$-\Delta u = \mu f(x,u) \text{ in } \mathbb{R}_N. \tag{3.14}$$

Equation (3.14) arises in many branches of mathematical physics (for example a study of vacuum stability, calculating lifetimes of metastable states) [40].

Throughout this and the next, section we impose the following conditions on f:

(a) $f \in C(\mathbb{R}_N \times \mathbb{R}, \mathbb{R})$.

(b) $\lim_{|x|\to\infty} f(x,t) = \bar{f}(t)$ uniformly for t in bounded subsets of $\mathbb{R}$.

(c) $\lim_{|t|\to\infty} \frac{f(x,t)}{|t|^{2^*-1}} = 0$ uniformly in $x \in \mathbb{R}_N$ and there exists $u \in C_0^\infty(\mathbb{R}_N)$ such that

$$\int_{\mathbb{R}_N} F(x,u)\,dx = \lambda_1 > 0.$$

(d) $-\infty < \liminf_{t\to 0} \frac{f(x,t)}{t} \le \limsup_{t\to 0} \frac{f(x,t)}{t} < 0$.

(e) There exists $\delta > 0$ such that

$$\lim_{|x|\to\infty} \frac{f(x,t) - \bar{f}(t)}{t} = 0$$

uniformly in $|t| < \delta$, $t \neq 0$.

To solve the problem (3.14) we consider the constrained minimization problem

$$(I_\mu) \qquad I_\mu = \inf\{I(u);\, u \in H^1(\mathbb{R}_N),\, \int_{\mathbb{R}_N} |\nabla u|^2\,dx = \mu\},$$

where $I(u) = -\int_{\mathbb{R}_N} F(x,t)\,dx$.

Letting $I^\infty(u) = -\int_{\mathbb{R}_N} \bar{F}(u)\,dx$, we define a variational problem at infinity

$$(I_\mu^\infty) \qquad I_\mu^\infty = \inf\{I^\infty(u); u \in H^1(\mathbb{R}_N); \int_{\mathbb{R}_N} |\nabla u|^2\,dx = \mu\}.$$

We present here an approach in which we avoid the use of the strict subadditivity property of I_μ. It is easy to check that both functions I_μ and I_μ^∞ are continuous on $[0,\infty)$.

Lemma 3.3 *We have $I_\mu^\infty < 0$ for all $\mu > 0$ and $I_\mu^\infty = \mu^{\frac{N}{N-2}} I_1^\infty$.*

Proof. It is sufficient to check the second relation. Setting $\sigma = \mu^{-\frac{1}{N-2}}$, $\mu > 0$, we obtain

$$\begin{aligned} I_\mu^\infty &= \inf\{-\int_{\mathbb{R}_N} \bar{F}(u)\,dx;\ \int_{\mathbb{R}_N} |\nabla u|^2\,dx = \mu\} \\ &= \inf\{-\int_{\mathbb{R}_N} \bar{F}(x)\,dx;\ \int_{\mathbb{R}_N} |\nabla u(\frac{x}{\sigma})|^2\,dx = 1\}. \end{aligned}$$

Letting $y = \sigma x$, we get

$$\begin{aligned} I_\mu^\infty &= \sigma^{-N} \inf\{-\int_{\mathbb{R}_N} \bar{F}(u(\frac{x}{\sigma}))\,dy;\ \int_{\mathbb{R}_N} |\nabla u(\frac{x}{\sigma})|^2\,dx = 1\} \\ &= \sigma^{-N} I_1^\infty = \mu^{\frac{N}{N-2}} I_1^\infty. \end{aligned}$$

□

Lemma 3.3 obviously implies that for every $\lambda > 0$

$$I_\lambda^\infty < I_\alpha^\infty + I_{\lambda-\alpha}^\infty \tag{3.15}$$

for all $\alpha \in (0,\lambda)$.

Lemma 3.4 *Problem (I_λ^∞) is solvable for each $\lambda > 0$. Each minimizing sequence for I_λ^∞ is relatively compact up to a translation in $H^1(\mathbb{R}_N)$.*

Proof. Let $\{u_m\}$ be a minimizing sequence for I_λ^∞. Since $\{u_m\}$ is bounded we may assume that $u_m \rightharpoonup u$ in $H^1(\mathbb{R}_N)$, $u_m \to u$ a. e. on $\mathbb{R}_N$ and $u_m \to u$ in $L^p_{\text{loc}}(\mathbb{R}_N)$, $2 \le p < 2^*$. Suppose that $u \not\equiv 0$, then $0 < \int_{\mathbb{R}_N} |\nabla u|^2\,dx \le \lambda$. We claim that $\int_{\mathbb{R}_N} |\nabla u|^2\,dx = \lambda$. In the contrary case $0 < \bar{\lambda} = \int_{\mathbb{R}_N} |\nabla u|^2\,dx < \lambda$. We now write

$$\int_{\mathbb{R}_N} |\nabla u_m|^2\,dx = \int_{\mathbb{R}_N} |\nabla u|^2\,dx + \int_{\mathbb{R}_N} |\nabla(u_m - u)|^2\,dx + o(1) \tag{3.16}$$

and since by Lemma 1.3

$$\int_{\mathbb{R}_N} \bar{F}(u_m)\,dx = \int_{\mathbb{R}_N} \bar{F}(u)\,dx + \int_{\mathbb{R}_N} \bar{F}(u_m - u)\,dx + o(1),$$

we deduce from (3.16) that

$$I^\infty(u_n) = I^\infty(u) + I^\infty(u_m - u) + o(1).$$

This yields

$$I^\infty(u_m) \geq I^\infty_{\bar\lambda} + I^\infty_{\lambda-\bar\lambda} + o(1).$$

Letting $m \to \infty$ we get

$$I^\infty_\lambda \geq I^\infty_{\bar\lambda} + I^\infty_{\lambda-\bar\lambda}$$

which contradicts (3.15). Since $\int_{\mathbb{R}_N} |\nabla u|^2\,dx = \lambda$, it follows from (3.16) that $\nabla u_m \to \nabla u$ in $L^2(\mathbb{R}_N)$ and the result follows in this case . It remains to consider the case $u \equiv 0$. By Lemma 1.1 either (a) $u_m \to 0$ in $L^\alpha(\mathbb{R}_N)$ for $2 < \alpha < 2^*$ or (b) there exists a sequence $\{y_m\} \subset \mathbb{R}_N$ such that $u_m(\cdot + y_m) \rightharpoonup v \not\equiv 0$ in $H^1(\mathbb{R}_N)$. If (a) occurs then by Corollary 1.1 $\lim_{m\to\infty} \int_{\mathbb{R}_N} \bar F(u_m)\,dx = 0$ and hence $I^\infty_\lambda = 0$ which is impossible. Therefore (b) holds. We let $v_m(\cdot) = u_m(\cdot + y_m)$ and because of translation invariance of I^∞ v_m is also a minimizing sequence for (I^∞_λ). We complete the proof by repeating the argument from the previous part. □

First we formulate the existence result which is parallel to Theorem 3.2 (Case (ii)).

Theorem 3.3 *Suppose that $I_\lambda < I^\infty_\lambda$ for some $\lambda > 0$. Then there exists $\lambda_\circ \in (0, \lambda]$ such that $I_{\lambda_\circ}$ is attained.*

Proof. Let $\{u_m\}$ be a minimizing sequence for I_λ. Assumptions (c) and (d) ensure that $\{u_m\}$ is bounded in $H^1(\mathbb{R}_N)$, $u_m \to u_\circ$ a.e. on $\mathbb{R}_N$. We claim that $u_\circ \not\equiv 0$. In the contrary case repeating the argument used in the proof of Theorem 3.2 we show that

$$\lim_{m\to\infty} \int_{\mathbb{R}_N} (\bar F(u_m) - F(x, u_m))\,dx = 0.$$

Then $I(u_m) = I^\infty(u_m) + o(1)$, that is, $I(u_m) \geq I^\infty_\lambda + o(1)$ and consequently $I_\lambda \geq I^\infty_\lambda$ which is impossible. Therefore $u_\circ \not\equiv 0$ and we write

$$\int_{\mathbb{R}_N} |\nabla u_m|^2\,dx = \int_{\mathbb{R}_N} |\nabla u_\circ|^2\,dx + \int_{\mathbb{R}_N} |\nabla(u_m - u_\circ)|^2\,dx + o(1).$$

Let $0 < \lambda_\circ = \int_{\mathbb{R}_N} |\nabla u_\circ|^2\,dx$. If $\lambda = \lambda_\circ$, then $\nabla u_m \to \nabla u_\circ$ in $L^2(\mathbb{R}_N)$ and the result follows. Therefore we have to consider the case $\lambda_\circ < \lambda$.

Application of Lemma 1.3 gives

$$\int_{\mathbb{R}_N} F(x, u_m)\, dx = \int_{\mathbb{R}_N} F(x, u_\circ)\, dx + \int_{\mathbb{R}_N} F(x, u_m - u_\circ)\, dx + o(1),$$

that is,

$$I(u_m) = I(u_\circ) + I(u_m - u_\circ) + o(1).$$

As in the proof of Theorem 3.2 we check that

$$I(u_m - u_\circ) = I^\infty(u_m - u_\circ) + o(1).$$

The last two relations imply that

$$I_\lambda = I(u_\circ) + I^\infty(u_m - u_\circ) + o(1) \tag{3.17}$$

as $m \to \infty$. On the other hand we always have

$$I_\lambda \leq I_{\lambda_\circ} + I^\infty_{\lambda-\lambda_\circ} \tag{3.18}$$

and

$$I(u_\circ) \geq I_{\lambda_\circ} \text{ and } I^\infty(u_m - u_\circ) \geq I^\infty_{\lambda-\lambda_\circ} + o(1). \tag{3.19}$$

Combining (3.17), (3.18) and (3.19) together we get $I(u_\circ) = I_{\lambda_\circ}$. □

Theorem 3.4 *Suppose $F(x,t) \geq \bar{F}(t)$ on $\mathbb{R}_N \times \mathbb{R}$. Then problem (3.14) (or variational problem (I_λ)) has a solution.*

Proof. According to (c) $\int_{\mathbb{R}_N} F(x,u)\, dx > 0$ for some $u \in H^1(\mathbb{R}_N)$. Hence $I_\lambda < 0$, where $\lambda = \int_{\mathbb{R}_N} |\nabla u|^2\, dx$. We now distinguish two cases: (i) $I_\lambda < I^\infty_\lambda$ and (ii) $I_\lambda = I^\infty_\lambda$.
Case (i). The solvability of problem (I_λ) is a consequence of Theorem 3.3.
Case (ii). By Lemma 3.4 problem (I^∞_λ) has a solution, that is, there exists $\bar{u} \in H^1(\mathbb{R}_N)$ such that $I^\infty_\lambda = I^\infty(\bar{u})$ and $\int_{\mathbb{R}_N} |\nabla \bar{u}|^2\, dx = \lambda$. Hence

$$I_\lambda \leq -\int_{\mathbb{R}_N} F(x, \bar{u})\, dx \leq -\int_{\mathbb{R}_N} \bar{F}(\bar{u})\, dx = I^\infty_\lambda,$$

and this means that $I_\lambda = I(\bar{u}) = I^\infty(\bar{u})$ and the result follows. □

3.3 Behaviour of Palais–Smale sequences

In this section we examine the nature of Palais–Smale sequences which are not relatively compact. We show that a Palais–Smale sequence either strongly converges to its weak limit or differs from it by at least one sequence which after suitable translation converges to a solution of a limit problem. Proposition 1.1 is a particular case of the main result, Theorem 3.5, of this section

To illustrate this phenomenon we consider the problem

$$\begin{cases} -\Delta u + a(x)u &= f(x,u) \text{ in } \mathbb{R}_N, \\ u \in H^1(\mathbb{R}_N). \end{cases} \tag{3.20}$$

We make the following assumptions:

(a_1) $f \in C(\mathbb{R}_N \times \mathbb{R}, \mathbb{R})$ and

$$\lim_{t\to 0} \frac{f(x,t)}{t} = 0 \text{ and } \lim_{|t|\to\infty} \frac{f(x,t)}{|t|^{\frac{N+2}{N-2}}} = 0$$

uniformly in $x \in \mathbb{R}_N$,

(a_2) $a \in C(\mathbb{R}_N, \mathbb{R})$, $a(x) \geqq 0$ on $\mathbb{R}_N$ and $\lim_{|x|\to\infty} a(x) = \bar{a} > 0$,

(a_3) $f'_t(x,t)$ exists and is continuous and satisfies

$$|f'_t(x,t)| \leq C\big(1 + |t|^{\frac{4}{N-2}}\big)$$

for all $(x,t) \in \mathbb{R}_N \times \mathbb{R}$ and some constant $C > 0$,

(a_4) There exist $A \in L^1(\mathbb{R}_N) \cap C(\mathbb{R}_N)$ with $A(x) \geqq 0$ on $\mathbb{R}_N$ and $\lim_{|x|\to\infty} A(x) = 0$ and a constant $\theta \in (0, \frac{1}{2})$ such that

$$A(x) + \theta u f(x,u) \geq F(x,u)$$

for all $(x,u) \in \mathbb{R}_N \times \mathbb{R}$, where $F(x,u) = \int_0^u f(x,t)\,dt$,

(a_5) $\lim_{|x|\to\infty} f(x,t) = \bar{f}(t)$ uniformly on bounded subsets of $\mathbb{R}$.

Let $I : H^1(\mathbb{R}_N) \to \mathbb{R}$ be a variational functional for problem (3.20) given by

$$I(u) = \frac{1}{2}\int_{\mathbb{R}_N} (|\nabla u|^2 + a(x)u^2)\,dx - \int_{\mathbb{R}_N} F(x,u)\,dx.$$

We associate with (3.20) problem at infinity

$$\begin{cases} -\Delta \quad u + \bar{a}u = \bar{f}(u) \text{ in } H^1(\mathbb{R}_N) \\ u \qquad \in H^1(\mathbb{R}_N). \end{cases} \tag{3.21}$$

A variational functional $I^\infty : H^1(\mathbb{R}_N) \to \mathbb{R}$ for (3.21) is given by

$$I^\infty(u) = \frac{1}{2}\int_{\mathbb{R}_N} (|\nabla u|^2 + \bar{u}^2)\, dx - \int_{\mathbb{R}_N} \bar{F}(u)\, dx,$$

where $\bar{F}(u) = \int_0^u \bar{f}(s)\, ds$.

Theorem 3.5 *Suppose that $(a_1), \ldots, (a_5)$ are satisfied. Let $\{u_m\} \subset H^1(\mathbb{R}_N)$ be a sequence such that*

$$I(u_m) \to 0 \text{ and } I'(u_m) \to 0 \text{ in } H^{-1}(\mathbb{R}_N)$$

as $m \to \infty$. Then either the sequence $\{u_m\}$ is relatively compact in $H^1(\mathbb{R}_N)$ or there exist an integer $k \geqq 1$, a solution u° of (3.20), nontrivial solutions $u^1, \ldots, u^k$ of (3.21) and sequences $\{x_m^1\}, \ldots, \{x_m^k\} \subset \mathbb{R}_N$ with $|x_m^i| \to \infty$ $(i = 1, \ldots, k)$ such that for some subsequence of $\{u_m\}$, denoted again by $\{u_m\}$, we have

$$u_m^\circ \equiv u_m \rightharpoonup u^\circ \text{ in } H^1(\mathbb{R}_N), \tag{3.22}$$

$$u_m^j(x) \equiv (u_m^{j-1} - u^{j-1})(x + x_m^j) \rightharpoonup u^j \text{ in } H^1(\mathbb{R}_N), \tag{3.23}$$

$j = 1, \ldots, k$. Furthermore,

$$\|u_m\|_{1,2}^2 \to \sum_{j=1}^k \|u^j\|_{1,2}^2 \tag{3.24}$$

and

$$c = I(u^\circ) + \sum_{j=1}^k I^\infty(u^j). \tag{3.25}$$

Proof. Writing

$$\begin{aligned} c + o(1) &= I(u_m) - \frac{1}{2}\langle I'(u_m), u_m\rangle \\ &= \frac{1}{2}\int_{\mathbb{R}_N} u_m f(x, u_m)\, dx - \int_{\mathbb{R}_N} F(x, u_m)\, dx \end{aligned}$$

$$\geq \left(\frac{1}{2}-\theta\right)\int_{\mathbb{R}_N} f(x,u_m)u_m\,dx - \int_{\mathbb{R}_N} A(x)\,dx,$$

we see that $\int_{\mathbb{R}_N} f(x,u_m)u_m\,dx$ and $\int_{\mathbb{R}_N} F(x,u_m)\,dx$ are bounded . This implies that $\{u_m\}$ is bounded in $H^1(\mathbb{R}_N)$. Therefore we may assume that $u_m \rightharpoonup u^\circ$ in $H^1(\mathbb{R}_N)$ and a.e. on $\mathbb{R}_N$. Since $I'(u_m) \to 0$ in $H^{-1}(\mathbb{R}_N)$, $u_\circ$ is a solution of problem (3.20). Setting $v_m^1 = u_m - u^\circ$ we have

$$\int_{\mathbb{R}_N} (|\nabla v_m^1|^2 + (v_m^1)^2)\,dx = \int_{\mathbb{R}_N} (|\nabla u_m|^2 + u_m^2)\,dx - \int_{\mathbb{R}_N} (|\nabla u^\circ|^2 + (u^\circ)^2)\,dx + o(1)$$

as $m \to \infty$ and by Lemma 1.3

$$\int_{\mathbb{R}_N} F(x,u_m)\,dx = \int_{\mathbb{R}_N} F(x,u^\circ)\,dx + \int_{\mathbb{R}_N} F(x,v_m^1)\,dx + o(1)$$

as $m \to \infty$. The last two relations imply that

$$I(v_m^1) = I(u_m) - I(u^\circ) + o(1). \tag{3.26}$$

Using assumptions (a_2) and (a_5) it is easy to check that

$$\lim_{m\to\infty} \int_{\mathbb{R}_N} (a(x) - \bar{a})(v_m^1)^2\,dx = 0$$

and

$$\lim_{m\to\infty} \int_{\mathbb{R}_N} (F(x,v_m^1) - \bar{F}(v_m^1))\,dx = 0.$$

This combined with (3.26) gives

$$I^\infty(v_m^1) = I(u_m) - I(u^\circ) + o(1). \tag{3.27}$$

Next we claim that

$$I^{\infty\prime}(v_m^1) = I'(u_m) - I'(u^\circ) + o(1) = o(1) \text{ in } H^{-1}(\mathbb{R}_N) \tag{3.28}$$

as $m \to \infty$. Indeed, for every $\phi \in H^1(\mathbb{R}_N)$ we have

$$\begin{aligned}\langle I'(u_m) \quad - \quad I'(u^\circ),\phi\rangle &= \int_{\mathbb{R}_N} (\nabla v_m^1 \nabla\phi + a(x)v_m^1\phi)\,dx \\ &- \int_{\mathbb{R}_N} (f(x,u_m) - f(x,u^\circ))\phi\,dx\end{aligned}$$

and

$$\langle I^{\infty\prime}(v_m^1), \phi\rangle = \int_{\mathbb{R}_n} (\nabla v_m^1 \nabla \phi + \bar{a} v_m^1 \phi)\, dx - \int_{\mathbb{R}_N} \bar{f}(v_m^1)\phi\, dx.$$

Therefore, to justify claim (3.28) we must show that

$$|\int_{\mathbb{R}_N} a(x) v_m^1 \phi\, dx - \int_{\mathbb{R}_N} \bar{a} v_m^1 \phi\, dx| \leqq o(1)\|\phi\|_{1,2} \tag{3.29}$$

and

$$|\int_{\mathbb{R}_N} (f(x, u_m) - f(x, u^\circ) - \bar{f}(v_m^1))\phi\, dx| \leqq o(1)\|\phi\|_{1,2}. \tag{3.30}$$

Since (3.29) is obvious we only check (3.30). For $R > 0$ we write using the mean value theorem and (a_3)

$$\begin{aligned}
&|\int_{\mathbb{R}_N} (f(x, u_m) - f(x, u^\circ) - \bar{f}(v_m^1))\phi\, dx \qquad (3.31)\\
&\leqq |\int_{B(0,R)} (f(x, u_m) - f(x, u^\circ) - \bar{f}(v_m^1))\phi\, dx| \\
&+ |\int_{\mathbb{R}_N - B(0,R)} (f(x, u_m) - f(x, u^\circ) - \bar{f}(v_m^1))\phi\, dx| \\
&\leqq C\left[\left(\int_{B(0,R)} |f(x, u_m) - f(x, u^\circ)|^{\frac{2N}{N+2}}\, dx\right)^{\frac{N+2}{2N}}\right. \\
&+ \left(\int_{B(0,R)} |\bar{f}(v_m^1)|^{\frac{2N}{N+2}}\, dx\right)^{\frac{N+2}{2N}} \\
&+ \left.\left(\int_{\mathbb{R}_N - B(0,R)} (u^\circ)^2\, dx\right)^{\frac{1}{2}} + \left(\int_{\mathbb{R}_N - B(0,R)} |u^\circ|^{\frac{2N}{N-2}}\, dx\right)^{\frac{N+2}{2N}}\right] \|\phi\|_{1,2} \\
&\cdot|\int_{\mathbb{R}_N - B(0,R)} f_u{}'(x, v_m^1 + \theta u^\circ) u^\circ \phi\, dx| \\
&+ |\int_{\mathbb{R}_N - B(0,R)} (f(x, v_m^1) - \bar{f}(v_m^1))\phi\, dx|
\end{aligned}$$

for some constant $C > 0$ and $\theta(x) \in (0, 1)$. By virtue of assumption (a_3)

we have

$$\left|\int_{\mathbb{R}_N - B(0,R)} f_u{}'(x, v_m^1 + \theta u^\circ) u^\circ \phi \, dx\right| \tag{3.32}$$

$$\leq C_1 \left[\left(\int_{\mathbb{R}_N - B(0,R)} (u^\circ)^2 \, dx \right)^{\frac{1}{2}} + \left(\int_{\mathbb{R}_N - B(0,R)} |u^\circ|^{\frac{2N}{N-2}} \right)^{\frac{N-2}{2N}} \right] \|\phi\|_{1,2}$$

for some constant $C_1 > 0$ independent of n and R. Introducing the following quantities

$$\epsilon_1(\delta) = \sup_{0<|t|\leq\delta} \frac{|f(x,t) - \bar{f}(t)|}{|t|}, \quad \delta > 0,$$

$$\epsilon_2(\epsilon) = \sup_{|t|>\frac{1}{\delta}} \frac{|f(x,t) - \bar{f}(t)|}{|t|^{\frac{N+2}{N-2}}}$$

and

$$\epsilon(R) = \sup_{|x|\geq R, \delta\leq|t|\leq\frac{1}{\delta}} |f(x,t) - \bar{f}(t)|,$$

we get the estimate

$$|\int_{\mathbb{R}_N - B(0,R)} (f(x, v_m^1) - \bar{f}(v_m^1)) \phi \, dx| \tag{3.33}$$

$$\leqq \int_{|x|\geq R, |v_m^1|<\delta} + \int_{|x|\geq R, \frac{1}{\delta}\leq|v_m^1|\leq\delta} + \int_{|x|\geq R, |v_m^1|\geq\frac{1}{\delta}} |f(x, v_m^1) - \bar{f}(v_m^1)| \phi \, dx$$

$$\leq \left[\left(\epsilon_1(\delta) + \frac{\epsilon(R)}{\delta} \right) \left(\int_{\mathbb{R}_N} (v_m^1)^2 \, dx \right)^{\frac{1}{2}} + \epsilon_2(\delta) \left(\int_{\mathbb{R}_N} |v_m^1|^{\frac{2N}{N-2}} \right)^{\frac{N+2}{2N}} \right] \|\phi\|_{1,2}.$$

Combining (3.33), (3.32) and (3.31) then letting $m \to \infty$, $R \to \infty$ and finally $\delta \to 0$ we get (3.30). This justifies our claim (3.28). If $\|v_m^1\|_{1,2} \to 0$ as $m \to \infty$, then we are done. So we may assume that

$$\lim_{m\to\infty} \|v_m^1\|_{1,2}^2 = l_1 > 0.$$

If $\liminf_{m\to\infty} \sup_{y\in\mathbb{R}_N} \int_{B(y,R)} (v_m^1)^2 \, dx = 0$, then by Lemma 1.1 $v_m^1 \to 0$ in $L^p(\mathbb{R}_N)$ for $2 < p < 2^*$. This yields

$$\lim_{m\to\infty} \int_{\mathbb{R}_N} \bar{f}(v_m^1) v_m^1 \, dx = 0$$

and by (3.28) we have $v_m^1 \to 0$ in $H^1(\mathbb{R}_N)$ which is impossible. Hence for some $R > 0$ there is $\alpha > 0$ such that $\liminf_{m\to\infty} \sup_{y\in\mathbb{R}_N} \int_{B(y,R)} (v_m^1)^2\, dx \geq \alpha$. Therefore, there exists a sequence $\{y_m\} \subset \mathbb{R}_N$ such that

$$\liminf_{m\to\infty} \int_{B(y_m^1,R)} (v_m^1)^2\, dx \geq \frac{\alpha}{2}. \tag{3.34}$$

Since $v_m^1 \rightharpoonup 0$ in $H^1(\mathbb{R}_N)$, the sequence $\{y_m^1\}$ must be unbounded. Indeed, in the contrary case we choose $R_\circ > 0$ so that $\sup_m(|y_m^1| + R) \leq R_\circ$. Let $\phi \in C_\circ^1(\mathbb{R}_N)$ be such that $0 \leqq \phi(x) \leqq 1$, supp $\phi \subset B(0, 2R_\circ)$ and $\phi(x) = 1$ on $B(0, R_\circ)$. Taking $v_m^1\phi$ as a test function in (3.28) we get

$$\int_{\mathbb{R}_N} (\nabla v_m^1 \nabla(\phi v_m^1) + \bar{a}\phi(v_m^1)^2)\, dx = \int_{\mathbb{R}_N} \bar{f}(v_m^1) v_m^1 \phi\, dx + o(1).$$

Thus

$$\begin{aligned}
&\int_{B(y_m^1,R)} (|\nabla v_m^1|^2 + \bar{a}(v_m^1)^2)\, dx \\
&\quad\leqq \int_{B(0,2R_\circ)} |\bar{f}(v_m^1)v_m^1|\, dx \\
&\quad + \sup_{\mathbb{R}_N} |\nabla\phi| \left(\int_{B(0,2R_\circ)} |\nabla v_m^1|^2\, dx\right)^{\frac{1}{2}} \left(\int_{B(0,2R_\circ)} |v_m^1|^2\, dx\right)^{\frac{1}{2}} + o(1)
\end{aligned}$$

and by the Sobolev embedding theorem and Lemma 1.3 we get

$$\lim_{m\to\infty} \int_{B(y_m^1,R)} (|\nabla v_m^1|^2 + (v_m^1)^2)\, dx = 0$$

which contradicts (3.34). We put

$$u_m^1(x) = v_m^1(x + y_m^1) = u_m(x + y_m^1) - u^\circ(x + y_m^1).$$

Since $\{u_m^1\}$ is bounded in $H^1(\mathbb{R}_N)$ we may assume that $u_m^1 \rightharpoonup u^1$ in $H^1(\mathbb{R}_N)$ and by (3.34) $u \not\equiv 0$. Obviously u^1 is a nontrivial solution of (3.21). We now set $v_m^2 = u_m^1 - u^1$. Then

$$\|v_m^2\|_{1,2}^2 = \|u_m^1\|_{1,2}^2 - \|u^1\|_{1,2}^2 + o(1)$$

as $m \to \infty$ and by Lemma 1.3

$$\int_{\mathbb{R}_N} \bar{F}(u_m^1)\, dx = \int_{\mathbb{R}_N} \bar{F}(u^1)\, dx + \int_{\mathbb{R}_N} \bar{F}(v_m^2)\, dx + o(1)$$

as $m \to \infty$. From the last two relations together with (3.27) we deduce that

$$\begin{aligned} I^\infty(v_m^2) &= I^\infty(u_m^1) - I^\infty(u^1) + o(1) \\ &= I^\infty(v_m^1) - I^\infty(u^1) + o(1) \\ &= I(u_m) - I(u^\circ) - I^\infty(u^1) + o(1). \end{aligned} \tag{3.35}$$

If $\|v_m^2\|_{1,2}^2 \to 0$ we can terminate the proof. In the contrary case $\lim_{m\to\infty} \|v_m^2\|_{1,2}^2 = l_2 > 0$. Repeating the previous argument we show the existence of a sequence $\{y_m^2\} \subset \mathbb{R}_N$ with $|y_m^2| \to \infty$ and a subsequence of $\{v_m^2\}$, relabelled again by $\{v_m^2\}$, such that

$$u_m^2(x) = v_m^2(x + y_m^2) \rightharpoonup u^2 \not\equiv 0$$

in $H^1(\mathbb{R}_N)$. Since

$$I^{\infty\prime}(v_m^2) = I^{\infty\prime}(u_m^1) - I^{\infty\prime}(u^1) + o(1) = o(1),$$

u^2 is a nontrivial solution of (3.21). Iterating this procedure we construct sequences

$$v_m^j = u_m^{j-1} - u^{j-1}, \quad j \geqq 2,$$

and $\{y_m^j\} \subset \mathbb{R}_N$ with $|y_m^j| \to \infty$ such that

$$u_m^j(x) = v_m^j(x + y_m^j) \rightharpoonup u^j(x) \not\equiv 0$$

in $H^1(\mathbb{R}_N)$, where u^j are solutions of (3.21). A simple iteration leads to the following relations

$$\begin{aligned} \|u_m^j\|_{1,2}^2 &= \|v_m^j\|_{1,2}^2 = \|u_m^{j-1}\|_{1,2}^2 - \|u^{j-1}\|_{1,2}^2 + o(1) \\ &= \|u_m\|_{1,2}^2 - \sum_{i=0}^{j-1} \|u^i\|_{1,2}^2 + o(1) \end{aligned}$$

$$\begin{aligned} I^\infty(u_m^j) &= I^\infty(v_m^j) = I^\infty(u^{j-1}) - I^\infty(u^{j-1}) + o(1) \\ &= I(u_m) - I(u^\circ) - \sum_{i=1}^{j-1} I^\infty(u^i) + o(1) \\ &= c - I(u^\circ) - \sum_{i=1}^{j-1} I^\infty(u^i) + o(1). \end{aligned}$$

On the other hand it follows from (a_1) and (a_5) that for every $\epsilon > 0$ there exist a constants $K_1 > 0$ and $K_2 > 0$ such that

$$\bar{f}(t)t \leqq \epsilon t^2 + K_1|t|^p + K_2|t|^{2^*}$$

for all $t \in \mathbb{R}$, where $2 < p < 2^*$. If $v \in H^1(\mathbb{R}_N)$ is a nontrivial solution of (3.21) then by the Sobolev embedding theorem

$$\begin{aligned}\int_{\mathbb{R}_N} (|\nabla v|^2 \quad + \quad v^2)\, dx &= \int_{\mathbb{R}_N} \bar{f}(v)v\, dx \leq \epsilon \int_{\mathbb{R}_N} v^2\, dx + K_1 \int_{\mathbb{R}_N} |v|^p\, dx \\ &+ \int_{\mathbb{R}_N} |v|^{2^*}\, dx \leqq \epsilon \int_{\mathbb{R}_N} v^2\, dx + C_1\|v\|_{1,2}^p + C_2\|v\|_{1,2}^{2^*}\end{aligned}$$

for some constants $C_1 > 0$ and $C_2 > 0$. Choosing $\epsilon < 1$ we see that $\|v\|_{1,2} \geqq C_3 > 0$, where $C_3 > 0$ is a constant independent of v. This shows that our construction must terminate after a final number of steps and this completes the proof. □

3.4 Bibliographical notes

The role of subadditivity conditions for thc constrained minimization has been first noted by P.L. Lions [73], [74], where these conditions were derived using the concentration-compactness principle [73], [74]. In our approach we stress on the derivation of the subadditivity condition on Lemma 1.2 and the vanishing Lemma 1.1. This approach is taken from Gongbao Li–Shusen Yan [69]. The proof of the global compactness theorem 3.5 is taken from Zhu–Cao [119].

Chapter 4
Artificial constraints

The purpose of this chapter is to extend some existence results of Chapter 2 to variational problems with more general nonlinearities. The existence of minimizers will be obtained by solving a minimization problem subject to an artificial constraint. The main point here is to determine the range of a variational functional for which the Palais–Smale condition holds. The important feature of the method presented here is a thourough examination of loss of mass at infinity of sequence of gradients of a Palais–Smale sequence.

4.1 Palais–Smale condition

The method of *artificial constraints* is based on the following observation. If $F : X \to \mathbb{R}$ is a C^1-functional on a Banach space X and if $F'(u_\circ) = 0$, then $\langle F'(u_\circ), u_\circ\rangle = 0$. Therefore it is natural to minimize the functional F on a manifold $\{u \in X - \{0\}; \langle F'(u), u\rangle = 0\}$. In the literature on variational methods this minimizing problem is often refered to as a minimizing problem with an artificial constraint (see [38], [39] and [79]).

We consider the following problem: find $u \in H^1(\mathbb{R}_N)$, $u \not\equiv 0$ on $\mathbb{R}_N$ satisfying the equation

$$-\Delta u + u = f(x, u) \text{ in } \mathbb{R}_N. \tag{4.1}$$

Equation (4.1) is a nonlinear (scalar) field equation of the mathematical physics and is often refered to as a *positive mass case.* A field equation representing a *zero mass case* is obtained by replacing in (4.1) the operator $-\Delta u + u$ by $-\Delta u$. A suitable Sobolev space for the zero mass case is the space $D^{1,2}(\mathbb{R}_N)$ which is a completion of $C_0^\infty(\mathbb{R}_N)$ (see Section 1.1) with respect to the norm

$$\|u\|_{D^{1,2}} = \left(\int_{\mathbb{R}_N} |\nabla u|^2\, dx\right)^{\frac{1}{2}}.$$

A variational approach to (4.1) through the concentration-compactness principle at infinity presented in this chapter can also be extended to a zero mass case. It is assumed that the nonlinearity $f \in C(\mathbb{R}_N \times \mathbb{R}, \mathbb{R})$ satisfies:

(i) $|f(x,t)| \leq A(1+|t|^s)$ for $(x,t) \in \mathbb{R}_N \times \mathbb{R}$, where $1 < s < 2^* - 1$ and $A > 0$ are constants.
(ii) $\lim_{t\to 0} \frac{f(x,t)}{t} = 0$ and $\lim_{|t|\to\infty} \frac{f(x,t)}{t} = \infty$ uniformly in $x \in \mathbb{R}_N$.
(iii) There exists $\mu \in (0, \frac{1}{2})$ such that $\mu t f(x,t) \geq F(x,t)$ for $(x,t) \in \mathbb{R}_N \times \mathbb{R}$, where $F(x,t) = \int_0^t f(x,s)\, ds$.
(iv) $\lim_{|x|\to\infty} f(x,t) = \bar{f}(t)$ uniformly for t in bounded subsets of $\mathbb{R}$.

Let $I : \mathbb{R}_N \to \mathbb{R}$ be a variational functional given by

$$I(u) = \frac{1}{2}\int_{\mathbb{R}_N} (|\nabla u|^2 + u^2)\, dx - \int_{\mathbb{R}_N} F(x,u)\, dx.$$

Critical points of I will be found with the aid of the *mountain pass theorem* (see [2]). The main difficulty is to find the range of level sets of I for which the Palais–Smale condition holds. To overcome this difficulty we introduce a functional $I^\infty : H^1(\mathbb{R}_N) \to \mathbb{R}$ defined by

$$I^\infty(u) = \frac{1}{2}\int_{\mathbb{R}_N} (|\nabla u|^2 + u^2)\, dx - \int_{\mathbb{R}_N} \bar{F}(u)\, dx,$$

where $\bar{F}(u) = \int_0^u \bar{f}(s)\, ds$ and consider the variational problem

$$J^\infty = \inf\{I^\infty(u);\ u \in H^1(\mathbb{R}_N), u \neq 0,\ \langle I^{\infty\prime}(u), u\rangle = 0\}.$$

The main objective of this section is to show that I satisfies the $(PS)_c$ condition for $c \in (0, J^\infty)$. This will be achieved by examining loss of mass at infinity of a weakly convergent minimizing sequence. Let $u_m \rightharpoonup u$ in $H^1(\mathbb{R}_N)$ and as in Chapter 1 we set

$$\beta_\infty = \lim_{R\to\infty} \limsup_{m\to\infty} \int_{|x|\geq R} |\nabla u_m|^2\, dx.$$

The proof of the fact that the $(PS)_c$ condition holds for $c \in (0, J^\infty)$ consists of two parts. We first show that $\beta_\infty = 0$. In the second step the application of Gagliardo–Nirenberg's inequality implies the convergence of $\{u_m\}$ in $L^r(\mathbb{R}_N)$ for $2 < r < 2^*$. This will allow us to deduce that up to a subsequence $u_m \to u$ in $H^1(\mathbb{R}_N)$.

Let $g : \mathbb{R}_N \times \mathbb{R} \to \mathbb{R}$ be a continuous function such that

$$|g(x,u)| \le b_1 |u|^p \text{ for } |u| \ge a_1 \text{ and } x \in \mathbb{R}_N,$$

$$|g(x,u)| \le |u|^q \text{ for } |u| \le a_2 \text{ and } x \in \mathbb{R}_N,$$

where $a_i > 0$, $b_i > 0$, $i = 1, 2$, $2 \le p, q < 2^*$ are constants.

Lemma 4.1 *Suppose that $u_m \rightharpoonup u$ in $H^1(\mathbb{R}_N)$ and $u_m \to u$ in $L^r_{\text{loc}}(\mathbb{R}_N)$ for $2 \le r < 2^*$ and a.e. on $\mathbb{R}_N$. Then there exist sequences $R_n \to \infty$ and $m_n \to \infty$ such that*

$$\begin{aligned} \lim_{n\to\infty} \int_{|x|\ge R_n} |\nabla u_{m_n}|^2 \, dx &= \lim_{n\to\infty} \int_{\mathbb{R}_N} |\nabla u_{m_n}|^2 \phi^2_{R_n} \, dx \\ &= \lim_{n\to\infty} \int_{\mathbb{R}_N} |\nabla (u_{m_n} \phi_{R_n})|^2 \, dx = \beta_\infty, \end{aligned} \tag{4.2}$$

$$\lim_{n\to\infty} \int_{\mathbb{R}_N} u^2_{m_n} |\nabla \phi_{R_n}|^2 \, dx = 0 \tag{4.3}$$

and

$$\lim_{n\to\infty} \int_{R_n \le |x| \le R_n+1} g(x, u_{m_n}) \, dx = 0. \tag{4.4}$$

Proof. We only consider the case $\beta_\infty > 0$. It is clear that

$$\lim_{R\to\infty} \lim_{m\to\infty} \int_{R\le|x|\le R+1} |g(x,u_m)| \, dx = 0.$$

Let $s_1 > 1$ be an integer and choose $R_1 > 0$ such that

$$\begin{aligned} \beta_\infty - \frac{1}{s_1} &\le \limsup_{m\to\infty} \int_{|x|\ge R_1+1} |\nabla u_m|^2 \, dx \\ &\le \limsup_{m\to\infty} \int_{|x|\ge R_1} |\nabla (\phi_{R_1} u_m)|^2 \, dx \le \beta_\infty + \frac{1}{s_1} \end{aligned}$$

and

$$\lim_{m\to\infty}\int_{R_1\le|x|\le R_1+1}|g(x,u_m)|\,dx\le\frac{1}{s_1}.$$

We now choose an integer $m_1>1$ so that

$$\beta_\infty-\frac{2}{s_1}\le\int_{|x|\ge R_1+1}|\nabla u_{m_1}|^2\,dx\le\int_{|x|\ge R_1}|\nabla(u_{m_1}\phi_{R_1})|^2\,dx\le\beta_\infty+\frac{2}{s_1}$$

and

$$\int_{R_1\le|x|\le R_1+1}|g(x,u_{m_1})|\,dx\le\frac{2}{s_1}.$$

Let $s_2>s_1$, we choose $R_2>R_1+2$ such that

$$\begin{aligned}\beta_\infty-\frac{1}{s_2}&\le\limsup_{m\to\infty}\int_{|x|\ge R_2+1}|\nabla u_m|^2\,dx\\&\le\limsup_{m\to\infty}\int_{|x|\ge R_2}|\nabla(u_m\phi_{R_2})|^2\,dx\le\beta_\infty+\frac{1}{s_2}\end{aligned}$$

and

$$\lim_{m\to\infty}\int_{R_2\le|x|\le R_2+1}|g(x,u_m)|\,dx\le\frac{1}{s_2}.$$

Next we choose $m_2>m_1$ so that

$$\beta_\infty-\frac{2}{s_2}\le\int_{|x|\ge R_2+1}|\nabla u_{m_2}|^2\,dx\le\int_{|x|\ge R_2}|\nabla(u_{m_2}\phi_{R_2})|^2\,dx\le\beta_\infty+\frac{2}{s_2}$$

and

$$\int_{R_2\le|x|\le R_2+1}|g(x,u_{m_2})|\,dx\le\frac{2}{s_2}.$$

A simple induction completes the proof of (4.2) and (4.4). We now observe that

$$\begin{aligned}\int_{|x|\ge R_k+1}|\nabla u_{m_k}|^2\,dx&=\int_{|x|\ge R_k+1}|\nabla u_{m_k}|^2\phi_{R_k}^2\,dx\\&\le\int_{|x|\ge R_k}|\nabla u_{m_k}|^2\phi_{R_k}^2\,dx\le\int_{|x|\ge R}|\nabla u_{m_k}|^2\,dx\end{aligned}$$

provided $0 < R \leq R_k$. We may assume that $\int_{|x|\geq R_k} |\nabla u_{m_k}|^2 \phi_{R_k}^2\, dx$ is convergent as $k \to \infty$, since otherwise we can replace $\{R_k\}$ and $\{m_k\}$ by suitable subsequences. Consequently letting $k \to \infty$ we get

$$\begin{aligned} \beta_\infty &= \lim_{k\to\infty} \int_{|x|\geq R_k+1} |\nabla u_{m_k}|^2\, dx = \lim_{k\to\infty} \int_{|x|>R_k+1} |\nabla u_{m_k}|^2 \phi_{R_k}^2\, dx \\ &\leq \lim_{k\to\infty} \int_{|x|\geq R_k} |\nabla u_{m_k}|^2 \phi_{R_k}^2\, dx \leq \limsup_{k\to\infty} \int_{|x|\geq R} |\nabla u_{m_k}|^2\, dx. \end{aligned}$$

Letting $R \to \infty$ we obtain

$$\lim_{R\to\infty} \limsup_{k\to\infty} \int_{|x|\geq R} |\nabla u_{m_k}|^2\, dx \leq \beta_\infty$$

and this implies that

$$\lim_{k\to\infty} \int_{|x|\geq R_k+1} |\nabla u_{m_k}|^2\, dx = \lim_{k\to\infty} \int_{|x|\geq R_k} |\nabla u_{m_k}|^2 \phi_{R_k}^2\, dx = \beta_\infty. \tag{4.5}$$

It follows from the Hölder inequality that

$$\begin{aligned} &|\int_{R_k\leq|x|\leq R_k+1} u_{m_k} \phi_{R_k} \nabla \phi_{R_k} \nabla u_{m_k}\, dx| \\ &\leq \left(\int_{|x|\geq R_k} |\nabla u_{m_k}|^2 \phi_{R_k}^2\, dx - \int_{|x|\geq R_k+1} |\nabla u_{m_k}| \phi_{R_k}^2\, dx\right)^{\frac{1}{2}} \\ &\times \left(\int_{R_k\leq|x|\leq R_k+1} u_{m_k}^2 \nabla \phi_{R_k}|^2\, dx\right)^{\frac{1}{2}}. \end{aligned}$$

From this and (4.5) we deduce that

$$\lim_{k\to\infty} \int_{R_k\leq|x|\leq R_k+1} u_{m_k} \phi_{m_k} \nabla \phi_{R_k} \nabla u_{m_k}\, dx = 0. \tag{4.6}$$

Writing

$$\begin{aligned} \int_{\mathbb{R}_N} |\nabla(u_{m_k}\phi_{R_k})|^2\, dx &= \int_{R_N} u_{m_k}^2 |\nabla \phi_{R_k}|^2\, dx \\ &+ 2\int_{\mathbb{R}_N} u_{m_k} \phi_{R_k} \nabla u_{m_k} \nabla \phi_{R_k}\, dx + \int_{\mathbb{R}_N} |\nabla u_{m_k}|^2 \phi_{R_k}^2\, dx \end{aligned}$$

we deduce (4.3) from (4.5) and (4.6). □

Proposition 4.1 *The functional I satisfies the $(PS)_c$ condition for $c \in (0, J^\infty)$.*

Proof. Let $\{u_m\} \subset H^1(\mathbb{R}_N)$ be such that $I(u_m) \to c \in (0, J^\infty)$ and $I'(u_m) \to 0$ in $H^{-1}(\mathbb{R}_N)$. It is easy to check that $\{u_m\}$ is bounded in $H^1(\mathbb{R}_N)$ and we may assume that $u_m \rightharpoonup u$ in $H^1(\mathbb{R}_N)$, $u_m \to u$ in $L^{s+1}_{loc}(\mathbb{R}_N)$ and a.e. on $\mathbb{R}_N$. We set $\theta_R = 1 - \psi_R(x)$ for $x \in \mathbb{R}_N$. First we estimate the difference $I(u_m) - I(\phi_R u_m)$. Writing

$$\begin{aligned} I(u_m) \quad &- \quad I(\phi_R u_m) \qquad (4.7) \\ &= \frac{1}{2}\int_{\mathbb{R}_N} \left(|\nabla u_m|^2 + u_m^2 - |\nabla(\phi_R u_m)|^2 - (\phi_R u_m)^2\right) dx \\ &- \int_{\mathbb{R}_N} F(x, u_m)\, dx + \int_{\mathbb{R}_N} F(x, u_m \phi_R)\, dx, \end{aligned}$$

we have

$$\begin{aligned} \int_{|x|\le R} &|\nabla u_m|^2\, dx + \int_{R\le|x|\le R+1} \left(|\nabla u_m|^2 - |\nabla(\phi_R u_m)|^2\right) dx \qquad (4.8) \\ &= \int_{|x|\le R} |\nabla u_m|^2\, dx + \int_{R\le|x|\le R+1} |\nabla u_m|^2 \theta_R^2\, dx \\ &- 2\int_{\mathbb{R}_N} |\nabla u_m|^2 \phi_R^2\, dx \\ &+ 2\int_{\mathbb{R}_N} |\nabla u_m|^2 \phi_R\, dx - 2\int_{\mathbb{R}_N} u_m \phi_R \nabla\phi_R \nabla u_m\, dx \\ &- \int_{\mathbb{R}_N} u_m^2 |\nabla\phi_R|^2\, dx \\ &= \int_{\mathbb{R}_N} |\nabla u_m|^2 \theta_R^2\, dx - 2\int_{\mathbb{R}_N} u_m \phi_R \nabla\phi_R \nabla u_m\, dx \\ &- \int_{\mathbb{R}_N} u_m^2 |\nabla\phi_R|^2\, dx + F^1_{R,m} \\ &= \int_{\mathbb{R}_N} |\nabla(u_m\theta_R)|^2\, dx - 2\int_{\mathbb{R}_N} u_m \phi_R \nabla\phi_R \nabla u_m\, dx \\ &- 2\int_{\mathbb{R}_N} u_m \theta_R \nabla u_m \nabla\theta_R\, dx - \int_{\mathbb{R}_N} u_m^2 |\nabla\phi_R|^2\, dx + F^1_{R,m}, \end{aligned}$$

where

$$F^1_{R,m} = 2\int_{\mathbb{R}_N} |\nabla u_m|^2 (\phi_R - \phi_R^2)\, dx,$$

and

$$\int_{\mathbb{R}_N} u_m^2\,dx - \int_{\mathbb{R}_N} \phi_R^2 u_m^2\,dx \tag{4.9}$$
$$= \int_{|x|\le R} u_m^2\,dx + \int_{R\le|x|\le R+1} (u_m^2 - \phi_R^2 u_m^2)\,dx$$
$$= \int_{|x|\le R} u_m^2\,dx + \int_{R\le|x|\le R+1} u_m^2(1+\phi_R^2 - 2\phi_R)\,dx$$
$$+ \; 2\int_{\mathbb{R}_N} (\phi_R - \phi_R^2)u_m^2\,dx = \int_{\mathbb{R}_N} u_m^2\theta_R^2\,dx + F_{R,m}^2,$$

where

$$F_{R,m}^2 = 2\int_{\mathbb{R}_N} (\phi_R - \phi_R^2)u_m^2\,dx.$$

Furthermore, we have

$$\int_{\mathbb{R}_N} F(x,u_m)\,dx - \int_{\mathbb{R}_N} F(x,\phi_R u_m)\,dx = \int_{\mathbb{R}_N} \theta_R F(x,u_m)\,dx \tag{4.10}$$
$$+ \int_{\mathbb{R}_N} \phi_R F(x,u_m)\,dx - \int_{\mathbb{R}_N} F(x,\phi_R u_m)\,dx$$
$$= \int_{\mathbb{R}_N} F(x,\theta_R u_m)\,dx + \int_{\mathbb{R}_N} \big(\theta_R F(x,u_m) - F(x,\theta_R u_m)\big)\,dx$$
$$+ \int_{\mathbb{R}_N} \big(\phi_R F(x,u_m) - F(x,\phi_R u_m)\big)\,dx.$$

Inserting (4.8), (4.9) and (4.10) into (4.7) and using the fact that $F_{R,m}^j \ge 0$, $j = 1, 2$, we obtain

$$I(u_m) - I(u_m\phi_R) \ge I(u_m\theta_R) - \int_{\mathbb{R}_N} u_m^2|\nabla\phi_R|^2\,dx \tag{4.11}$$
$$- \int_{\mathbb{R}_N} u_m\phi_R\nabla\phi_R\nabla u_m\,dx - \int_{\mathbb{R}_N} u_m\theta_R\nabla u_m\nabla\theta_R\,dx$$
$$- \; F_{R,m}^3 - F_{R,m}^4,$$

where

$$F_{R,m}^3 = \int_{\mathbb{R}_N} \big(\theta_R F(x,u_m) - F(x,\theta_R u_m)\big)\,dx$$

and

$$F^4_{R,m} = \int_{\mathbb{R}_N} (\phi_R F(x,u_m) - F(x,\phi_R u_m))\, dx.$$

Since $\lim_{m\to\infty}\langle I'(u_m), u_m\theta_R^2\rangle = 0$ uniformly in $R \geq 1$, we have

$$\begin{aligned}\int_{\mathbb{R}_N} & |\nabla(u_m\theta_R)|^2\, dx + \int_{\mathbb{R}_N} (u_m\theta_R)^2\, dx \\ &= \int_{\mathbb{R}_N} u_m^2 |\nabla\theta_R|^2\, dx + \int_{\mathbb{R}_N} f(x,u_m)u_m\theta_R^2\, dx + o(1).\end{aligned}$$

From this we derive that

$$\begin{aligned}\int_{\mathbb{R}_N} |\nabla(u_m\theta_R)|^2 &+ \int_{\mathbb{R}_N} (u_m\theta_R)^2 - \int_{\mathbb{R}_N} f(x,u_m\theta_R)u_m\theta_R\, dx \quad (4.12)\\ &= \int_{\mathbb{R}_N} u_m^2|\nabla\theta_R|^2\, dx + F^5_{R,m} + o(1),\end{aligned}$$

where

$$F^5_{R,m} = \int_{\mathbb{R}_N} \left(f(x,u_m)u_m\theta_R^2 - f(x,u_m\theta_R)u_m\theta_R\right) dx$$

Combining (4.12) with (4.11) we obtain

$$\begin{aligned} I(u_m) &- I(u_m\phi_R) \geq \frac{1}{2}\int_{\mathbb{R}_N} (|\nabla(u_m\theta_R)|^2 + (u_m\theta_R)^2) \qquad (4.13)\\ &- \mu\int_{\mathbb{R}_N} f(x,u_m\theta_R)u_m\theta_R\, dx \\ &- \int_{\mathbb{R}_N} u_m^2|\nabla\phi_R|^2\, dx - \int_{\mathbb{R}_N} u_m\phi_R\nabla\phi_R\nabla u_m\, dx \\ &- \int_{\mathbb{R}_N} u_m\theta_R\nabla u_m\nabla\theta_R\, dx - F^3_{R,m} - F^4_{R,m} \\ &\geq (\frac{1}{2}-\mu)\int_{\mathbb{R}_N} (|\nabla(u_m\theta_R)|^2 + (u_m\theta_R)^2)\, dx - \int_{\mathbb{R}_N} u_m^2|\nabla\phi_R|^2\, dx \\ &- \int_{\mathbb{R}_N} u_m\phi_R\nabla\phi_R\nabla u_m\, dx - \int_{\mathbb{R}_N} u_m\theta_R\nabla u_m\nabla\theta_R\, dx \\ &+ \mu\int_{\mathbb{R}_N} u_m^2|\nabla\theta_R|^2\, dx - F^3_{R,m} - F^4_{R,m} + \mu F^5_{R,m} + o(1) \\ &\geq -\int_{\mathbb{R}_N} u_m^2|\nabla\phi_R|^2\, dx - \int_{\mathbb{R}_N} u_m\phi_R\nabla\phi_R\nabla u_m\, dx \end{aligned}$$

$$- \int_{\mathbb{R}_N} u_m\theta_R \nabla u_m \nabla\theta_R \, dx - F^3_{R,m} - F^4_{R,m} + \mu F^5_{R,m} + o(1).$$

Writing

$$I(\phi_R u_m) - I^\infty(\phi_R u_m) = \int_{\mathbb{R}_N} \big(\bar{F}(\phi_R u_m) - F(x, \phi_R u_m)\big)\, dx$$

and setting

$$\begin{aligned} G^1_{R,m} &= -\int_{\mathbb{R}_N} u_m^2 |\nabla\phi_R|^2 \, dx - \int_{\mathbb{R}_N} u_m\phi_R \nabla\phi_R \nabla u_m \, dx \\ &- \int_{\mathbb{R}_N} u_m\theta_R \nabla u_m \nabla\theta_R \, dx - F^3_{R,m} - F^4_{R,m} + \mu F^5_{R,m} + o(1) \end{aligned}$$

and

$$G^2_{R,m} = \int_{\mathbb{R}_N} \big(\bar{F}(\phi_R u_m) - F(x, \phi_R u_m)\big)\, dx$$

we derive from (4.13) the estimate

$$I(u_m) - I^\infty(\phi_R u_m) \geq G^1_{R,m} + G^2_{R,m} + o(1). \tag{4.14}$$

Let $v_m(x) = \big(\phi_R u_m\big)(\sigma x)$, $\sigma > 0$, and we have

$$\begin{aligned} \langle I^{\infty\prime}(v_m), v_m\rangle &= \sigma^{2-N} \int_{\mathbb{R}_N} |\nabla(\phi_R u_m)|^2 \, dx \\ &+ \sigma^{-N} \int_{\mathbb{R}_N} (\phi_R u_m)^2 \, dx \\ &- \sigma^{-N} \int_{\mathbb{R}_N} \bar{f}(u_m\phi_R) u_m\phi_R \, dx \\ &= \sigma^{-N}(\sigma^2 - 1) \int_{\mathbb{R}_N} |\nabla(\phi_R u_m)|^2 \, dx \\ &+ \sigma^{-N} \langle I^{\infty\prime}(u_m\phi_R), u_m\phi_R\rangle. \end{aligned} \tag{4.15}$$

Since $\langle I'(u_m), u_m\phi_R^2\rangle = o(1)$ uniformly for $R \geq 1$, we have

$$\begin{aligned} \langle I^{\infty\prime}(u_m\phi_R), u_m\phi_R\rangle &= \int_{\mathbb{R}_N} u_m^2 |\nabla\phi_R|^2 \, dx \\ &+ \int_{\mathbb{R}_N} \big(f(x, u_m) u_m \phi_R^2 - \bar{f}(\phi_R u_m)\phi_R u_m\big)\, dx + o(1). \end{aligned} \tag{4.16}$$

Our objective is to show that

$$\beta_\infty = \lim_{R\to\infty} \limsup_{m\to\infty} \int_{\mathbb{R}_N} |\nabla(\phi_R u_m)|^2 \, dx = 0.$$

Arguing indirectly we assume that $\beta_\infty > 0$. It follows from Lemma 4.1 that there exist sequences $\{R_n\}$ and $\{u_{m_n}\}$ such that (4.2) and (4.3) are satisfied. According to (4.4) of Lemma4.1 $\{R_n\}$ and $\{u_{m_n}\}$ can be chosen so that $\lim_{n\to\infty} F^j_{R_n,m_n} = 0$, $j = 3,4,5$. Hence also $\lim_{n\to\infty} G^1_{R_n,m_n} = 0$. We now show that

$$\lim_{R\to\infty} G^2_{R,m_n} = 0 \tag{4.17}$$

uniformly in m. By virtue of (i) and (iv) for every $\epsilon > 0$ there corresponds a constant $T > 0$ such that

$$|\bar{F}(u)|, |F(x,u)| \leq \frac{\epsilon}{2}|u|^{2^*}$$

for $|u| \geq T$. Writing

$$|G^2_{R,m}| \leq \int_{|u_m|\leq T} |\bar{F}(\phi_R u_m) - F(x, \phi_R u_m)| \, dx + \frac{\epsilon}{2} \sup_{m\geq 1} \int_{\mathbb{R}_N} |u_m|^{2^*} \, dx$$

we deduce from (iv) that (4.17) holds. In a similar manner we show that

$$\lim_{n\to\infty} \int_{\mathbb{R}_N} \left(f(x, u_{m_n}) u_{m_n} \phi^2_{R_n} - \bar{f}(\phi_{R_n} u_{m_n}) \phi_{R_n} u_{m_n}\right) dx = 0. \tag{4.18}$$

We now choose for each n, σ_n so that

$$(\sigma_n^2 - 1) \int_{\mathbb{R}_N} |\nabla(\phi_{R_n} u_{m_n})|^2 \, dx = -\langle {I^\infty}'(u_{m_n}\phi_{R_n}), u_{m_n}\phi_{R_n}\rangle \tag{4.19}$$

Since $\beta_\infty > 0$ we deduce from (4.16),(4.18) and (4.3) that $\lim_{n\to\infty} \sigma_n = 1$. With this choice of σ_n by (4.15) we have

$$\langle {I^\infty}'(v_{m_n}), v_{m_n}\rangle = 0 \tag{4.20}$$

for each n. Next we evaluate the value of I^∞ at $v_{m_n}(x) = \left(\phi_{R_n} u_{m_n}\right)(\sigma_n x)$. Using the fact that $\lim_{n\to\infty} \sigma_m = 1$ we get

$$\begin{aligned} I^\infty(v_{m_n}) &= \frac{1}{2}\sigma_n^{2-N} \int_{\mathbb{R}_N} |\nabla(\phi_{R_n} u_{m_n})|^2 \, dx \\ &+ \frac{\sigma_n^{-N}}{2} \int_{\mathbb{R}_N} \left(u_{m_n}\phi_{R_n}\right)^2 dx - \sigma_n^{-N} \int_{\mathbb{R}_N} \bar{F}(u_{m_n}\phi_{R_n}) \, dx \end{aligned} \tag{4.21}$$

$$
\begin{aligned}
&= \frac{1}{2}\sigma_n^{-N}(\sigma_n^2 - 1)\int_{\mathbb{R}_N} |\nabla(\phi_{R_n} u_{m_n})|^2\,dx \\
&+ (\sigma_n^{-N} - 1) I^\infty(\phi_{R_n} u_{m_n}) \\
&+ I^\infty(\phi_{R_n} u_{m_n}) = I^\infty(\phi_{R_n} u_{m_n}) + o(1).
\end{aligned}
$$

Letting $n \to \infty$ we deduce from (4.14), (4.20) and (4.21) that $c \geq J_\infty$ and we have arrived at a contradiction. Therefore $\beta_\infty = 0$. To proceed further we set

$$\alpha_\infty = \lim_{R\to\infty} \limsup_{m\to\infty} \int_{\mathbb{R}_N} |u_m \phi_R|^{s+1}\,dx.$$

By the Sobolev embedding theorem we have

$$\limsup_{m\to\infty} \int_{\mathbb{R}_N} |u_m|^{s+1}\,dx = \int_{\mathbb{R}_N} |u|^{s+1}\,dx + \alpha_\infty.$$

Let us now recall the Gagliardo - Nirenberg inequality: for each $u \in H^1(\mathbb{R}_N)$ we have

$$\|u\|_{s+1} \leq C\|\nabla u\|_2^\alpha \|u\|_2^{1-\alpha}$$

with $0 < \alpha < 1$ and $C > 0$ being constants independent of u. We apply this inequality to $\phi_R u_m$ and deduce that $\alpha_\infty = 0$. Therefore $u_m \to u$ in $L^{s+1}(\mathbb{R}_N)$ and by the dominated convergence theorem we get

$$\lim_{n\to\infty} \int_{\mathbb{R}_N} f(x, u_{m_n}) u_{m_n}\,dx = \int_{\mathbb{R}_N} f(x,u)u\,dx.$$

On the other hand, u satisfies equation (4.1) and this implies

$$
\begin{aligned}
\lim_{n\to\infty} \int_{\mathbb{R}_N} \left(|\nabla u_{m_n}|^2 + u_{m_n}^2\right) dx &= \lim_{n\to\infty} \int_{\mathbb{R}_N} f(x, u_{m_n}) u_{m_n}\,dx \\
&= \int_{\mathbb{R}_N} \left(|\nabla u|^2 + u^2\right) dx
\end{aligned}
$$

and the result easily follows. □

Remark 4.1 If $\bar{f} \not\equiv 0$, then $J^\infty > 0$.

Indeed, we have

$$\|u\|_{1,2}^2 = \int_{\mathbb{R}_N} u\bar{f}(u)\,dx \leqq \epsilon\|u\|_{1,2}^2 + C_\epsilon \|u\|_{1,2}^{2^*}.$$

Selecting $0 < \epsilon < 1$ we get $(1-\epsilon)\|u\|_{1,2}^2 \leqq C_\epsilon \|u\|_{1,2}^{2^*}$. Hence we can find $c > 0$ such that $c \leqq \|u\|_{1,2}$ for each $u \not\equiv 0$ satisfying $\langle I^{\infty\prime}(u), u\rangle = 0$. On the other hand by (iii) we have

$$\begin{aligned} I^\infty(u) &= \frac{1}{2}\|u\|_{1,2}^2 - \int_{\mathbb{R}_N} \bar{F}(u)\,dx = \frac{1}{2}\int_{\mathbb{R}_N} u\bar{f}(u)\,dx - \int_{\mathbb{R}_N} \bar{F}(u)\,dx \\ &\geqq \left(\frac{1}{2}-\mu\right)\int_{\mathbb{R}_N} u\bar{f}(u)\,dx = \left(\frac{1}{2}-\mu\right)\|u\|_{1,2}^2 \geqq \left(\frac{1}{2}-\mu\right)c^2 > 0. \end{aligned}$$

Proposition 4.2 *If f is independent of x, that is, $f(x,u) = \bar{f}(u)$ and $\bar{f} \not\equiv 0$, then each sequence $\{u_m\} \subset H^1(\mathbb{R}_N)$ satisfying $I^\infty(u_m) \to J^\infty$ and $I^{\infty\prime}(u_m) \to 0$ in $H^{-1}(\mathbb{R}_N)$ is relatively compact up to a translation.*

Proof. Since $\{u_m\}$ is bounded in $H^1(\mathbb{R}_N)$ we may assume that $u_m \rightharpoonup u$ in $H^1(\mathbb{R}_N)$. We now use Lemma 1.1 to show that $u \not\equiv 0$. In the contrary case $u_m \rightharpoonup 0$ in $H^{-1}(\mathbb{R}_N)$ and either (i)

$$\liminf_{m\to\infty} \sup_{y\in\mathbb{R}_N} \int_{B(0,1)} |u_m|^2\,dx = 0$$

or (ii)

$$\liminf_{m\to\infty} \sup_{y\in\mathbb{R}_N} \int_{B(y,1)} |u_m|^2\,dx > 0.$$

If (i) occurs then by Lemma 1.1 $u_m \to 0$ in $L^r(\mathbb{R}_N)$ for all $2 < r < 2^*$. Using the fact that $I^{\infty\prime}(u_m) \to 0$ in $H^{-1}(\mathbb{R}_N)$ we see that $u_m \to 0$ in $H^1(\mathbb{R}_N)$ and this is impossible. Hence the case (ii) prevails, which implies the existence of a sequence $\{y_m\} \subset \mathbb{R}_N$ such that $u_m(\cdot + y_m) \rightharpoonup u \not\equiv 0$ in $H^1(\mathbb{R}_N)$. Due to the translation invariance of I^∞, $\{u_m(\cdot + y_m)\}$ is still the $(PS)_c$ sequence for I^∞. Let

$$\beta_\infty = \lim_{R\to\infty} \limsup_{m\to\infty} \int_{|x|>R} \left(|\nabla u_m|^2 + u_m^2\right) dx.$$

We now show that $\beta_\infty = 0$. Arguing indirectly we assume that $\beta_\infty > 0$. The limit function u is obviously a solution of equation (4.1). Hence $\langle I^{\infty\prime}(u), u\rangle = 0$ and consequently

$$J^\infty \leqq \frac{1}{2}\int_{\mathbb{R}_N} \left(|\nabla u|^2 + u^2\right) dx - \int_{\mathbb{R}_N} \bar{F}(u)\,dx. \tag{4.22}$$

Since $\langle I^{\infty\prime}(u_m), u_m\phi_R\rangle \to 0$ as $m \to \infty$, uniformly in $R \geqq 1$, we get

$$\int_{\mathbb{R}_N} (\nabla u_m u_m \nabla\phi_R + |\nabla u_m|^2\phi_R + u_m^2\phi_R)\, dx = \int_{\mathbb{R}_N} u_m \bar{f}(u_m)\phi_R\, dx + o(1),$$

which yields

$$\beta_\infty = \lim_{R\to\infty} \limsup_{m\to\infty} \int_{\mathbb{R}_N} u_m \bar{f}(u_m)\phi_R\, dx. \tag{4.23}$$

On the other hand we deduce from the relation

$$J^\infty + \int_{\mathbb{R}_N} \bar{F}(u_m)\, dx + o(1) = \frac{1}{2}\int_{\mathbb{R}_N} (|\nabla u_m|^2 + u_m^2)\, dx$$

that

$$\begin{aligned} J^\infty \quad &+ \int_{\mathbb{R}_m} \bar{F}(u)\, dx + \lim_{R\to\infty} \limsup_{m\to\infty} \int_{|x|\geqq R} \bar{F}(u_m)\, dx \\ &\geq \frac{1}{2}\int_{\mathbb{R}_N} (|\nabla u|^2 + u^2)\, dx + \beta_\infty. \end{aligned}$$

Using (4.22) we obtain

$$\lim_{R\to\infty} \limsup_{m\to\infty} \int_{|x|\geqq R} \bar{F}(u_m)\, dx \geqq \beta_\infty. \tag{4.24}$$

Combining (4.23), (4.24) and assumption (iii) we get

$$\begin{aligned} \beta_\infty \quad &\leqq \lim_{R\to\infty} \limsup_{m\to\infty} \int_{|x|\geqq R} \bar{F}(u_m)\, dx \leqq \mu \lim_{R\to\infty} \limsup_{m\to\infty} \int_{|x|\geqq R} u_m \bar{f}(u_m)\, dx \\ &< \lim_{R\to\infty} \limsup_{m\to\infty} \int_{|x|\geqq R} u_m \bar{f}(u_m)\, dx = \beta_\infty, \end{aligned}$$

which is impossible. To complete the proof we repeat the final part of the proof of Proposition 4.1. □

4.2 Existence result

The existence result for problem (4.1) is based on the mountain pass theorem. In this section we maintain the assumptions (i) - (iv) of Section 4.1.

Lemma 4.2 *There exist constants $\alpha > 0$ and $\rho > 0$ such that*

$$I(u) \geq \alpha \text{ for } \|u\| = \rho.$$

Proof. By virtue of assumptions (i) and (ii) given $\epsilon > 0$ we can find $C_\epsilon > 0$ such that

$$F(x, u) \leq \epsilon u^2 + C_\epsilon |u|^{2^*}$$

for $(x, u) \in \mathbb{R}_N \times \mathbb{R}$. Thus we have by the Sobolev embedding theorem

$$I(u) \geq \frac{1}{2}\|u\|^2 - \epsilon\|u\|^2 - C_\epsilon S^{-\frac{2^*}{2}} \|u\|^{2^*}.$$

Selecting $\epsilon > \frac{1}{2}$ the result follows. □

Lemma 4.3 *Let $u \in H^1(\mathbb{R}_N)$ be a function with compact support ω such that $u \geq 0$ on $\mathbb{R}_N$ and $\|u\|_{1,2} = 1$. Then there exists $R_o > 0$ such that*

$$I(Ru) < 0 \text{ for } R \geq R_o.$$

Proof. By assumption (ii) $\lim_{t\to\infty} \frac{f(x,t)}{t} = \infty$ uniformly in $x \in \mathbb{R}_N$. Therefore to every $k > 0$ there corresponds $M = M(k) > 0$ such that $f(x, s) \geq ks$ for $s \geq M$. Thus

$$F(x, s) \geq \frac{ks^2}{2} + \int_0^M f(x, t)\, dt - \frac{k}{2}M^2$$

for $s \geq M$. Using this we derive the following estimate

$$\begin{aligned} \int_{\mathbb{R}_N} F(x, Ru)\, dx &= \int_{x\in\omega, u\geq \frac{M}{R}} F(x, Ru)\, dx \\ &+ \int_{x\in\omega, u\leq \frac{M}{R}} F(x, Ru)\, dx \\ &\geq \frac{kR^2}{2} \int_{x\in\omega, u\geq \frac{M}{R}} u^2\, dx + \int_{x\in\omega, u\geq \frac{M}{R}} \int_0^M f(x, t)\, dt\, dx \\ &- C + \int_{x\in\omega, u\leq \frac{M}{R}} F(x, Ru)\, dx \end{aligned} \tag{4.25}$$

where $C > 0$ is a constant depending on $M > 0$. Next, there exists a constant $C_1 > 0$ depending on M such that

$$|\int_{x\in\omega, u\geq \frac{M}{R}} \int_0^M f(x,t)\,dt\,dx| \leq C_1 \text{ and } |\int_{x\in\omega, u\leq \frac{M}{R}} F(x,Ru)\,dx| \leq C_1. \tag{4.26}$$

Combining (4.25) and (4.26) we get

$$\int_{\mathbb{R}_N} F(x,Ru)\,dx \geq \frac{kR^2}{2} \int_{x\in\omega, u\geq \frac{M}{R}} u^2\,dx - C_2$$

for some constant $C_2 > 0$ depending on M. We next fix $R_o > 0$ so that

$$\int_{x\in\omega, u\geq \frac{M}{R}} u^2\,dx \geq \frac{1}{2} \int_{\mathbb{R}_N} u^2\,dx$$

for $R \geq R_o$. Choosing $k = 4\left(\int_{\mathbb{R}_N} u^2\,dx\right)^{-1}$ we obtain the following estimate for $I(Ru)$

$$I(Ru) \leq -R^2 + C_2$$

for $R \geq R_o$. This obviously implies the assertion of our lemma. □

We now define a mountain pass level for I. By Lemma 4.3 there exists $e \in H^1(\mathbb{R}_N)$ such that $I(e) < 0$ and $\|e\|_{1,2} > \rho$. Let

$$\Gamma = \{\gamma \in C([0,1], H^1(\mathbb{R}_N));\ \gamma(0) = 0,\ \gamma(1) = e\}$$

and set

$$c = \inf_{\gamma\in\Gamma} \max_{t\in[0,1]} I(\gamma(t)).$$

The following existence result is a consequence of Proposition 4.1.

Theorem 4.1 *Suppose that there exists a path $\gamma_o \in \Gamma$ such that*

$$\max_{t\in[0,1]} I(\gamma_o(t)) < J^\infty. \tag{4.27}$$

Then problem (4.1) has at least one nontrivial solution.

4.3 Critical points

We denote by c_∞ the mountain pass level of I^∞, that is,

$$c_\infty = \inf_{\gamma\in\Gamma}\max_{t\in[0,1]} I^\infty(\gamma(t)),$$

where

$$\Gamma = \{\gamma \in C([0,1], H^1(\mathbb{R}_N)),\ \gamma(0) = 0,\ \gamma(1) = e\}$$

with $I^\infty(e) < 0$.

Proposition 4.3 *Suppose that f is independent of x, that is, $f(x,u) = \bar{f}(u)$, $\bar{f} \not\equiv 0$ and that $\frac{\bar{f}(t)}{t}$ is nondecreasing for $|t| > 0$. Then there exists $u \in H^1(\mathbb{R}_N)$ such that $I^\infty(u) = J^\infty$, that is, problem (4.1) has a nontrivial solution.*

Proof. Let $\{u_m\}$ be a minimizing sequence for J^∞. We set

$$g(t) = I^\infty(tu_m) = \frac{t^2}{2}\int_{\mathbb{R}_N}(|\nabla u_m|^2 + u_m^2)\,dx - \int_{\mathbb{R}_N}\bar{F}(tu_m)\,dx$$

and we have

$$\begin{aligned} g'(t) &= t\int_{\mathbb{R}_N}(|\nabla u_m|^2 + u_m^2)\,dx - \int_{\mathbb{R}_N} u_m\bar{f}(tu_m)\,dx \\ &= t\int_{\mathbb{R}_N} u_m^2\left(\frac{\bar{f}(u_m)}{u_m} - \frac{\bar{f}(tu_m)}{tu_m}\right)dx. \end{aligned}$$

If $t \in (0,1)$, then $g'(t) \geq 0$, while for $t \geqq 1$, $g'(t) \leqq 0$. Thus $g(1) = I^\infty(u_m) = \max_{t\geqq 0} I^\infty(tu_m)$. Hence we may construct a path γ_{u_m} for each u_m such that (see [46])

$$c_\infty \leqq \max_{\gamma_{u_m}} I(u) = I(u_m),$$

and therefore $J^\infty \geqq c_\infty$. By the mountain pass (see [2]) theorem, there exists $\{u_m\} \subset H^1(\mathbb{R}_N)$ such that $I^{\infty\prime}(u_m) \to 0$ and $I^\infty(u_m) \to c_\infty$. Since $c_\infty \leqq J^\infty$, it follows from Proposition 4.2 that $\{u_m\}$ is relatively compact up to a translation and this completes the proof. □

Proposition 4.4 *Suppose that (i) - (iv) hold, $f(x,t) \geqq 0$ for $t \geq 0$, $f(x,t) \geq \bar{f}(t)$ for $(x,t) \in \mathbb{R}_N \times \mathbb{R}$, $a(x) \leq \bar{a}$ for $x \in \mathbb{R}_N$ and that $\frac{f(x,t)}{t}$ is nondecrasing for $|t| > 0$. Then problem (4.1) has nontrivial solution.*

Proof. We may assume that $F(x,u) \not\equiv \bar{F}(u)$ and that $f(x,u) = 0$ for $t \leqq 0$. By Proposition 4.3 there exists $u_\circ$ such that $J^\infty = I^\infty(u_\circ)$. Since $I^\infty(u_\circ) = I^\infty(|u_\circ|)$, by the maximum principle we may assume that $u_\circ(x) > 0$ on $\mathbb{R}_N$. By virtue of (iii) $I(tu_\circ) < 0$ for t sufficiently large. As in Proposition 4.3 we show that $I^\infty(u_\circ) = \max_{t \geqq 0} I^\infty(tu_\circ)$. Then

$$J^\infty = I^\infty(u_\circ) \geqq I^\infty(t_\circ u_\circ) > I(t_\circ u_\circ) = \sup_{t \geqq 0} I(tu_\circ).$$

We can now construct a path $\gamma_{u_\circ}$ such that $c \leqq \max_{\gamma_{u_\circ}} I(u) < J^\infty$. This shows that condition (4.27) is satisfied. □

In Propositions 4.3 and 4.4 we have verified condition (4.27) assuming that $\frac{\bar{f}(t)}{|t|}$ and $\frac{f(x,t)}{|t|}$ are nondecreasing for $t \neq 0$. We now give an example of a nonlinearity f satisfying (i) - (iv) for which the above monotonicity conditions do not hold, nonetheless (4.27) is satisfied.

Proposition 4.5 *Let*

$$f(x,t) = Q_1(x)|t|^{q_1-2}t - Q_2(x)|t|^{q_2-2}t,$$

where $2 < q_2 < q_1 < 2^*$. *We suppose that* Q_1 *and* Q_2 *are nonnegative continuous functions satisfying* $Q_1(x) \geqq Q_1(\infty) = \lim_{|x|\to\infty} Q_1(x)$ *and* $Q_2(x) \leqq Q_2(\infty) = \lim_{|x|\to\infty} Q_2(x)$ *and one of these inequalities is strict on a set of positive measure. Then problem (4.1) has a nontrivial solution.*

Proof. The nonlinearity obviously satisfies conditions (i) - (iv). In particular condition (iii) holds with $\mu = \frac{1}{q_2}$. We now check that if $\langle I^{\infty\prime}(v), v\rangle = 0$, $v \not\equiv 0$, then

$$I^\infty(v) = \sup_{t \geqq 0} I^\infty(tv).$$

It is easy to see that $I^\infty(tv) \to -\infty$ as $t \to \infty$ and $I^\infty(tv) > 0$ for $t > 0$ sufficiently small. Hence there exists $t' > 0$ such that

$$I^\infty(t'v) = \sup_{t \geqq 0} I^\infty(tv) \ \text{ and } \ \frac{d}{dt} I^\infty(tv)\,|_{t=t'} = 0.$$

Setting

$$k = \int_{\mathbb{R}_N} (|\nabla v|^2 + v^2)\,dx,\ k_i = \int_{\mathbb{R}_N} Q_i(\infty)|v|^{q_i-1}\,dx, \quad i = 1, 2,$$

we write

$$\begin{aligned}\frac{d}{dt}I^\infty(tv) &= t\int_{\mathbb{R}_N}(|\nabla v|^2+v^2)\,dx + t^{q_2-1}\int_{\mathbb{R}_N}Q_2(\infty)|v|^{q_2-1}\,dx \\ &- t^{q_1-1}\int_{\mathbb{R}_N}Q_1(\infty)|v|^{q_1-1}\,dx = tk+t^{q_2-1}k_2-t^{q_1-1}k_1=0,\end{aligned}$$

which is equivalent to

$$k+t^{q_2-2}k_2-t^{q_1-2}k_1=0,\quad t>0. \tag{4.28}$$

We show that this equation has a unique solution. To show this we put $g_1(t)=k_1t^{q_1-2}$, $g_2(t)=k+k_2t^{q_2-2}$. Then

$$(g_1-g_2)'(t)=(q_1-2)k_1t^{q_1-3}-(q_2-2)k_2t^{q_2-3}$$

and we see that

$$(g_1-g_2)'(t)>0 \text{ for } t>s=\left(\frac{k_2(q_2-2)}{k_1(q_1-2)}\right)^{\frac{1}{q_1-q_2}}.$$

If the graph of g_1 intersects the graph of g_2 at $\bar{t}$, then $k_1\bar{t}^{q_1-2}=k+k_2\bar{t}^{q_2-2}$, so $\bar{t}>\left(\frac{k_2}{k_1}\right)^{\frac{1}{q_1-q_2}}>s$. On the other hand g_1-g_2 is strictly increasing on (s,∞), so $\bar{t}$ is a unique intersection point. Since $\frac{d}{dt}I^\infty(tv)\mid_{t=t'}=0$ and $\langle I^{\infty\prime}(v),v\rangle=0$, we have

$$\frac{d}{dt}I^\infty(tv)\mid_{t=1}=\int_{\mathbb{R}_N}(|\nabla v|^2+v^2+Q_2(\infty)|v|^{q_2-1}-Q_1(\infty)v|^{q_1-1})\,dx=0.$$

Since equation (4.28) has a unique root we have $t'=1$, so $I^\infty(tv)\leqq I^\infty(v)$ for $t>0$. We can construct a path γ_v such that $I^\infty(v)=\max_{\gamma_v}I^\infty(u)$. This implies that

$$J^\infty\geqq c_\infty=\inf_{\gamma\in\Gamma}\max_{\gamma}I^\infty(u).$$

By Propositions 4.1 and 4.2 the $(PS)_c$-condition holds for $c\leqq J^\infty$. The mountain pass theorem implies that there exists $u_\circ\in H^1(\mathbb{R}_N)$ such that $I^\infty(u_\circ)=c_\infty$, $I^{\infty\prime}(u_\circ)=0$. Hence $c_\infty=I^\infty(u_\circ)\geqq J^\infty$ and we have $I^\infty(u_\circ)=J^\infty$. Similarly, there exists $t^*>0$ such that

$$I(t^*u_\circ)=\sup_{t\geqq 0}I(tu_\circ).$$

Thus

$$\sup_{t \geqq 0} I(tu_\circ) = I(t^* u_\circ) \leqq I^\infty(t^* u_\circ) \leqq I^\infty(u_\circ) = J^\infty.$$

Finally, we can construct a path $\gamma_{u_\circ}$ such that

$$c \leqq \max_{u \in \gamma_{u_\circ}} I(u) = I(t^* u_\circ) < J^\infty$$

and this completes the proof. □

4.4 Unbounded domains

Interesting variants of the existence results can be obtained for the equation 4.1 on unbounded domains. We consider the problem

$$\begin{cases} -\Delta u + u = f(u) & \text{in } \Omega, \\ u = 0 \text{ on } \partial\Omega, u > 0 & \text{on } \Omega, \end{cases} \tag{4.29}$$

where $\Omega \subset \mathbb{R}_N$ is unbounded domain.

We shall assume throughout that the nonlinearity $f : [0, \infty) \to \mathbb{R}$ is of class C^1 and satisfies

(a_1) $f(t) = o(t)$ for $t \to 0$.
(a_2) There exists $1 < p < \frac{N+2}{N-2}$ such that $\lim_{t\to\infty} \frac{f(t)}{t^p} = 0$.
(a_3) There exists $p \geq \theta > 2$ such that

$$0 < \theta F(t) \leq f(t)t$$

for all $t > 0$, where $F(t) = \int_0^t f(s)\, dx$.
(a_4) The function $t \to \frac{f(t)}{t}$ is nondecreasing on $(0, \infty)$.

To formulate assumptions on Ω, we write for $1 \leq l \leq N-1$, $\mathbb{R}_N = \mathbb{R}_l \times \mathbb{R}_{N-l}$. For a point $z \in \mathbb{R}_N$ we have $z = (t, x)$, where $t \in \mathbb{R}_l$ and $x \in \mathbb{R}_{N-l}$. Let $t \in \mathbb{R}_l$. We denote by $\Omega^t \subset \mathbb{R}_{N-l}$ the projection of Ω onto $\mathbb{R}_{N-l}$, that is,

$$\Omega^t = \{x \in \mathbb{R}_{N-l};\ (t, x) \in \Omega\}.$$

We impose the following restrictions on the domain Ω:

(b_1) Ω is a smooth subset of $\mathbb{R}_N$ and the projections Ω^t are bounded uniformly in $t \in \mathbb{R}_l$.

(b_2) There exists a nonempty closed set $F \subset \mathbb{R}_{N-l}$ such that $F \subset \Omega^t$ for all $t \in \mathbb{R}_l$.

(b_3) For each $\delta > 0$ there exists an $M > 0$ such that

$$\Omega^t \subset \{x \in \mathbb{R}_{N-l};\ \text{dist}(x, F) < \delta\}$$

for $|t| \geq M$.

Typical examples of domains satisfying (b_1), (b_2) and (b_3) are

$$\Omega = \{(t, x) \in \mathbb{R} \times \mathbb{R}_{N-1};\ |x| < g(t)\}$$

and

$$\Omega = \{(t, x) \in \mathbb{R}_{N-1} \times \mathbb{R};\ 0 < x < h(t)\},$$

where $g : \mathbb{R} \to (0, \infty)$ and $h : \mathbb{R}_{N-1} \to (0, \infty)$ are smooth functions satisfying

$$g(t) > \lim_{|s| \to \infty} g(s) \text{ for all } t \in \mathbb{R}$$

and

$$h(t) > \lim_{|s| \to \infty} h(s) \text{ for all } t \in \mathbb{R}_{N-1}.$$

Solutions to problem (4.29) will be found as critical points of the functional $J : H_0^1(\Omega) \to \mathbb{R}$ given by

$$J(u) = \frac{1}{2} \int_\Omega \left(|\nabla u^2 + u^2\right) dx - \int_\Omega F(u)\, dx,$$

where f is extended by 0 for $s < 0$.

A starting point for the existence result is to find the range of level sets of J for which the (PS)-condition holds. It is easy to check that J has a mountain pass geometry.

Lemma 4.4 *There exist constants $\alpha > 0$ and $\rho > 0$ such that*

$$J(u) \geq \alpha \textit{ for all } \|u\|_{1,2} = \rho.$$

Moreover, for every $u \in H_0^1(\Omega)$, with $u > 0$ on a set of positive measure, there exists $\tau = \tau(u) > 0$ such that $\|u\|_{1,2} > \rho$ and $J(\tau u) < 0$.

Let

$$\mathcal{P} = \{\gamma \in C([0,1], H_0^1(\Omega));\ \gamma(0) = 0 \text{ and } J(\gamma(1)) < 0\}.$$

According to Lemma 4.4, $\mathcal{P} \not\equiv 0$ and we can define the mountain pass level

$$c(\Omega) = \inf_{\gamma \in \mathcal{P}} \sup_{t \in [0,1]} J(\gamma(t)).$$

It is clear that $\alpha \leq c(\Omega) < \infty$.

Lemma 4.5 *Suppose that* $(a_1) - -(a_4)$ *hold. Then*

$$c(\Omega) = \inf_{u \in H_0^1(\Omega), u \not\equiv 0} \sup_{t \geq 0} J(tu).$$

The proof of this lemma is postponed to Section 6.3.

To formulate a criterion for the existence of a critical point of the functional J we define sets

$$\Omega_k = \{(t,x) \in \Omega;\ |t| < k\},\ k > 0.$$

It is obvious that Ω_k is a bounded set for each k. Let $\chi \in C^1(\mathbb{R})$ be a function such that $\chi(t) = 0$ for $t \leq 0$, $\chi(t) = 1$ for $t \geq 1$ and $0 \leq \chi(t) \leq 1$. We set $\eta_k(t,x) = \chi(|t| - k)$.

By straightforward calculations we can check that

$$\int_{\Omega_{k+1} - \Omega_k} \left(|\nabla \eta_k u|^2 + |\eta_k u|^2\right) dz \leq C \int_{\Omega_{k+1} - \Omega_k} \left(|\nabla u|^2 + u^2\right) dz$$

for each $u \in H_0^1(\Omega)$ and each k, where $C > 0$ is a constant independent of k and u.

For each $k > 0$ we set $H_k = H_0^1(\Omega_k)$ and $\tilde{H}_k = H_0^1(\Omega - \bar{\Omega}_k)$. The functional J restricted to subspaces H_k and $\tilde{H}_k$ has also a mountain pass structure and we set

$$c_k = \inf_{\gamma \in \mathcal{P}_k} \sup_{t \in [0,1]} J(\gamma(t))$$

and

$$\tilde{c}_k = \inf_{\gamma \in \tilde{\mathcal{P}}_k} \sup_{t \in [0,1]} J(\gamma(t)),$$

where

$$\mathcal{P}_k = \{\gamma \in \mathcal{P};\ \gamma(s) \in H_k \text{ for all } s \in [0,1]\}$$

and

$$\tilde{\mathcal{P}}_k = \{\gamma \in \mathcal{P};\ \gamma(s) \in \tilde{H}_k \ \text{ for all } \ s \in [0,1]\}.$$

It is very easy to check that if $k < l$, then $H_k \subset H_l$ and $\tilde{H}_k \supset \tilde{H}_l$. Therefore both sequences $\{c_k\}$ and $\{\tilde{c}_k\}$ are monotone and moreover $\lim_{k\to\infty} c_k = c(\Omega)$.

We now extend Proposition 4.1 to unbounded domains Ω satisfying (b_1), (b_2) and (b_3).

Proposition 4.6 *Suppose that*

$$c(\Omega) < \lim_{k\to\infty} \tilde{c}_k.$$

Then there exists a critical point of J such that $J(u) = c(\Omega)$.

Proof. By virtue of the mountain pass theorem without the Palais–Smale condition there exists a sequence $\{u_n\} \subset H_0^{1,2}(\Omega)$ such that $J(u_n) \to c(\Omega)$ and $J'(u_n) \to 0$ in $H^{-1}(\Omega)$ as $n \to \infty$. It follows from (a_3) that

$$J(u_n) - \frac{1}{\theta}\langle J'(u_n), u_n\rangle \geq \left(\frac{1}{2} - \frac{1}{\theta}\right)\int_\Omega \left(|\nabla u_n|^2 + u_n^2\right) dx$$

and this implies that $\{u_n\}$ is bounded in $H_0^{1,2}(\Omega)$. Let

$$\gamma_\infty = \lim_{k\to\infty} \limsup_{n\to\infty} \int_{\Omega_k} \left(|\nabla u_n|^2 + u_n^2\right) dx.$$

We claim that $\gamma_\infty > 0$. If we assume, contrary to the claim, that $\gamma_\infty = 0$, then by an obvious modification of Lemma 4.1 we can find sequences $k_s \to \infty$ and $m_s \to \infty$ as $s \to \infty$ such that

$$\int_{\Omega_{k_s+1}} \left(|\nabla u_{m_s}|^2 + u_{m_s}^2\right) dx \leq \frac{1}{s}. \tag{4.30}$$

We set $w_s(z) = \eta_{k_s}(z) u_{m_s}(z)$ for $z \in \Omega$. Then $w_s \in \tilde{H}_{k_s}$. We now estimate $J(tw_s) - J(tu_{m_s})$ for $t \geq 0$. Towards this end we write

$$\begin{aligned} J(tw_s) \quad - \quad J(tu_{m_s}) &= -\frac{t^2}{2}\int_{\Omega_{k_s+1}} \left(|\nabla u_{m_s}|^2 + u_{m_s}^2\right) dz \\ &+ \int_{\Omega_{k_s+1}} F(tu_{m_s})\, dz \\ &+ \frac{t^2}{2}\int_{\Omega_{k_s+1}-\Omega_{k_s}} \left(|\nabla w_s|^2 + w_s^2\right) dz - \int_{\Omega_{k_s+1}-\Omega_{k_s}} F(tw_s)\, dz. \end{aligned} \tag{4.31}$$

It follows from (4.30) that

$$\int_{\Omega_{k_s+1}-\Omega_{k_s}} \left(|\nabla w_s|^2 + w_s^2\right) dx \le \frac{C_1}{s} \tag{4.32}$$

for some constant $C_1 > 0$ independent of s. We next determine the estimates for the integrals involving nonlinear terms. By (a_1), (a_2) and the Sobolev embedding theorem we have

$$\begin{aligned} \int_{\Omega_{k_s+1}} F(tu_{m_s})\, dx &\le C \int_{\Omega_{k_s+1}} \left(t^2 u_{m_s}^2 + t^{p+1}|u_{m_s}|^{p+1}\right) dx \\ &\le C_2\left(\frac{t^2}{s} + \frac{t^{p+1}}{s^{\frac{p+1}{2}}}\right) \end{aligned} \tag{4.33}$$

and

$$\int_{\Omega_{k_s+1}} F(tw_s)\, dx \le C_2\left(\frac{t^2}{s} + \frac{t^{p+1}}{s^{\frac{p+1}{2}}}\right), \tag{4.34}$$

where $C_2 > 0$ is a constant independent of s. Combining the estimates (4.31), (4.32), (4.33) and (4.34) we get

$$|J(tw_s) - J(tu_{m_s})| \le C_3\left(\frac{t^2}{s} + \frac{t^{p+1}}{s^{\frac{p+1}{2}}}\right) \tag{4.35}$$

for some constant $C_3 > 0$ independent of s. Since $J(u_n) = c(\Omega) + o(1)$ and $\langle J'(u_n), u_n\rangle = o(1)$ we may assume that $u_{m_s} > 0$ on a set of positive measure in $\mathbb{R}_N$ for each s. Therefore for each s there exists $\tau_s > 0$ such that $J(\tau_s u_{m_s}) = -1$. We now show that the sequence $\{\tau_s\}$ is bounded. Indeed, using (a_1), (a_2) and (a_3) we derive for every $\epsilon > 0$ and k sufficiently large the following estimate

$$\begin{aligned} \frac{c(\Omega)}{2} &\le J(u_{m_s}) - \frac{1}{2}\langle J'(u_{m_s}), u_{m_s}\rangle \\ &= -\int_\Omega F(u_{m_s})\, dx + \frac{1}{2}\int_\Omega f(u_{m_s})u_{m_s}\, dx \\ &\le \epsilon \int_\Omega (u_{m_s}^+)^2\, dx + K\int_\Omega (u_{m_s}^+)^{p+1}\, dx, \end{aligned}$$

where $K > 0$ is a constant depending on ϵ. Since $\{u_{m_s}\}$ is bounded in $L^2(\Omega)$ we can choose $\epsilon > 0$ so that

$$\frac{c(\Omega)}{4} \le K\int_\Omega (u_{m_s}^+)^{p+1}\, dx. \tag{4.36}$$

Further, applying the Hölder inequality we obtain

$$\int_\Omega |u_{m_s}^+|^{p+1}\,dz \le \left(\int_\Omega |u_{m_s}^+|^{\theta+1}\,dz\right)^\alpha \left(\int_\Omega |u_{m_s}^+|^{2^*}\,dz\right)^{1-\alpha}, \tag{4.37}$$

where $\alpha = \frac{2^*-(p+1)}{2^*-(\theta+1)}$. It therefore follows from (4.36), (4.37), the Sobolev embedding theorem and the boundedness of $\{u_n\}$ in $H_\circ^1(\Omega)$ that

$$\int_\Omega |u_{m_s}|^{p+1}\,dz \ge \beta, \tag{4.38}$$

for some constant $\beta > 0$ independent of s. Finally, using (a_3) we can estimate τ_s in the following way

$$-1 = J(\tau_s u_{m_s}) \le C\frac{\tau_s^2}{2}\left(\int_\Omega (|\nabla u_{m_s}|^2 + u_{m_s}^2)\,dz - B\tau_s^{\theta+1}\int_\Omega |u_{m_s}^+|^\theta\,dz\right.$$

for some constant $B > 0$. It is obvious that the last inequality and (4.38) imply that $\{\tau_s\}$ is a bounded sequence, that is, $\tau_s \le \bar\tau$ for all s. Therefore by (4.35) we have

$$\sup_{t\ge 0} J(tw_s) = \sup_{t\in[0,\bar\tau]} J(tw_s) \le \sup_{t\in[0,\bar\tau]} J(tu_{m_s}) + \frac{M}{s}$$

for some constant $M > 0$ independent of s. From this inequality we deduce that

$$\lim_{k\to\infty} \tilde c_k \le c(\Omega)$$

which is impossible. Since $\gamma_\infty > 0$ we may assume that there exists a subsequence of $\{u_m\}$, denoted again by $\{u_m\}$, such that $u_m \rightharpoonup u \not\equiv 0$ in $H_\circ^1(\Omega)$. Repeating the final part of the proof of Proposition 4.2 we show that $\{u_m\}$ is relatively compact in $H_\circ^1(\Omega)$. □

To proceed futher we need to show the strict monotonicity of the mountain pass levels with respect of the inclusion of sets. To establish this we introduce some notation. Let $A \subset \mathbb{R}_N$. We denote by $H(A)$ the closed subspace of $H^1(\mathbb{R}_N)$ consisting of functions in $H^1(\mathbb{R}_N)$ vanishing a.e. on $\mathbb{R}_N - A$. If $A = \bar\Sigma$, where Σ is a domain with a Lipschitz boundary, then $H_\circ^1(\Sigma) = H(A)$. We now note that if $H(A) \ne \emptyset$, then the restriction of J to $H(A)$ has the mountain pass structure and we can define the mountain pass level in the usual way which will be denoted by $c(A)$. If $H(A) = \emptyset$, we set $c(A) = \infty$.

Lemma 4.6 *Suppose that* $(a_1) - -(a_4)$ *hold. Let* $\Sigma \subset \mathbb{R}_N$ *be a domain with a smooth boundary and let* A *be a closed subset of* Σ. *If* $c(A)$ *is a critical value of* J *restricted to* $H(A)$, *then*

$$c(\Sigma) < c(A).$$

Proof. Let $u_\circ \in H(A)$ be a critical point of J corresponding to the value $c(A)$, that is, $J(u_\circ) = c(A)$. Since $c(A) > 0$, $u_\circ > 0$ on a set of positive measure in $\mathbb{R}_N$. It follows from the unique continuation theorem [92] that $u_\circ$ is not a solution of the Dirichlet problem

$$\begin{aligned} -\Delta u + u &= f(u) \text{ in } \Sigma, \\ u &= 0 \text{ on } \partial\Sigma. \end{aligned}$$

Consequently, $u_\circ$ is not a critical point of J in $H^1_\circ(\Sigma)$. We set $J_\circ = J \mid_{H^1_\circ(\Sigma)}$. To complete the proof we construct a path $\gamma \in C([0,1], H^1_\circ(\Sigma))$ such that $\gamma(0) = 0$, $J_\circ(\gamma(1)) < 0$ and

$$\max_{t\in[0,1]} J_\circ(\gamma(t)) < c(A). \tag{4.39}$$

First we choose $\delta > 0$ and $\rho > 0$ such that

$$||J_\circ{}'|| \geq \rho \text{ for } u \in H_\circ 1(\Sigma), \quad \text{with } ||u - u_\circ||_{1,2} < \delta. \tag{4.40}$$

Let $K = \{u \in H^1_\circ(\Sigma);\ J_\circ{}'(u) = 0\}$. There exists a locally Lipschitz pseudogradient $V : H^1_\circ(\Sigma) - K \to H^1_\circ(\Sigma)$ such that for all $u \in B(u_\circ, \delta)$

$$||V(u)||_{1,2} \leq 2||J_\circ{}'(u)|| \tag{4.41}$$

and

$$\langle J_\circ{}'(u), V(u)\rangle \geq ||J_\circ{}'(u)||^2. \tag{4.42}$$

We now consider the initial value problem (the pseudogradient flow)

$$\eta' = -V(\eta), \quad \eta(0,u) = u, \tag{4.43}$$

where $u \in H^1_\circ(\Sigma)$. We now observe that for each compact set $\mathcal{K} \subset H^1_\circ(\Sigma) - K$, there exists a constant $\sigma_\circ > 0$ such that $\eta(t,u)$ is defined on $[0, \sigma_\circ]$ for all $u \in \mathcal{K}$. As $\mathcal{K}$ we choose $\mathcal{K} = \{\tau u_\circ;\ \tau \in [0, \tau_\circ]\}$, where $\tau_\circ = \tau_\circ(u_\circ)$ is a constant determined by Lemma 4.4. It then follows from (4.40) and (4.41) that $\frac{d}{dt}J_\circ(\eta(t,u)) \le 0$ for all $u \in \mathcal{K}$ and $t \in [0,\sigma_\circ]$ and

$$\frac{d}{dt}J_\circ(\eta(t,u)) \le -\rho^2 \text{ for all } u \in \mathcal{K} \cap B(u_\circ, \frac{\delta}{2}) \tag{4.44}$$

and $\sigma \in [0,\sigma_1]$, with $\sigma_1 > 0$ sufficiently small. Integrating (4.44) we get

$$J_\circ(\eta(\sigma_1,u)) \le J_\circ(u) - \rho^2\sigma_1 \text{ for all } u \in B(u_\circ, \frac{\delta}{2}) \cap \mathcal{K}. \tag{4.45}$$

We now observe that $J_\circ(u_\circ) = \max_{t\ge 0} J_\circ(tu_\circ)$ since $u_\circ$ is a critical point of $J_\circ$ on $H(A)$. We choose τ_1 so that $\tau u_\circ \in B(u_\circ, \frac{\delta}{2})$ for all $\tau \in (1-\tau_1, 1+\tau_1)$. It then follows from (4.45) that

$$J_\circ(\eta(\sigma_1,\tau u_\circ)) \le J_\circ(u_\circ) - \rho^2\sigma_1$$

for all $\tau \in (1-\tau_1, 1+\tau_1)$ and

$$J_\circ(\eta(\sigma_1,\tau u_\circ)) \le J_\circ(\tau u_\circ) < J_\circ(u_\circ)$$

for all $\tau \in [0,\tau_\circ] - (1-\tau_1, 1+\tau_1)$. We now define a path $\gamma \in C([0,1], H^1_\circ(\Sigma))$ by $\gamma(t) = \eta(\sigma_1, t\tau_\circ u_\circ)$ for $t \in [0,1]$. The last two relations show that (4.39) holds and this completes the proof. □

We digress here momentarily to remark that problem (4.29) has a solution on a cylindrical set. By assumption (b_1) $\hat{F} = \mathbb{R}_l \times F \subset \Omega$. Let $c(\hat{F})$ be the mountain pass level of J restricted to $H(\hat{F})$. Repeating a slightly modified argument of Remark 2.1 of Chapter 2 we can show that $c(\hat{F})$ is a critical value of J restricted to $H(\hat{F})$. It therefore follows from Lemma 4.5 that

$$c(\Omega) < c(\hat{F}). \tag{4.46}$$

For a bounded set $S \subset \mathbb{R}_{N-l}$ we set

$$S_\delta = \{x \in \mathbb{R}_{N-l};\ \text{dist}(x,S) < \delta\}.$$

Lemma 4.7 *If $S \subset \mathbb{R}_{N-l}$ is a closed, nonempty and bounded set, then*

$$\lim_{\delta\to 0} c(\hat{S}_\delta) = c(\hat{S}).$$

Proof. Since we always have $c(\hat{S}_\delta) \le c(\hat{S})$, we only need to show that

$$a = \liminf_{\delta\to 0} c(\hat{S}_\delta) \ge c(\hat{S}). \tag{4.47}$$

Let $\delta_n \to 0$ be such that $c(\hat{S}_{\delta_n}) \to a$. For each δ_n, there exists a critical point $u_n \in H(\hat{S}_{\delta_n})$ of the restriction of J to $H(\hat{S}_{\delta_n})$ such that $c(\hat{S}_{\delta_n}) = J(u_n)$. Since $\langle J'(u_n), u_n\rangle = 0$, we check, using (a_3), that

$$\int_{\mathbb{R}_N} (|\nabla u_n|^2 + u_n^2)\, dx \leq \frac{2\theta}{\theta - 2} a < \infty.$$

This shows that $\{u_n\}$ is bounded in $H^1(\mathbb{R}_N)$. Therefore we may assume that $u_n \rightharpoonup u$ in $H^1(\mathbb{R}_N)$ and $u_n \to u$ in L^s_{loc}, $2 < s < 2^*$. Since S is closed we see that $u \in H(\hat{S})$ and u is a critical point of $J\mid_{H(\hat{S})}$. We may always assume that $u \not\equiv 0$. Since otherwise, using Lemma 1.1, we can find a sequence $z_n = (t_n, x_n)$ such that

$$\int_{B(z_n,R)} u_n^2\, dz \geq d > 0$$

for all n and some constants $R > 0$ and $d > 0$. Since $\{x_n\}$ is bounded and J is translation invariant in t, we can assume that $z_n = 0$ which justifies our claim. We now observe that since u is a nonzero critical point of $J\mid_{H(\hat{S})}$, we must have $J(u) = \max_{t\geq 0} J(tu)$. Hence $J(u) \geq c(\hat{S})$. Thus (4.47) will follow if we can show that

$$\liminf_{n\to\infty} J(u_n) \geq J(u).$$

Let η be a C^1-function with a compact support in $\mathbb{R}_N$. Then $\langle J'(u_n), u_n\eta\rangle = o(1)$ and we have

$$\begin{aligned}\int_{\mathbb{R}_N} \eta|\nabla u_n|^2\, dz &= -\textstyle\int_{\mathbb{R}_N} (u_n\nabla\eta\nabla u_n + u_n^2\eta)\, dz \\ &\quad + \textstyle\int_{\mathbb{R}_N} f(u_n)u_n\eta\, dz + o(1).\end{aligned}$$

Since $u_n \rightharpoonup u$ in $H^1(\mathbb{R}_N)$ we get that

$$\begin{aligned}\lim_{n\to\infty}\int_{\mathbb{R}_N} |\nabla u_n|^2\eta\, dz &= -\int_{\mathbb{R}_N} (u\nabla u\nabla\eta + u^2)\, dz \\ &\quad + \int_{\mathbb{R}_N} f(u)u\eta\, dz = \int_{\mathbb{R}_N} \eta|\nabla u|^2\, dz.\end{aligned}$$

This shows that $u_n \to u$ in H^1 on bounded subsets of $\mathbb{R}_N$. Let ϕ_R be a cutoff function from Proposition 1.2. For each $\delta > 0$ we can find $R > 0$

such that

$$\lim_{n\to\infty}\left(\frac{1}{2}\int_{B(0,R+1)}\left(|\nabla u_n|^2+u_n^2\right)dz-\int_{B(0,R+1)}F(u_n)\,dx\right) \tag{4.48}$$
$$=\frac{1}{2}\int_{B(0,R+1)}\left(|\nabla u|^2+u^2\right)dz-\int_{B(0,R+1)}F(u)\,dz\geq J(u)-\delta.$$

Taking $u_n\phi_R$ as a test function we get

$$\begin{aligned}0=\langle J'(u_n),u_n\phi_R\rangle &= \int_{\mathbb{R}_N-B(0,R+1)}\left(|\nabla u_n|^2+u_n^2\right)dz\\ &- \int_{\mathbb{R}_N-B(0,R+1)} f(u_n)u_n\,dz+E_n,\end{aligned}$$

where

$$\begin{aligned}E_n &= \int_{B(0,R+1)-B(0,R)}\left(\nabla u_n\nabla(u_n\phi_R)+u_n^2\phi_R\right)dz\\ &- \int_{B(0,R+1)-B(0,R)} f(u_n)u_n\phi_R\,dz.\end{aligned}$$

Choosing R larger, if necessary, we have $|E_n|<\delta$. This and (4.48) imply that

$$\begin{aligned}\frac{1}{2}\int_{\mathbb{R}_N-B(0,R+1)}&\left(|\nabla u_n|^2+u_n^2\right)dz-\int_{\mathbb{R}_N-B(0,R+1)}F(u_n)\,dz\\ &= \int_{\mathbb{R}_N-B(0,R+1)}\left(\frac{1}{2}f(u_n)u_n-F(u_n)\right)dz\\ &- \frac{1}{2}E_n\geq-\frac{1}{2}E_n.\end{aligned}$$

Combining (4.48) with the last two estimates we get $\lim_{n\to\infty}J(u_n)\geq J(u)-2\delta$ and the result easily follows. □

We use Lemma 4.6 in the proof of the following existence result.

Theorem 4.2 *Problem (4.29) admits a positive solution.*

Proof. It follows from assumptions (b_1) and (b_2) that $\hat{F}=\mathbb{R}_l\times F\subset\Omega$. Since $\lim_{k\to\infty}c_k=c(\Omega)$, we deduce from (4.47) that for some $c<c(\hat{F})$, we have $c_k<c$ for sufficiently large k. Hence it suffices to show that $\tilde{c}_k\geq c$ for sufficiently large k. According to (b_3) for each $\delta>0$ we have $\Omega-B(0,k)\subset\mathbb{R}_l\times F_\delta$ provided k is sufficiently large. Hence by Lemma 4.6

$c(\hat{F}_\delta) \leq \tilde{c}_k$ for k large. The desired inequality follows from Lemma 4.7 if we choose δ sufficiently small. □

4.5 Bibliographical notes

The method of the artificial constraints originates in the works of Nehari [79] and Coffman [38], [39]. The results presented in Sections 4.1–4.3 are due to Chabrowski–Yang [33] and Yang–Zhu [112], [113]. The existence results on unbounded domains in Section 4.4 are taken from the papers of del Pino–Felmer [88], [89].

Chapter 5

Inverse power method

A minimizer of problem $(\mathcal{M})$ with $b \equiv 1$ can be taken as radially symmetric function due to the Schwarz symmetrization. In this chapter we consider a nonlinear elliptic problem with coefficients and nonlinearities independent of x. Using the inverse power method we develop an algorithm leading to nonradial solutions. In the final part of this chapter we consider the case where coefficient and nonlinearity satisfy some symmetry condition in x.

5.1 Iterative procedure

The aim of this section is to introduce an iterative procedure for the eigenvalue problem

$$-\Delta u + a(x)u = \lambda f(x, u) \quad \text{in } \mathbb{R}_N. \tag{5.1}$$

This procedure is a generalization of the *inverse power* method for computing the eigenvalues of a symmetric matrix. We follow here the ideas from the papers [49], [50] and [52].

Throughout this section we assume that:

(A) a is a continuous, nonnegative bounded function such that $a(x) \geqq a_0 > 0$ for $|x| \geqq r > 0$.

(B) The nonlinearity $f(x, u)$ is continuous on $\mathbb{R}_N \times \mathbb{R}$ and nondecreasing in u and satisfies

$$|f(x, u)| \leqq C|t|^{q-1} \quad \text{on } \mathbb{R}_N \times \mathbb{R},$$

where $C > 0$ is a constant and $2 < q < 2^*$.

Let $B : H^1(\mathbb{R}_N) \times H^1(\mathbb{R}_N) \to \mathbb{R}$ be a bilinear form given by

$$B(u,v) = \int_{\mathbb{R}_N} (\nabla u \nabla v + a(x)uv)\, dx.$$

We define a norm $\|u\|_a = B(u,u)^{\frac{1}{2}}$.

Lemma 5.1 *The norms* $\|\cdot\|_a$ *and* $\|\cdot\|_{1,2}$ *are equivalent.*

Proof. We obviously have

$$\int_{\mathbb{R}_N} (|\nabla u|^2 + a(x)u^2)\, dx \leqq \max(1, \|a\|_\infty)\|u\|_{1,2}^2.$$

It follows from (A) that

$$\int_{|x|\geqq r} (|\nabla u|^2 + u^2)\, dx \leqq \max\left(1, \frac{1}{a_0}\right) \int_{|x|\geqq r} (|\nabla u|^2 + a(x)u^2)\, dx.$$

On the other hand by the Hölder and Sobolev inequalities we have

$$\begin{aligned} \int_{|x|\leqq r} (|\nabla u|^2 + u^2)\, dx &\leqq \int_{|x|\leqq r} |\nabla u|^2\, dx + |B(0,r)|^{\frac{2}{N}} \left(\int_{|x|\leqq r} |u|^{2^*}\, dx\right)^{\frac{2}{2^*}} \\ &\leqq \left(1 + S^{-1}|B(0,r)|^{\frac{2}{N}}\right) \int_{\mathbb{R}_N} |\nabla u|^2\, dx. \end{aligned}$$

□

For each $R > 0$ we set

$$S_R = \{u \in H^1(\mathbb{R}_N);\ \|u\|_a = R\}.$$

Lemma 5.2 *For each* $v \in S_R$*, there exists a unique solution* $u \in H^1(\mathbb{R}_N)$ *of the equation*

$$-\Delta u + a(x)u = f(x,v) \ \text{ in } \mathbb{R}_N. \tag{5.2}$$

Proof. For a fixed $v \in H^1(\mathbb{R}_N)$ the functional $u \in H^1(\mathbb{R}_N) \to \int_{\mathbb{R}_N} f(x,v)u\, dx$ is linear and continuous on $H^1(\mathbb{R}_N)$. The result follows from the Lax–Milgram theorem. □

We now define a mapping $T : S_R \to S_R$ by $u = Tv$, where $u = \frac{R}{\|w\|_a} w$ and w is a unique solution of (5.2). The function $u = Tv$ satisfies the equation

$$-\Delta u + a(x)u = \lambda f(x,v) \text{ in } \mathbb{R}_N \tag{5.3}$$

with $\lambda = \frac{R}{\|w\|_a}$. Iterating this procedure we obtain a sequence $\{u_j\} \subset S_R$ such that

$$-\Delta u_{j+1} + a(x)u_{j+1} = \lambda_{j+1} f(x, u_j) \text{ in } \mathbb{R}_N, \tag{5.4}$$

where $u_{j+1} = \frac{R}{\|w_{j+1}\|_a} w_{j+1}$ and $\lambda_{j+1} = \frac{R}{\|w_{j+1}\|_a}$ with w_{j+1} satisfying

$$-\Delta w_{j+1} + a(x)w_{j+1} = f(u_j) \text{ in } \mathbb{R}_N. \tag{5.5}$$

To examine the convergence of the sequence $\{u_j\}$ we define a functional

$$\Phi(u) = \int_{\mathbb{R}_N} F(x, u)\, dx \text{ for } u \in H^1(\mathbb{R}_N),$$

where $F(x,t) = \int_0^t f(x,s)\, ds$.

Lemma 5.3 *If $v \in S_R$ and $u = Tv$, then*

$$\Phi(u) \geq \Phi(v)$$

and $\Phi(u) = \Phi(v)$ if and only if $u = v$. In particular $\lim_{j\to\infty} \Phi(u_j)$ exists.

Proof. Since u satisfies (5.3) we see that

$$B(u,u) = \lambda \int_{\mathbb{R}_N} uf(x,v)\, dx \text{ and } B(u,v) = \lambda \int_{\mathbb{R}_N} vf(x,v)\, dx.$$

On the other hand

$$B(u,v) \leqq \frac{B(u,u) + B(v,v)}{2} = B(u,u)$$

and hence

$$\int_{\mathbb{R}_N} (u - v) f(x,v)\, dx \geqq 0$$

with equality if and only if $u = v$. By the convexity of Φ we get

$$\Phi(u) \geqq \Phi(v) + \int_{\mathbb{R}_N} (u - v) f(x,v)\, dx \tag{5.6}$$

and the result follows. □

Lemma 5.4 *For each $j \geqq 1$*

$$\frac{1}{2}\|u_{j+1} - u_j\|_a^2 \leqq \lambda_{j+1} |\Phi(u_{j+1} - \Phi(u_j))|.$$

Proof. Let $u = u_{j+1}$, $v = u_j$ and $\lambda = \lambda_{j+1}$. Since $||u||_a^2 = ||v||_a^2 = R^2$, we see that

$$\frac{1}{2}||u - v||_a^2 = B(u, u - v) = \lambda \int_{\mathbb{R}_N} (u - v) f(x, v)\, dx$$

and the result follows from (5.6). □

Lemma 5.5 *If $\Phi(u_\circ) > 0$, then $0 < \lambda_{j+1} \leqq \frac{R^2}{\Phi(u_\circ)}$ for all $j \geqq 1$.*

Proof. It follows from (5.4) that

$$\lambda_{j+1} = \frac{B(u_{j+1}, u_j)}{\int_{\mathbb{R}_N} u_j f(x, u_j)\, dx} \leqq \frac{R^2}{\int_{\mathbb{R}_N} u_j f(x, u_j)\, dx}$$

and applying (5.6) with $u = 0$ and $v = u_j$ we get

$$\int_{\mathbb{R}_N} u_j f(x, u_j)\, dx \geqq \Phi(u_j) \geqq \Phi(u_\circ).$$

We easily deduce from this inequality the upper bound for λ_j. □

As in Proposition 1.2 we introduce the quantities α_∞ and β_∞ controlling the loss of mass of $\{u_m\}$ at infinity

$$\alpha_\infty = \lim_{R\to\infty} \lim_{m\to\infty} \int_{|x|\geq R} |u_m|^q\, dx$$

and

$$\beta_\infty = \lim_{R\to\infty} \lim_{m\to\infty} \int_{|x|\geq R} \left(|\nabla u_m|^2 + a u_m^2\right) dx.$$

Proposition 5.1 *Let $\Phi(u_\circ) > 0$ and suppose that f satisfies (B). If $\alpha_\infty = 0$, then $\beta_\infty = 0$ and consequently the sequence $\{u_j\}$ is relatively compact in $H^1(\mathbb{R}_N)$. (This means that the sequence $\{u_j\}$ has a subsequence converging in $H^1(\mathbb{R}_N)$ to a function u satisfying (5.1) with $\lambda > 0$).*

Proof. Let ϕ_R be a function defined in the proof of Proposition 1.2. Since u_{j+1} is a weak solution of (5.4) we get

$$\int_{\mathbb{R}_N} \left[\nabla u_{j+1} \nabla (u_{j+1}\phi_R) + a u_{j+1}^2 \phi_R\right] dx = \lambda_{j+1} \int_{\mathbb{R}_N} u_{j+1}\phi_R f(x, u_j)\, dx. \tag{5.7}$$

By virtue of the Sobolev inequality we get

$$\begin{aligned} \left|\int_{\mathbb{R}_N} u_{j+1}\phi_R f(x,u_j)\,dx\right| &\leqq C\|u_{j+1}\|_q \left(\int_{|x|\geqq R} |u_j|^q\,dx\right)^{\frac{q-1}{q}} \\ &\leqq C_1 R \left(\int_{|x|\geqq R} |u_j|^q\,dx\right)^{\frac{q-1}{q}}. \end{aligned}$$

Since $\alpha_\infty = 0$,

$$\lim_{R\to\infty}\limsup_{j\to\infty} \int_{|x|\geqq R} u_{j+1}\phi_R f(x,u_j)\,dx = 0.$$

We may assume that $u_j \rightharpoonup u$ in $H^1(\mathbb{R}_N)$, $u_j \to u$ in $L^2_{\text{loc}}(\mathbb{R}_N)$ and $u_j \to u$ a.e. on $\mathbb{R}_N$. Hence

$$\lim_{R\to\infty}\limsup_{j\to\infty} \left|\int_{\mathbb{R}_N} \nabla u_{j+1} u_{j+1} \nabla\phi_R\,dx\right| = 0.$$

Letting $j \to \infty$ and then $R \to \infty$ we deduce from (5.7) that $\beta_\infty = 0$ and result follows. □

5.2 Nonradial solutions

The inverse power method will be now applied to the problem

$$-\Delta u + u = f(u) \text{ in } \mathbb{R}_N, \tag{5.8}$$

where $f(u) = C|u|^{q-1}$ and $C > 0$ is a constant.

Let

$$r^2 = x_1^2 + \ldots + x_m^2, \quad s^2 = x_{m+1}^2 + \ldots + x_N^2$$

and

$$H^1_{r,s}(\mathbb{R}_N) = \{u \in H^1(\mathbb{R}_N);\ u(x) = u(r,s)\}.$$

As in Section 5.1 we introduce a sequence $\{u_j\}$ for equation (5.8).

Lemma 5.6 *If $u_o \in H^1_{r,s}(\mathbb{R}_N)$, then $u_j \in H^1_{r,s}(\mathbb{R}_N)$.*

Proof. Let

$$T = \begin{pmatrix} R & 0 \\ 0 & S \end{pmatrix},$$

where R and S are $m \times m$ and $N - m \times N - m$ orthogonal matrices, respectively. Setting $x^T = Tx$ we have $\left[\frac{\partial x^T}{\partial x}\right] = T$. Since $u_\circ \in H^1_{r,s}(\mathbb{R}_N)$, we see that $u_\circ(x) = u_\circ(x^T)$. Let w_1 be a solution in $H^1(\mathbb{R}_N)$ of the equation

$$-\Delta w_1 + w_1 = f(u_\circ) \text{ in } \mathbb{R}_N.$$

Then for every $\zeta \in H^1(\mathbb{R}_N)$ we have

$$\int_{\mathbb{R}_N} [\nabla_x w_1 \nabla_x \zeta + w_1 \zeta] \, dx = \int_{\mathbb{R}_N} f(u_\circ)\zeta \, dx. \tag{5.9}$$

Since $\nabla_x w_1(x^T) = \nabla_{x^T} w_1(x^T)T$, $\nabla_x \zeta(x) = \nabla_{x^T}\zeta(x)T$, $dx = dx^T$, where $\zeta(x) = \zeta_1(x^T)$, $\zeta_1 \in H^1(\mathbb{R}_N)$, we obtain applying (5.9) with $\zeta(x) = \zeta_1(x^T)$ that

$$\begin{aligned} &\int_{\mathbb{R}_N} \left(\nabla_x w_1(x^T)\nabla_x \zeta(x) + w_1(x^T)\zeta(x)\right) dx \\ = &\int_{\mathbb{R}_N} \left(\nabla_{x^T} w_1(x^T)\nabla_{x^T}\zeta_1(x^T) + w_1(x^T)\zeta_1(x^T)\right) dx^T \\ = &\int_{\mathbb{R}_N} f(u_\circ(x^T))\zeta_1(x^T)\, dx^T = \int_{\mathbb{R}_N} f(u_\circ(x))\zeta(x)\, dx. \end{aligned} \tag{5.10}$$

It follows from (5.9) and (5.10)

$$\int_{\mathbb{R}_N} \left[\nabla\left(w_1(x) - w_1(x^T)\right)\nabla\zeta + \left(w_1(x) - w_1(x^T)\right)\zeta\right] dx = 0.$$

In particular, if $\zeta(x) = w_1(x) - w_1(x^T)$ we obtain

$$|\nabla w_1(x) - \nabla w_1(x^T)| = 0 \text{ and } \zeta(x) = w_1(x) - w_1(x^T) = 0.$$

Hence $w_1(x) = w_1(x^T)$ which yields $u_1(x) = u_1(x^T) = u_1(r,s)$. A simple induction completes the proof. □

Lemma 5.7 *Let $m > 1$ and $N - m > 1$ and let $\{u_j\}$ be a sequence of successive approximations (obtained by the inverse power method). Then $\alpha_\infty = 0$.*

Proof. We show that to every $\epsilon > 0$ there corresponds $K > 0$ such that

$$\int_K^\infty dr \int_{\mathbb{R}_{N-m}} |u_j(r, x_{m+1}, \ldots, x_N)|^q r^{m-1}\, dx_{m+1} \ldots dx_N < \epsilon \tag{5.11}$$

and

$$\int_{\mathbb{R}_m} dx_1 \ldots dx_m \int_K^\infty |u(x_1, \ldots, x_m, s)|^q s^{N-m+1}\, ds < \epsilon. \tag{5.12}$$

Indeed, letting

$$A(u) = \left(\frac{\partial u}{\partial r}\right)^2 + \sum_{i=m+1}^{N} \left(\frac{\partial u}{\partial x_i}\right)^2 + u^2,$$

we have

$$||u_j||_{1,2} = \omega_N \int_0^\infty dr \int_{\mathbb{R}_{N-m}} A(u_j) r^{m-1}\, dx_{m+1} \ldots dx_N < \infty.$$

Letting $w_j = r^{\frac{m-1}{2}} u_j$ we write

$$\frac{\partial w_j}{\partial r} = \frac{m-1}{2} r^{\frac{m-3}{2}} u_j + r^{\frac{m-1}{2}} \frac{\partial u_j}{\partial r}, \quad i = 1, \ldots, m,$$

$$\frac{\partial w_j}{\partial x_i} = r^{\frac{m-1}{2}} \frac{\partial u_j}{\partial x_i}, \quad i = m+1, \ldots, N.$$

Hence

$$\begin{aligned}\int_0^\infty dr & \int_{\mathbb{R}_{N-m}} |w_j(r, x_{m+1}, \ldots, x_N)|^2\, dx_{m+1} \ldots dx_N \\ &= \int_0^\infty dr \int_{\mathbb{R}_{N-m}} |u_j(r, x_{m+1}, \ldots, x_N)|^2 r^{m-1}\, dx_{m+1} \ldots dx_N < \infty\end{aligned}$$

and

$$\begin{aligned}\int_1^\infty dr & \int_{\mathbb{R}_{N-m}} |\nabla w_j(r, x_{m+1}, \ldots, x_N)|^2\, dx_{m+1} \ldots dx_N \\ &= \int_1^\infty dr \int_{\mathbb{R}_{N-m}} A(w_j)\, dx_{m+1} \ldots dx_N,\end{aligned}$$

$$\left(\int_1^\infty dr \int_{\mathbb{R}_{N-m}} \left(\frac{\partial w_j}{\partial r}\right)^2 dx_{m+1},\ldots dx_N\right)^{\frac{1}{2}}$$
$$\leqq \left(\frac{m-1}{2}\int_1^\infty dr \int_{\mathbb{R}_{N-m}} u_j^2 r^{m-3}\, dx_{m+1}\ldots dx_N\right)^{\frac{1}{2}}$$
$$+ \left(\int_1^\infty dr \int_{\gg ath_{N-m}} \left(\frac{\partial u_j}{\partial r}\right)^{\frac{1}{2}} r^{m-1}\, dx_{m+1}\ldots dx_N\right)^{\frac{1}{2}} < \infty$$

and

$$\int_1^\infty dr \int_{\mathbb{R}_{N-m}} \sum_{i=m+1}^N \left(\frac{\partial w_j}{\partial x_i}\right)^2 dx_{m+1}\ldots dx_N$$
$$= \int_1^\infty dr \int_{\mathbb{R}_{N-m}} \sum_{i=m+1}^N \left(\frac{\partial u_j}{\partial x_i}\right)^2 r^{m-1}\, dx_{m+1}\ldots dx_N.$$

Consequently $w_j \in H^1((1,\infty)\times \mathbb{R}_{N-m})$ and by the Sobolev embedding theorem $w_j \in L^q((1,\infty)\times\mathbb{R}_{N-m})$, that is,

$$\int_{K_1}^\infty dr \int_{\mathbb{R}_{N-m}} |u_j(r, x_{m+1},\ldots,x_N)|^q r^{m-1}\, dx_{m+1}\ldots dx_N$$
$$= \omega_m \int_{K_1}^\infty dr \int_{\mathbb{R}_{N-m}} |w_j(r, x_{m+1},\ldots,x_N)|^q r^{-\frac{(m-1)q}{2}}\, dx_{m+1}\ldots x_N$$
$$\leqq \omega_m K_1^{-\frac{(m-1)q}{2}} C$$

for some constant $C > 0$. Taking K_1 sufficiently large (5.11) easily follows. To show (5.12) we set $w_j = s^{\frac{N-m-1}{2}} u_j$. In a similar manner we show that $w_j \in H^1(\mathbb{R}_m \times (1,\infty))$. Therefore we can find $K_2 > 0$ such that

$$\int_{\mathbb{R}_m} dx_1 \ldots dx_m \int_{K_2}^\infty |u_j(x_1,\ldots,x_m,s)|^q s^{\frac{N-m-1}{2}}\, ds < \epsilon.$$

Letting $K_\circ = \max(K_1, K_2)$ we get

$$\int_{r^2+s^2 \geqq K_\circ} |u_j|^q\, dx < \epsilon$$

and the assertion follows. □

We are now in a position to establish the existence of nonradial solutions.

Theorem 5.1 *Problem (5.8) admits a nonradial solutions for* $N \neq 3, 5$.

Proof. Case $N = 2m$.
The function $u_j(x) = u_j(r,s)$ satisfies the equation

$$-\Delta u_j + u_j = \lambda_j f(u_{j-1}),$$

which we write in the form

$$-\frac{\partial^2 u_j}{\partial r^2} - \frac{m-1}{r}\frac{\partial u_j}{\partial r} - \frac{\partial^2 u_j}{\partial s^2} - \frac{m-1}{s}\frac{\partial u_j}{\partial s} + u_j = \lambda_j f(u_{j-1}).$$

Since u_j are even functions in x_i, we have

$$\frac{\partial u(0,s)}{\partial r} = 0 \text{ and } \frac{\partial u(r,0)}{\partial s} = 0.$$

We introduce new variables $\xi = r+s$, $\eta = r-s$, $\xi \in (0,\infty)$, $\eta \in (-\infty,\infty)$. Then

$$u_j(r,s) = u_j(\frac{\xi+\eta}{2}, \frac{\xi-\eta}{2}) = \bar{u}_j(\xi,\eta),$$

$\bar{u}_j \in C^2(0,\infty) \times (-\infty,\infty))$ and satisfies

$$\begin{cases} -2\left(\frac{\partial^2 \bar{u}}{\partial \xi^2} + \frac{\partial^2 \bar{u}_j}{\partial \eta^2}\right) - \frac{2(m-1)}{\xi+\eta}\left(\frac{\partial \bar{u}}{\partial \xi} + \frac{\partial \bar{u}}{\partial \eta}\right) - \frac{2(m-1)}{\xi-\eta}\left(\frac{\partial \bar{u}}{\partial \eta} - \frac{\partial \bar{u}}{\partial \eta}\right) \\ +\bar{u}_j(\xi,\eta) = \lambda_j f(\bar{u}_{j-1}) \\ \frac{\partial \bar{u}_j}{\partial \xi} = \frac{\partial \bar{u}_j}{\partial \eta} \text{ on } \xi = \eta \\ \frac{\partial \bar{u}_j}{\partial \xi} = -\frac{\partial \bar{u}_j}{\partial \eta} \text{ on } \xi = -\eta. \end{cases} \tag{5.13}$$

We now show that if $\bar{u}_{j-1}$ is an odd function in η then $\bar{u}_j$ is also an odd function in η. Towards this end we set $v_j(\xi,\eta) = \bar{u}_j(\xi,-\eta)$. We then get

$$\begin{aligned} -2&\left(\frac{\partial^2 v_j(\xi,\eta)}{\partial \xi^2} + \frac{\partial^2 v_j}{\partial \eta^2}\right) - \frac{2(m-1)}{\xi+\eta}\left(\frac{\partial v_j}{\partial \xi} + \frac{\partial v_j}{\partial \eta}\right) \\ -& \quad \frac{2(m-1)}{\xi-\eta}\left(\frac{\partial v_j}{\partial \xi} - \frac{\partial v_j}{\partial \eta}\right) + v_j(\xi,\eta) \\ =& \quad -2\left(\frac{\partial^2 \bar{u}_j(\xi,-\eta)}{\partial \xi^2} + \frac{\partial^2 \bar{u}_j(\xi,-\eta)}{\partial(-\eta)^2}\right) \\ -& \quad \frac{2(m-1)}{\xi-(-\eta)}\left(\frac{\partial \bar{u}_j(\xi,-\eta)}{\partial \xi} - \frac{\partial \bar{u}_j(\xi,-\eta)}{\partial(-\eta)}\right) \\ -& \quad \frac{2(m-1)}{\xi+(-\eta)}\left(\frac{\partial \bar{u}_j(\xi,-\eta)}{\partial \xi} + \frac{\partial u_j(\xi,-\eta)}{\partial(-\eta)}\right) + \bar{u}_j(\xi,-\eta) \\ =& \quad \lambda_j f(\bar{u}_{j-1}(\xi,-\eta)) = -\lambda_j f(\bar{u}_{j-1}(\xi,\eta)), \end{aligned}$$

that is, $-v_j$ also satisfies (5.13). By virtue of the uniqueness of solution of (5.4) we have $\bar{u}_j(\xi,\eta) = -\bar{u}_j(\xi,-\eta)$, that is, $\bar{u}_j(\xi,\eta)$ is odd in η. By Proposition 5.1 $\{u_j\}$ and $\{\lambda_j\}$ have subsequences convergent to u and $\lambda > 0$, respectively, satisfying

$$-\Delta u + u = \lambda f(u) \quad \text{in } \mathbb{R}_N.$$

Since $f(t)$ is independent of x, by a change of variables, we tranform this to the equation (5.8). It is clear that u satisfies $u(r,s) = -u(s,r)$.
Case $N = 2m+3$, $m > 1$.
We set $r^2 = x_1^2 + \dots x_m^2$, $s^2 = x_{m+1}^2 + \dots + x_{2m}^2$ and $t = x_{2m+1}^2 + x_{2m+2}^2 + x_{3m+3}^2$. As in Lemma 5.6 we show that $\alpha_\infty = 0$ and that equation (5.8) has a solution. Furthermore u_j satisfies the equation

$$-\frac{\partial u_j^2}{\partial r^2} - \frac{m-1}{r}\frac{\partial u_j}{\partial r} - \frac{\partial^2 u_j}{\partial s^2} - \frac{m-1}{s}\frac{\partial u_j}{\partial s} - \frac{\partial^2 u_j}{\partial t^2} - \frac{2}{t}\frac{\partial u_j}{\partial t} + u_j = \lambda_j f(u_{j-1}),$$

with

$$\frac{\partial^2 u_j(0,s,t)}{\partial r} = 0, \quad \frac{\partial u_j(r,0,t)}{\partial s} = 0, \quad \frac{\partial u_j(r,s,0)}{\partial t} = 0.$$

Introducing new variables $\xi = r+s$, $\eta = r-s$, we set $u_j(r,s,t) = \bar{u}_j(\xi,\eta,t)$. As in the previous case we show that if $\bar{u}_{j-1}$ is an odd function in η, so does $\bar{u}_j$. Therefore $u(r,s,t) = \bar{u}(\xi,\eta,t)$ is an odd function in η and $u(r,s,t)$ satisfies $u(r,s,t) = -u(s,r,t)$ which means that (5.8) has a nontrivial solution. □

Example 5.1 As an first approximation we can take

$$\bar{u}_o(\xi,\eta) = \frac{\eta}{(1+\xi^2+\eta^2)^M} \quad \text{if } N = 2m,$$

and

$$\bar{u}_o(\xi,\eta,t) = \frac{\eta}{(1+\xi^2+\eta^2+t^2)^M} \quad \text{if } N = 2m+3$$

with M sufficiently large to ensure that $u_o \in H^1(\mathbb{R}_N)$.

Theorem 5.1 can be extended to equation (5.1) with $a(x) = a(r,s)$ and $f(x,u) = f(r,s,u)$ satisfying the symmetry conditions $a(r,s) = a(r,s)$ and $f(r,s,u) = f(s,r,u)$.

5.3 Equation with non-constant coefficient

In this section we show how to construct nonradial solution of (5.1) in the case where $a(x)$ is not constant. We consider the equation

$$-\Delta u + a(x)u = \lambda b(x)|u|^{q-2}u \text{ in } \mathbb{R}_N. \tag{5.14}$$

We assume that a and b are continuous and positive functions satisfying

$$\lim_{|x|\to\infty} a(x) = a(\infty) \text{ and } \lim_{|x|\to\infty} b(x) = b(\infty).$$

Applying the iteration procedure from Section 5.1 we obtain a sequence $\{u_j\} \subset H^1(\mathbb{R}_N)$ satisfying

$$-\Delta u_{j+1} + a(x)u_{j+1} = \lambda_{j+1} b(x)|u_j|^{q-2}u_j \text{ in } \mathbb{R}_N \text{ and } \|u_j\| = R.$$

Therefore we may assume that there exists a subsequence $\{u_{j_k}\} \subset H^1(\mathbb{R}_N)$ such that $u_{j_k} \rightharpoonup u$ in $H^1(\mathbb{R}_N)$, $u_{j_k} \to u$ a.e. on $\mathbb{R}_N$ and $\lambda_{j_k} \to \lambda$. For the ease of notations we denote this sequence again by $\{u_j\}$.

We define variational functionals

$$I(u) = \frac{1}{2}\int_{\mathbb{R}_N} (|\nabla u|^2 + a(x)u^2)\,dx - \frac{\lambda}{q}\int_{\mathbb{R}_N} b(x)|u|^q\,dx$$

and

$$\bar{I}(u) = \frac{1}{2}\int_{\mathbb{R}_N} (|\nabla u|^2 + a(\infty)u^2)\,dx - \frac{\lambda}{q}\int_{\mathbb{R}_N} b(\infty)|u|^q\,dx.$$

We set

$$\alpha = \inf\{\bar{I}(u);\, u \in H^1(\mathbb{R}_N) - \{0\}, \langle \bar{I}'(u), u\rangle = 0\}.$$

We commence by showing that $\{u_j\}$ is a Palais–Smale sequence for I.

Proposition 5.2 *The sequence $\{u_j\}$ is a Palais–Smale sequence for the functional I, that is,*

$$I(u_j) \to \left(\frac{1}{2} - \frac{1}{q}\right)R^2 \text{ and } I'(u_j) \to 0$$

as $j \to \infty$.

Proof. By straightforward calculations we check that

$$\begin{aligned} I(u_j) &= \frac{1}{2}\int_{\mathbb{R}_N}\left(|\nabla u_j|^2 + a(x)u_j^2\right)dx - \frac{\lambda}{q}\int_{\mathbb{R}_N} b(x)|u_j|^q\,dx \\ &= \frac{R^2}{2} - \frac{1}{q}\left(\lim_{j\to\infty}\frac{R^2}{\int_{\mathbb{R}_N} b(x)|u_{j-1}|^{q-2}u_{j-1}u_j\,dx}\right)\int_{\mathbb{R}_N} b(x)|u_j|^q\,dx. \end{aligned}$$

Since

$$\lim_{j\to\infty}\left(\int_{\mathbb{R}_N} b(x)|u_{j-1}|^{q-2}u_{j-1}u_j\,dx - \int_{\mathbb{R}_N} b(x)|u_j|^q\,dx\right) = 0$$

we have

$$\lim_{j\to\infty}\frac{\int_{\mathbb{R}_N} b(x)|u_{j-1}|^{q-2}u_{j-1}u_j\,dx}{\int_{\mathbb{R}_N} b(x)|u_j|^q\,dx} = 1.$$

Hence

$$\lim_{j\to\infty} I(u_j) = \left(\frac{1}{2} - \frac{1}{q}\right)R^2.$$

Let $\phi \in H^1(\mathbb{R}_N)$ with $\|\phi\|_a = 1$. It follows from

$$\int_{\mathbb{R}_N}\left(\nabla u_j \nabla\phi + a(x)u_j\phi\right)dx - \lambda_j\int_{\mathbb{R}_N} b(x)|u_{j-1}|^{q-2}u_{j-1}\phi\,dx$$

that

$$\begin{aligned} \langle I'(u_j),\phi\rangle &= \lambda_j\int_{\mathbb{R}_N} b(x)|u_{j-1}|^{q-2}u_{j-1}\phi\,dx \\ &- \lambda\int_{\mathbb{R}_N} b(x)|u_j|^{q-2}u_j\phi\,dx \\ &= (\lambda_j-\lambda)\int_{\mathbb{R}_N} b(x)|u_{j-1}|^{q-2}u_{j-1}\phi\,dx \\ &+ \lambda\int_{\mathbb{R}_N} b(x)\left(|u_{j-1}|^{q-2}u_{j-1} - |u_j|^{q-2}u_j\right)\phi\,dx. \end{aligned} \tag{5.15}$$

By the Hölder and Sobolev inequalities we have

$$\begin{aligned} &|\int_{\mathbb{R}_N} b(x)|u_{j-1}|^{q-2}u_{j-1}\phi\,dx| \\ &\leqq C_1\left(\int_{\mathbb{R}_N}|u_{j-1}|^q\,dx\right)^{\frac{q-1}{q}}\left(\int_{\mathbb{R}_N}|\phi|^q\,dx\right)^{\frac{1}{q}} \end{aligned} \tag{5.16}$$

$$\leq \quad C_2 R^{q-1} \|\phi\|_a = C_2 R^q$$

and

$$\begin{aligned}
& | \int_{\mathbb{R}_N} b(x) \left(|u|_{j-1}^{q-2} u_{j-1} - |u_j|^{q-2} u_j \right) \phi \, dx | \qquad (5.17) \\
= & \; | \int_{\mathbb{R}_N} \phi(x) \int_0^1 b(x) \frac{d}{dt} |tu_{j-1} + (1-t)u_j|^{q-2} \left(tu_{j-1} + (1-t)u_j \right) dx | \\
\leq & \; C_3(q-1) \int_{\mathbb{R}_N} |\phi| \left(\int_0^1 |tu_{j-1} + (1-t)u_j|^{q-2} |u_{j-1} - u_j| \, dt \right) dx \\
\leq & \; C_3(q-1) \int_0^1 \|\phi\|_q \|u_{j-1} - u_j\|_q^2 \|tu_{j-1} + (1-t)u_j\|_q^{q-2} \, dt \\
\leq & \; C_4 \|u_{j-1} - u_j\|_q,
\end{aligned}$$

for some constants $C_i > 0$, $i = 1, 2, 3, 4$. Applying again the Sobolev inequality we get

$$\begin{aligned}
\|u_{j-1} - u_j\|_q & \leq \; C_5 \|u_j - u_{j-1}\|_a \\
& \leq \; C_5 \lambda_j^{\frac{1}{2}} \left(\int_{\mathbb{R}_N} b(x) |u_j|^q \, dx - \int_{\mathbb{R}_N} b(x) |u_{j-1}|^q \, dx \right)^{\frac{1}{2}}.
\end{aligned}$$

Since the sequence $\Phi(u_j) = \int_{\mathbb{R}_N} b(x) |u_j|^q \, dx$ is increasing and bounded we see that $\lim_{j\to\infty} \|u_{j-1} - u_j\|_q = 0$. Combining this with (5.16) and (5.17) and the fact that $\lim_{j\to\infty} \lambda_j = \lambda$ we deduce from (5.15) that

$$\sup_{\|\phi\|_a = 1} |\langle I'(u_j), \phi \rangle| \to 0 \text{ as } j \to \infty.$$

□

An obvious modification of Proposition 4.1 leads to the following.

Proposition 5.3 *If $\{u_m\}$ is a $(PS)_c$-sequence for I with $c \in (0, \alpha)$, then $\{u_m\}$ is relatively compact in $H^1(\mathbb{R}_N)$.*

5.4 Even and odd solutions

We now show that under suitable assumptions on $b(x)$ the approximating sequence $\{u_j\}$ satisfies the $(PS)_c$-condition with $c \in (0, \alpha)$.

Let $\bar{r} = (r, 0, \ldots, 0) \in \mathbb{R}_N$, $r > 0$, and set

$$K_r = \min_{x \in B(\bar{r},r)} \frac{b(x)}{b(\infty)}, \quad \text{where } B(\bar{r}, r) = \{x \in \mathbb{R}_N;\ (x_1 - r)^2 + x_2^2 \ldots + x_N^2 < r^2\}.$$

As in Chapter 1 we introduce the notation

$$M_\infty = \inf\{\int_{\mathbb{R}_N} (|\nabla u|^2 + a(\infty)u^2)\, dx;\ u \in H^1(\mathbb{R}_N),\ b(\infty) \int_{\mathbb{R}_N} |u|^p\, dx = 1\}.$$

According to Theorem 2.1 there exists $\bar{u} \in H^1_r(\mathbb{R}_N)$ such that

$$M_\infty = \inf_{u \in H^1(\mathbb{R}_N) - \{0\}} \frac{\int_{\mathbb{R}_N} (|\nabla u|^2 + a(\infty)u^2)\, dx}{\left(\int_{\mathbb{R}_N} b(\infty)|u|^p\, dx\right)^{\frac{2}{p}}} = \frac{\int_{\mathbb{R}_N} (|\nabla \bar{u}|^2 + a(\infty)\bar{u}^2)\, dx}{\left(\int_{\mathbb{R}_N} b(\infty)|\bar{u}|^p\, dx\right)^{\frac{2}{p}}}.$$

It is also known that there exists $\bar{u}_r \in H^1_0(B(0,1)$ such that

$$\begin{aligned} M(B(0,r)) &= \inf_{u \in H^1_0(B(0,r)) - \{0\}} \frac{\int_{B(0,r)} (|\nabla u|^2 + a(\infty)u^2)\, dx}{\left(\int_{B(0,r)} b(\infty)|u|^p\, dx\right)^{\frac{2}{p}}} \\ &= \frac{\int_{B(0,r)} (|\nabla \bar{u}_r|^2 + a(\infty)\bar{u}_r^2)\, dx}{\left(\int_{B(0,r)} b(\infty)|\bar{u}_r|^p\, dx\right)^{\frac{2}{p}}}. \end{aligned}$$

If we set $\bar{u}_r^+(x) = \bar{u}^r(x_1 - r, x_2, \ldots, x_N)$ for $x \in B(\bar{r}, r)$, then

$$M(B(0,r)) = \frac{\int_{B(\bar{r},r)} \left(|\nabla \bar{u}_r^+|^2 + (\bar{u}_R^+)^2\right) dx}{\left(\int_{B(\bar{r},r)} b(\infty)|u_R^+|^2\, dx\right)^{\frac{2}{p}}}.$$

Lemma 5.8 *Suppose that $a(x) \geqq a(\infty)$ on $\mathbb{R}_N$ and that*

$$K_r > 2^{\frac{p-2}{2}} \left(\frac{M(B(0, \frac{r}{2}))}{M_\infty}\right)^{\frac{p}{2}} \tag{5.18}$$

for some $r > 0$, then

$$\alpha > (\frac{1}{2} - \frac{1}{p}) R^2. \tag{5.19}$$

Proof. If $u \in H^1(\mathbb{R}_N) - \{0\}$ and

$$t = \left(\frac{\int_{\mathbb{R}_N} (|\nabla u|^2 + a(\infty)u^2)\, dx}{\lambda b(\infty) \int_{\mathbb{R}_N} |u|^p\, dx} \right)^{\frac{1}{p-2}},$$

then

$$t^2 \int_{\mathbb{R}_N} (|\nabla u|^2 + a(\infty)u^2)\, dx = \lambda t^p \int_{\mathbb{R}_N} b(\infty)|u|^p\, dx.$$

This means that $\langle \bar{I}'(tu), tu \rangle = 0$. Therefore

$$\begin{aligned}
\alpha &= \inf\{\bar{I}(u);\ u \in H^1(\mathbb{R}_N) - \{0\},\ \langle \bar{I}'(u), u\rangle = 0\} \\
&= \inf_{u \in H^1(\mathbb{R}_N) - \{0\}} \left\{ \frac{t^2}{2} \int_{\mathbb{R}_N} (|\nabla u|^2 + a(\infty)u^2)\, dx - \frac{\lambda}{p} t^p \int_{\mathbb{R}_N} b(\infty)|u|^p\, dx \right\} \\
&= \inf_{u \in H^1(\mathbb{R}_N) - \{0\}} \left\{ \frac{\lambda t^p}{2} \int_{\mathbb{R}_N} b(\infty)|u|^p\, dx - \frac{\lambda t^p}{p} \int_{\mathbb{R}_N} b(\infty)|u|^p\, dx \right\} \\
&= \lambda^{-\frac{2}{p-2}} \left(\frac{1}{2} - \frac{1}{p}\right) \inf_{u \in H^1(\mathbb{R}_N) - \{0\}} \frac{\left(\int_{\mathbb{R}_N} (\nabla u|^2 + a(\infty)u^2)\, dx \right)^{\frac{p}{p-2}}}{\left(\int_{\mathbb{R}_N} b(\infty)|u|^p\, dx \right)^{\frac{2}{p-2}}} \\
&= \lambda^{-\frac{2}{p-2}} \left(\frac{1}{2} - \frac{1}{p}\right) M_\infty^{\frac{p}{p-2}}.
\end{aligned}$$

Since $\lambda \leqq \frac{R^2}{\int_{\mathbb{R}_N} b(x)|u_\circ|^p\, dx}$, we see that

$$\alpha \geqq \left(\frac{1}{2} - \frac{1}{p}\right) \left(\frac{\int_{\mathbb{R}_N} b(x)|u_\circ|^p\, dx}{R^p} \right)^{\frac{2}{p-2}} M_\infty^{\frac{p}{p-2}} R^2. \tag{5.20}$$

If $u_\circ$ is an odd function and $u_\circ = \bar{u}^+_{\frac{r}{2}}$ for $x_1 > 0$, then by (5.18) we have

$$\begin{aligned}
\frac{\int_{\mathbb{R}_N} b(x)|u_\circ|^p\, dx}{R^p} M_\infty^{\frac{p}{2}} &\geqq \frac{K_r \int_{\mathbb{R}_N} b(\infty)|u_\circ|^p\, dx}{\left(\int_{\mathbb{R}_N} (|\nabla u_\circ|^2 + a(\infty)u_\circ^2)\, dx \right)^{\frac{p}{2}}} M_\infty^{\frac{p}{2}} \\
&= K_r 2^{1-\frac{p}{2}} \left(\frac{M_\infty}{M(B(0, \frac{r}{2}))} \right)^{\frac{p}{2}} > 1.
\end{aligned}$$

This, combined with (5.20), shows that (5.19) is satisfied. □

As an immediate consequence of Lemma 5.7 and Proposition 5.3 we derive the existence of a nonradial solution.

Theorem 5.2 *Suppose that $a(x)$ and $b(x)$ are even with respect to $x_1 > 0$, $a(x) \leqq a(\infty)$ on $\mathbb{R}_N$ and that $b(x)$ satisfies (5.18). Then*

(i) if $u_\circ$ is even and $u_\circ(x) = u^{+}_{\frac{r}{2}}(x)$ for $x_1 > 0$, then there exists a positive solution u of (5.14) which is even in x_1

(ii) if $u_\circ$ is odd and $u_\circ(x) = \bar{u}^{+}_{\frac{r}{2}}$ for $x_1 > 0$, then there exists a solution u of (5.14) which is odd in x_1.

5.5 Bibliographical notes

The use of the inverse power method to construct an algorithm generating sequences convergent to solutions of nonlinear elliptic problems is due to Edeyland–Spruck [52]. The results of Sections 5.2, 5.3 and 5.4 are patterned after the papers of Yi Ding and Yi Ding–Xiaxi Ding [49], [50]. A construction of nonradial solutions based on Bartsch's fountain theorem [10] can be found in the work of Bartsch–Willem [12].

Chapter 6

Effect of topology

In the first part of this chapter we investigate how the number of solutions of the Dirichlet problem in exterior domains for elliptic equations involving a subcritical Sobolev exponent depends on the topology of the domain. The multiplicity of solutions can be expressed in terms of the Lusternik–Schnirelman category of the domain. We then relate the number of solutions to the graph topology of coefficients of the equation. In the second part, these ideas will be used to examine the behaviour of ground -states of equations with subcritical nonlinearity and involving a small positive parameter. This is motivated by a search for standing waves of the following nonlinear Schrödinger equation

$$ih\frac{\partial\psi}{\partial t} = -\frac{h^2}{2m}\Delta\psi + V(x)\psi - \gamma|\psi|^{p-2}\psi \text{ in } \mathbb{R}^N,$$

where h, m, γ and p are positive constants and $p > 2$. A standing wave is a solution of the form

$$\psi(x,t) = \exp\{i\frac{Et}{h}\}u(x),$$

where $E \in \mathbb{R}$ is a constant. Substituting ψ into the equation and assuming that $2m = 1$ we obtain

$$-h^2\Delta u + \big(V(x) - E\big)u = |u|^{p-2}u \text{ in } \mathbb{R}^N.$$

In this chapter we investigate in the case where $V(x) - E$ is bounded on $\mathbb{R}^N$ from below by a positive constant. Solutions of this equation exhibit concentration phenomena. In particular, we show that there exist for each

$h > o$ a positive solution with the property: $u(x_h) = \sup_{x\in\mathbb{R}^N} u(x)$ and $x_h \to x_o$ and $V(x_o) = \inf_{x\in\mathbb{R}^N} V(x)$.

6.1 Best Sobolev constants in exterior domains

We consider the Dirichlet problem

$$\begin{cases} -\Delta u + \lambda u & = u^{p-1} \text{ in } \Omega, \\ u(x) > 0 \text{ on } & \Omega,\ u(x) = 0 \text{ on } \partial\Omega, \end{cases} \tag{6.1}$$

where $2 < p < 2^*$, $n \geq 3$ and Ω is unbounded doman with smooth boundary $\partial\Omega \neq 0$ such that $\overline{\mathbb{R}^N - \Omega}$ is bounded. Since Ω is a large interior domain, the minimization problem

$$M(\lambda, p, \Omega) = \inf\{\int_\Omega (|\nabla u|^2 + \lambda u^2)\, dx;\ u \in H_o^1(\Omega),\ \int_\Omega |u|^p\, dx = 1\}$$

does not have a minimizer (see Proposition 2.1. Moreover, it follows from the proof of Proposition 2.1 that

$$M(\lambda, p, \Omega) = M(\lambda, p, \mathbb{R}^N),$$

where $M(\lambda, p, \mathbb{R}^N)$ is defined as above with Ω replaced by $\mathbb{R}^N$. Unlike in Chapters 1 and 2, we stress the dependence of M on λ, which plays a crucial role in finding a solution to problem (6.1). According to Corollary 2.2 every minimizing sequence $\{u_m\}$ for $M(\lambda, p, \Omega)$ admits the following decomposition

$$u_n(x) = w_n(x) + \bar{u}(x - y_n), \tag{6.2}$$

where $w_n \to 0$ in $H^1(\mathbb{R}^N)$, $|y_n| \to \infty$ and $\bar{u}$ is a ground state solution of the equation

$$-\Delta u + \lambda u = |u|^{p-2}u \text{ in } \mathbb{R}^N. \tag{6.3}$$

We set $M(\lambda, p, \rho) = M(\lambda, p, B(0, \rho))$. It is known that the minimization problem on the ball has a solution which is positive and radially symmetric. The quantity $M(\lambda, p, \rho)$ is independent of the centre of the ball and depends only on the radius of the ball, that is, $M(\lambda, p, \rho) = M(\lambda, p, B(x, \rho))$ (see Gidas, Ni & Nirenberg [58]). It follows from the dicussion in Section 2.1 that $M(\lambda, p, \rho)$ is increasing in ρ as $\rho \to 0$, that is,

$$M(\lambda, p, \rho_1) \geq M(\lambda, p, \rho_2) \text{ if } \rho_2 > \rho_1. \tag{6.4}$$

We also have

$$\lim_{\rho\to 0} M(\lambda, p, \rho) = \infty \tag{6.5}$$

and

$$\lim_{\rho\to\infty} M(\lambda, p, \rho) = M(\lambda, p, \mathbb{R}^N). \tag{6.6}$$

Indeed, let u_ρ be a minimizer for $M(\lambda, p, \rho)$. Then by the Hölder and Sobolev inequalities we have

$$\begin{aligned} 1 &= \int_{B(0,\rho)} |u_\rho|^p \, dx \le |B(0,\rho)|^{\frac{2^*-p}{2^*}} \left(\int_{B(0,\rho)} |u_\rho|^{2^*} \, dx \right)^{\frac{p}{2^*}} \\ &\le |B(0,\rho)|^{\frac{2^*-p}{2^*}} S^{-\frac{p}{2}} M(\lambda, p, \rho)^{\frac{p}{2}}. \end{aligned}$$

This inequality implies (6.5). To justify (6.6) let $\rho_n \to \infty$. We extend each minimizer u_{ρ_n} by zero outside $B(0, \rho_n)$. The sequence $\{u_{\rho_n}\}$ is bounded in $H_r(\mathbb{R}^N)$, where $H^1_r(\mathbb{R}^N)$ is a subspace of radially symmetric functions of $H^1(\mathbb{R}^N)$. Since $H^1_r(\mathbb{R}^N)$ is compactly embedded into $L^p(\mathbb{R}^N)$, $2 < p < 2^*$, we see that $M(\lambda, p, \rho_n) \to M(\lambda, p, \mathbb{R}^N)$.

To proceed further we rescale the quantities $M(\lambda, p, \mathbb{R}^N)$ and $M(\lambda, p, r)$ in the following way. Let $T_\lambda : H^1(\mathbb{R}^N) \to H^1(\mathbb{R}^N)$ be given by

$$T_\lambda(u) = \lambda^{-\frac{N}{2p}} u\left(\frac{x}{\sqrt{\lambda}}\right).$$

Then

$$||T_\lambda(u)||_{L^p(B(0,\sqrt{\lambda}r))} = ||u||_{L^p(B(0,r))}$$

and

$$\int_{B(0,\sqrt{\lambda}r)} \left(|\nabla T_\lambda(u)|^2 + T_\lambda(u)^2\right) dx = \frac{1}{\lambda^{1-\left(\frac{N}{2}-\frac{N}{p}\right)}} \int_{B(0,r)} \left(|\nabla u|^2 + \lambda u^2\right) dx.$$

Hence

$$M(\lambda, p, r) = \lambda^{1-\left(\frac{N}{2}-\frac{N}{p}\right)} M(1, p, \sqrt{\lambda}r)$$

and

$$M(\lambda, p, \mathbb{R}^N) = \lambda^{1-\left(\frac{N}{2}-\frac{N}{p}\right)} M(1, p, \mathbb{R}^N),$$

From this and (6.6) we easily deduce the following.

Lemma 6.1 *For every $p \in (2, 2^*)$ there exists $\bar{\lambda} = \bar{\lambda}(p)$ such that*

$$M(\lambda, p, r) < 2^{1-\frac{2}{p}} M(\lambda, p, \mathbb{R}^N)$$

for every $\lambda \geq \bar{\lambda}(p)$.

Clearly, $\bar{\lambda}(p)$ depends also on r. However, we do not emphasize this dependence, as r will be suitably chosen and kept fixed in the next section.

6.2 Role of barycentre

In what follows we assume that $0 \in \mathbb{R}^N - \bar{\Omega}$. We fix a number $r > 0$ such that the sets $\Omega^+ = \{x \in \mathbb{R}^N;\ \text{dist}\ (x, \Omega) \leq r\}$, $\Omega^- = \{x \in \Omega;\ \text{dist}\ (x, \partial\Omega) \geq 2r\}$ and Ω are homotopically equivalent. Let $\bar{\rho} = \inf\{\rho;\ \mathbb{R}^N - \Omega \subset B(0, \rho)\}$. By $\beta : H^1(\mathbb{R}^N) - \{0\} \to \mathbb{R}^N$ we denote a "barycentre" on $H^1(\mathbb{R}^N)$ given by

$$\beta(u) = \frac{\int_{\mathbb{R}^N} \chi(x)|u(x)|^p\, dx}{\int_{\mathbb{R}^N} |u(x)|^p\, dx},$$

where

$$\chi(x) = \begin{cases} x & \text{for } |x| \leq R \\ R\frac{x}{|x|} & \text{for } |x| > R, \end{cases}$$

and $R = 3\bar{\rho}$.

By $E_\lambda : H_0^1(\Omega) \to \mathbb{R}$ we define a functional given by

$$E_\lambda(u) = \int_\Omega \left(|\nabla u|^2 + \lambda u^2\right) dx.$$

Lemma 6.2 *Let $p \in (2, 2^*)$ and set*

$$\sigma(\lambda) = \inf\{E_\lambda(u);\ u \in H_0^1(\Omega),\ \int_\Omega |u|^p\, dx = 1,\ \beta(u) \notin \Omega^+\}.$$

Then

$$\liminf_{\lambda \to \infty} \frac{\sigma(\lambda)}{M(\lambda, p, r)} > 1. \tag{6.7}$$

Proof. We argue by contradiction. Assume that (6.7) does not hold. Then there exists a sequence $\{u_n\} \subset H_0^1(\Omega)$ such that $\|u_n\|_{L^p(\Omega)} = 1$,

$\beta(u_n) \notin \Omega^+$ for each n and

$$\lim_{n\to\infty} \frac{E_{\lambda_n}(u_n)}{M(\lambda_n, p, r)} = \alpha \leq 1.$$

Let T_{λ_n} be a rescaling defined in Section 6.1. Then

$$\lim_{n\to\infty} \frac{\int_{\sqrt{\lambda_n}\Omega} \left(|\nabla T_{\lambda_n}(u_n)|^2 + T_{\lambda_n}(u_n)^2\right) dx}{M(1, p, \sqrt{\lambda_n} r)} = \alpha \leq 1.$$

Since $\|T_{\lambda_n}(u_n)\|_{L^p(\mathbb{R}^N)} = \|T_{\lambda_n}(u_n)\|_{L^p(\sqrt{\lambda_n}\Omega)} = 1$, and

$$\begin{aligned} &\int_{\sqrt{\lambda_n}\Omega} \left(|\nabla T_{\lambda_n}(u_n)|^2 + T_{\lambda_n}(u_n)^2\right) dx \\ &\quad = \int_{\mathbb{R}^N} \left(|\nabla T_{\lambda_n}(u_n)|^2 + T_{\lambda_n}(u_n)^2\right) dx > M(1, p, \mathbb{R}^N) \end{aligned}$$

we see that $\alpha = 1$ and

$$\lim_{n\to\infty} \int_{\mathbb{R}^N} \left(|\nabla T_{\lambda_n}(u_n)|^2 + T_{\lambda_n}(u_n)^2\right) dx = M(1, p, \mathbb{R}^N).$$

We now observe that $T_{\lambda_n}(u_n) = 0$ on $\mathbb{R}^N - \sqrt{\lambda_n}\Omega$, so by (6.2) we have

$$T_{\lambda_n}(u_n) = w_n(x) + \bar{u}(x - y_n),$$

with $w_n \to 0$ in $H^1(\mathbb{R}^N)$; $|y_n| \to \infty$, where $\bar{u}$ is a ground state solution of the equation (6.3). In the next step we show that (i) dist $\left(\chi(\frac{y_n}{\sqrt{\lambda_n}}), \Omega\right) \to 0$ and (ii) $|\beta(u_n) - \chi(\frac{y_n}{\sqrt{\lambda_n}})| \to 0$ as $n \to \infty$. Let $\{\rho_n\} \subset \mathbb{R}$ be sequence such that $\rho_n \to \infty$ and $\frac{\rho_n}{\sqrt{\lambda_n}} \to 0$ as $n \to \infty$. Then

$$\int_{B\left(\frac{y_n}{\sqrt{\lambda_n}}, \frac{\rho_n}{\sqrt{\lambda_n}}\right)} |u_n|^p \, dx = \int_{B(y_n, \rho_n)} |T_{\lambda_n}(u_n)|^p \, dx \to 1$$

as $n \to \infty$. Since $u_n \in H_0^1(\Omega)$, we must have $B\left(\frac{y_n}{\sqrt{\lambda_n}}, \frac{\rho_n}{\sqrt{\lambda_n}}\right) \cap \Omega \neq \emptyset$ for n sufficiently large. From this we deduce that

$$\text{dist}\left(\frac{y_n}{\sqrt{\lambda_n}}, \Omega\right) \to 0 \tag{6.8}$$

as $n \to \infty$ because $\frac{\rho_n}{\sqrt{\lambda_n}} \to 0$. If $|\frac{y_n}{\sqrt{\lambda_n}}| \leq R$, then $\chi(\frac{y_n}{\sqrt{\lambda_n}}) = \frac{y_n}{\sqrt{\lambda_n}}$ and (i) follows from (6.8). Finally, if $|\frac{y_n}{\sqrt{\lambda_n}}| > R$, then $\chi(\frac{y_n}{\sqrt{\lambda_n}}) = R$ and consequently

$\chi(\frac{y_n}{\sqrt{\lambda_n}}) \in \Omega$ and dist $\left(\chi(\frac{y_n}{\sqrt{\lambda_n}}), \Omega\right) = 0$. To prove (ii) we write

$$\begin{aligned}
|\beta(u_n) &- \chi(\frac{y_n}{\sqrt{\lambda_n}})| = |\int_{\mathbb{R}^N} \chi(x)|u_n(x)|^p\,dx - \chi(\frac{y_n}{\sqrt{\lambda_n}})| \\
&= |\int_{\mathbb{R}^N} \left[\chi(x) - \chi(\frac{y_n}{\sqrt{\lambda_n}})\right] |u_n(x)|^p\,dx \\
&\leq \int_{B(\frac{y_n}{\sqrt{\lambda_n}}, \frac{\rho_n}{\sqrt{\lambda_n}})} |\chi(x) - \chi(\frac{y_n}{\sqrt{\lambda_n}})||u_n(x)|^p\,dx \\
&+ \int_{\mathbb{R}^N - B(\frac{y_n}{\sqrt{\lambda_n}}, \frac{\rho_n}{\sqrt{\lambda_n}})} |\chi(x) - \chi(\frac{y_n}{\sqrt{\lambda_n}})||u_n(x)|^p\,dx = i_n^1 + i_n^2.
\end{aligned}$$

Since $\int_{B(\frac{y_n}{\sqrt{\lambda_n}}, \frac{\rho_n}{\sqrt{\lambda_n}})} |u_n(x)|^p\,dx \leq 1$ and

$$\max\left\{|\chi(x) - \chi(\frac{y_n}{\sqrt{\lambda_n}}); \, x \in B(\frac{y_n}{\sqrt{\lambda_n}}, \frac{\rho_n}{\sqrt{\lambda_n}})\right\} \leq \frac{\rho_n}{\sqrt{\lambda_n}} \to 0$$

as $n \to \infty$, we see that $i_n^1 \to 0$ as $n \to \infty$. To estimate i_n^2 we notice that

$$|\int_{\mathbb{R}^N - B(\frac{y_n}{\sqrt{\lambda_n}}, \frac{\rho_n}{\sqrt{\lambda_n}})} |u_n(x)|^p\,dx = 1 - \int_{B(\frac{y_n}{\sqrt{\lambda_n}}, \frac{\rho_n}{\sqrt{\lambda_n}})} |u_n(x)|^p\,dx \to 0.$$

This, combined with $|\chi(x) - \chi(\frac{y_n}{\sqrt{\lambda_n}})| \leq 2R$, implies that $i_n^2 \to 0$ as $n \to \infty$. It now follows from (i) and (ii) that dist $(\beta(u_n), \Omega) \to 0$ as $n \to \infty$, which is impossible, since $\beta(u_n) \notin \Omega^+$. □

As an immediate consequence of Lemma 6.1 and (6.7) we obtain

Corollary 6.1 *For every $p \in (2, 2^*)$ there exists $\bar{\lambda} = \bar{\lambda}(p)$ such that for every $\lambda \geq \bar{\lambda}$ we have*

$$M(\lambda, p, r) < 2^{1-\frac{2}{p}} M(\lambda, p, \Omega)$$

and if $\int_\Omega |u(x)|^p\,dx = 1$ and $E_\lambda(u) \leq M(\lambda, p, r)$, then $\beta(u) \in \Omega^+$.

We now define the following energy level of the functional E_λ: for $p \in (2, 2^*)$, $\lambda > 0$ and $\rho > 0$ we set

$$\mu(\lambda, p, \rho) = \inf\{E_\lambda(u); \, u \in H_0^1(\Omega), \int_\Omega |u(x)|^p\,dx = 1, \, |\beta(u)| < \rho\}.$$

Lemma 6.3 *For every $\rho < R$ we have*

$$\mu(\lambda, p, \rho) > M(\lambda, p, \Omega). \tag{6.9}$$

Proof. If $\rho \in (0, R)$, then

$$\mu(\lambda, p, \rho) \geq M(\lambda, p, \Omega) = M(\lambda, p, \mathbb{R}^N).$$

Arguing by contradiction, assume that

$$\mu(\lambda, p, \rho) = M(\lambda, p, \mathbb{R}^N).$$

Then there exists a sequence $\{u_n\} \subset H^1_\circ(\Omega)$ such that $\int_\Omega |u_n(x)|^p\, dx = 1$, $|\beta(u_n)| < \rho$ and $E_\lambda(u_n) \to M(\lambda, p, \mathbb{R}^N)$ as $n \to \infty$. By the decomposition formula (6.2) we have

$$u_n(x) = w_n(x) + \bar{u}(x - y_n),$$

with $w_n \to 0$ in $H^1(\mathbb{R}^N)$ and $|y_n| \to \infty$ as $n \to \infty$. We claim that

$$|\beta(\bar{u}(x - y_n))| \to R \quad \text{as } n \to \infty. \tag{6.10}$$

Indeed, let $\rho_n = |y_n|^{\frac{1}{2}}$. We then have, for sufficiently large n,

$$\begin{aligned}
|\beta(\bar{u}(x - y_n)) &- R\frac{y_n}{|y_n|}| = |\int_{\mathbb{R}^N} (\chi(x) - R\frac{y_n}{|y_n|})|\bar{u}(x - y_n)|^p\, dx| \\
&\leq \int_{B(y_n,\rho_n)} R|\frac{x}{|x|} - \frac{y_n}{|y_n|}|\bar{u}(x - y_n)|^p\, dx \\
&+ \int_{\mathbb{R}^N - B(y_n,\rho_n)} |\chi(x) - R\frac{y_n}{|y_n|}||\bar{u}(x - y_n)|^p\, dx \\
&\leq \frac{2R}{|y_n|^{\frac{1}{2}}} \int_{B(y_n,\rho_n)} |\bar{u}(x - y_n|^p\, dx \\
&+ 2R\int_{\mathbb{R}^N - B(y_n,\rho_n)} |\bar{u}(x - y_n)|^p\, dx.
\end{aligned}$$

Since $|y_n| \to \infty$ and the ground state $\bar{u}$ decays exponentially at infinity, both integrals tend to 0 as $n \to \infty$. Since $w_n \to 0$ in $H^1(\mathbb{R}^N)$ and β is continuous on $H^1(\mathbb{R}^N)$, we deduce from (6.10) that $\beta(u_n) \to R$ as $n \to \infty$, which contradicts the assumption $|\beta(u_n)| < \rho < R$. □

Corollary 6.2 *For a given $p \in (2, 2^*)$ let $\bar{\lambda}(p)$ be chosen as in Corollary 6.1. Then for every $\lambda \geq \bar{\lambda}(p)$ there exists a number $\tau(\lambda) > 2r$ such that for every $\tau \geq \tau(\lambda)$ the following holds: if $\int_\Omega |u(x)|^p\, dx = 1$, $E_\lambda(u) \leq M(\lambda, p, \tau)$, then $\beta(u) \in \mathbb{R}^N - B(0, 2\bar{\rho})$.*

Proof. By the definition of R, we have $2\bar{\rho} < R$. Hence $\mu(\lambda, p, 2\bar{\rho}) > M(\lambda, p, \Omega) = M(\lambda, p, \mathbb{R}_N)$. The assertion follows from (6.6). □

To proceed further we introduce a manifold $\Sigma = \{u \in H_o^1(\Omega); \int_\Omega |u(x)|^p = 1\}$. It is easy to check that any critical point v of E_λ on Σ such that $E_\lambda(v) \leq 2^{1-\frac{2}{p}} M(\lambda, p, \Omega)$ does not change sign on Ω. Indeed, let v be critical point of $E_\lambda |_\Sigma$ satisfying $E_\lambda(v) = c < 2^{1-\frac{2}{p}} M(\lambda, p, \Omega)$. Since

$$\int_\Omega (\nabla v \nabla \phi + \lambda v \phi)\, dx = c \int_\Omega |v|^{p-2} v \phi \, dx$$

for each $\phi \in H_o^1(\Omega)$, we see that

$$\int_\Omega (|\nabla v^\pm|^2 + \lambda (v^\pm)^2)\, dx = c \int_\Omega |v^\pm|^p \, dx.$$

Assuming that $v^+ \neq 0$ and $v^- \neq 0$, we obtain

$$M(\lambda, p, \Omega) \|v^\pm\|_{L^p}^2 \leq \int_\Omega (|\nabla v^\pm|^2 + \lambda (v^\pm)^2)\, dx = c \int_\Omega |v^\pm|^p \, dx.$$

From this we deduce that $\|v^\pm\|_{L^p}^p \geq \left[\frac{M(\lambda,p,\Omega)}{c}\right]^{\frac{p}{p-2}}$ and hence $\|v\|_{L^p}^p \geq 2\left[\frac{M(\lambda,p,\Omega)}{c}\right]^{\frac{p}{p-2}}$, that is, $c \geq 2^{1-\frac{2}{p}} M(\lambda, p, \Omega)$, which is impossible.

6.3 Multiplicity in terms of category

To establish the existence of multiple solutions, we use the notion of the relative category [118].

Let X be a topological space and X_1, X_2 closed subsets of X such that $X_2 \subseteq X_1$. We say that the relative category in X of X_1 with respect to X_2 is n and we use notation $\mathrm{cat}_X[X_1, X_2] = n$, if and only if n is the least nonnegative integer such that $X_1 = \bigcup_0^n A_i$, $X_2 \subset A_o$, where A_i, $i = 0, \ldots, n$, are closed and there exist $h_i \in C([0,1] \times A_i, X)$, $i = 0, \ldots, n$ such that

(1) $h_i(0, x) = x$ for all $x \in A_i$, $i = 0, \ldots, n$,
(2) for each $i = 1, \ldots, n$ there exists $p_i \in X$ such that $h_i(1, x) = p_i$ for all $x \in A_i$,
(3) $h_o(1, x) \in X_2$ for all $x \in A_o$, $h_o(t, x) \in X_2$ for all $x \in A_o \cap X_2$ and all $t \in [0, 1]$.

Theorem 6.1 *Let M be a complete Riemann manifold and $f \in C^1(M, \mathbb{R})$. We set*

$$f^c = \{u \in M;\ f(u) \le c\},$$

$$K^c = \{u \in M;\ f(u) = c,\ \nabla f\mid_M (u) = 0.\}$$

Let $-\infty < a' < a < b < b' < \infty$ and suppose that f satisfies the Palais–Smale condition on the set $\{u \in M;\ f(u) \in (a', b')\}$ and moreover $K^a = K^b = \emptyset$. Then

$$\#\{u \in M;\ f(u) \in [a, b],\ \nabla f\mid_M (u) = 0\} \ge \operatorname{cat}_{f^b}[f^b, f^a].$$

Theorem 6.2 *There exists a function $\bar{\lambda} : (2, 2^*) \to (0, \infty)$ such that for every $\lambda \ge \bar{\lambda}(p)$ problem (6.1) admits at least* $\operatorname{cat}_{\bar{\Omega}}[\bar{\Omega}, \mathbb{R}^N - B(0, \bar{\rho})]$ *distinct solutions.*

Proof. As $\bar{\lambda}$ we take a function chosen in Corollary 6.1. Let $\lambda \ge \bar{\lambda}(p)$ and we choose $\tau(\lambda) \ge 2r$ as in Corollary 6.2. We now select $\hat{\tau} > \tau(\lambda)$ such that

$$\{u \in \Sigma;\ E_\lambda(u) = M(\lambda, p, \hat{\tau}),\ \nabla E_\lambda\mid_\Sigma = 0\} = \emptyset.$$

In the contrary case we would have infinitely many critical levels $M(\lambda, p, \tau)$ of E_λ on Σ satisfying

$$M(\lambda, p, \tau) < M(\lambda, p, 2r) < 2^{1-\frac{2}{p}} M(\lambda, p, \Omega).$$

In a similar manner we can choose $\hat{r} \in [r, 2r)$ such that

$$\{u \in \Sigma;\ E_\lambda(u) = M(\lambda, p, \hat{r}),\ \nabla E_\lambda\mid_\Sigma (u)\} = \emptyset.$$

According to Proposition 2.4 the functional E_λ satisfies the Palais–Smale condition on the set

$$\{u \in \Sigma;\ M(\lambda, p, \Omega) < E_\lambda(u) < 2^{1-\frac{2}{p}} M(\lambda, p, \Omega)\}$$

Since $M(\lambda, p, \Omega) < M(\lambda, p, \hat{\tau}) < M(\lambda, p, \hat{r}) < 2^{1-\frac{2}{p}} M(\lambda, p, \Omega)$, by Theorem 6.1 the functional $E_\lambda\mid_\Sigma$ has at least $\operatorname{cat}_{E_\lambda^{M(\lambda,p,\hat{r})}}\left[E_\lambda^{M(\lambda,p,\hat{r})}, E_\lambda^{M(\lambda,p,\hat{\tau})}\right]$ critical points below the level $M(\lambda, p, \hat{r})$. We now show that for $\bar{R} \ge R$ sufficiently large

$$\operatorname{cat}_{E_\lambda^{M(\lambda,p,\hat{r})}}\left[E_\lambda^{M(\lambda,p,\hat{r})}, E_\lambda^{M(\lambda,p,\tau)}\right] \ge \operatorname{cat}_{\Omega^+}[\Omega^-, \mathbb{R}^N - B(0, \bar{R} + \hat{\tau})]. \tag{6.11}$$

This and homotopy equivalence between Ω^+, Ω^- and Ω and between $\mathbb{R}^N - B(0,\bar\rho)$ and $\mathbb{R}^N - B(0,\bar R+\hat\tau)$ can be used to conclude that

$$\operatorname{cat}_{E_\lambda^{M(\lambda,p,\hat r)}}\left[E_\lambda^{M(\lambda,p,\hat r)}, E_\lambda^{M(\lambda,p,\hat\tau)}\right] \geq \operatorname{cat}_{\bar\Omega}[\bar\Omega, \mathbb{R}^N - B(0,\bar\rho)],$$

which completes the proof. Therefore it remains to prove (6.11). Toward this end we define the mapping $\Phi : \bar\Omega \to H_0^1(\Omega)$ by

$$\Phi(y)(x) = \begin{cases} u_{\lambda,p,\hat r}(x-y) & \text{for } y \in \Omega^- \cap B(0,\bar R+\hat r) \\ u_{\lambda,p,\sigma}(x-y) & \text{for } y \in B(0,\bar R+\hat\tau) - B(0,\bar R+\hat r) \\ u_{\lambda,p,\hat\tau}(x-y) & \text{for } y \in \mathbb{R}^N - B(0,\bar R+\hat\tau), \end{cases}$$

where $\bar R \geq R$ and $\sigma = |y| - \bar R$ and $u_{\lambda,p,s}$ is a radially symmetric minimizer for $M(\lambda,p,s)$. The mapping Φ is continuous and satisfies

$$\begin{cases} \Phi(\bar\Omega^-) \subset E_\lambda^{M(\lambda,p,\hat r)} &, \Phi(\mathbb{R}^N - B(0,\bar R+\hat\tau)) \subset E_\lambda^{M(\lambda,p,\hat\tau)} \\ (\beta\circ\Phi)(y) = y & \text{for all } y \in \bar\Omega \cap B(0,R-\hat r) \\ |(\beta\circ\Phi)(y) > 2\bar\rho & \text{for all } y \in \mathbb{R}^N - B(0,R-\hat r), \end{cases} \tag{6.12}$$

for $\bar R$ sufficiently large. Let $\eta : \Omega^+ \to \Omega^+$ be a continuous mapping defined by

$$\eta(x) = \begin{cases} x & \text{if } |x| \leq \bar\rho \text{ or } |x| \geq \bar R+\hat\tau \\ \frac{x}{|x|}\left(\bar\rho + \frac{|x|-\bar\rho}{\bar\rho}(\bar R - \bar\rho + \hat\tau)\right) & \text{if } \bar\rho \leq |x| \leq 2\bar\rho \\ (\bar R+\hat\tau)\frac{x}{|x|} & \text{if } 2\bar\rho \leq |x| \leq \bar R+\hat\tau. \end{cases}$$

Then

$$(\eta\circ\beta\circ\Phi(y) = y \text{ for } y \in \bar\Omega^- \cap B(0,\bar\rho),$$

$$|(\eta\circ\beta\circ\Phi)(y)| \geq \bar\rho \text{ for } |y| \geq \bar\rho,$$

and

$$|(\eta\circ\beta\circ\Phi)(y)| \geq \bar R+\hat\tau \text{ for } |y| \geq \bar R+\hat\tau.$$

If we define the homotopy $\mathcal{K} \in C([0,1]\times\Omega^-, \Omega^-)$ by

$$\mathcal{K}(t,y) = t(\eta\circ\beta\circ\Phi)(y) + (1-t)y,$$

we see that the mapping $\eta \circ \beta \circ \Phi : \Omega^- \to \Omega^-$ is homotopically equivalent to the identity map. To prove (6.11) suppose that

$$\mathrm{cat}_{E_\lambda^{M(\lambda,p,\hat r)}}\left[E_\lambda^{M(\lambda,p,\hat r)}, E_\lambda^{M(\lambda,p,\hat\tau)}\right] = n.$$

This means that n is the least integer such that

$$E_\lambda^{M(\lambda,p,\hat r)} \subset \bigcup_{i=1}^n A_i, \quad E_\lambda^{M(\lambda,p,\hat\tau)} \subset A_o,$$

where A_i, $i = 0, \ldots, n$, are closed and there exist $h_i \in C\big([0,1] \times A_i, E_\lambda^{M(\lambda,p,\hat r)}\big)$, $i = 0, \ldots, n$ such that

(1) $h_i(0,u) = u$ for all $u \in A_i$, $i = 0, \ldots, n$,
(2) for every $i = 1, \ldots, n$ there exists $p_i \in E_\lambda^{M(\lambda,p,\hat r)}$ such that $h_i(1,u) = p_i$ for all $u \in A_i$,
(3) $h_o(1,u) \in E_\lambda^{M(\lambda,p,\hat\tau)}$ for all $u \in A_o$ and $h_o(t,u) \in E_\lambda^{M(\lambda,p,\hat\tau)}$ for all $u \in A_o \cap E_\lambda^{M(\lambda,p,\hat\tau)}$ and $t \in [0,1]$.

We now set $K_i = \Phi^{-1}(A_i)$, $i = 0, \ldots, n$. It follows from (6.12) that $\Omega^- \subset \bigcup_{i=0}^n K_i$ and $\mathbb{R}^N - B(0, \bar R + \hat\tau) \subset K_o$. We then define $g_i \in C([0,1] \times K_i, \Omega^+)$, $i = 0, \ldots, n$, by

$$g_i(t,x) = \begin{cases} \mathcal{K}(2t, x) & \text{for } 0 \le t \le \frac{1}{2} \\ \eta \circ \beta \circ h_i(2t-1, \cdot) \circ \Phi(x) & \text{for } \frac{1}{2} \le t \le 1. \end{cases}$$

It is easy to check that the sets K_i and mappings g_i satisfy the conditions (i), (ii), (iii) of the definition of the relative category in Ω^+ of Ω^- with respect to $\mathbb{R}^N - B(0, \bar R + \hat\tau)$. Hence $\mathrm{cat}_{\Omega^+}[\Omega^- - B(0, \bar R + \hat\tau)] \le n$. □

6.4 Schrödinger equation

The problem of the existence of standing waves for nonlinear Schrödinger equation can be reduced to finding solutions of

$$(P_\epsilon) \qquad \begin{cases} -\epsilon^2 \Delta u + V(x) u &= |u|^{p-2} u \text{ in } \mathbb{R}^N \\ u(x) > 0 \text{ on } & \mathbb{R}^N, \end{cases}$$

where $1 < p < 2^*$, $N \ge 3$. We assume that the function $V(x)$ is continuous and bounded on $\mathbb{R}^N$ and

(V_1) $0 < V_o = \inf_{x \in \mathbb{R}^N} V(x)$,

(V_2) $\lim_{|x|\to\infty} V(x) > V_\circ$.

The multiplicity result for the problem (P_ϵ) will be described in terms of relative category of the following sets

$$\mathcal{V} = \{x \in \mathbb{R}^N;\ V(x) = V_\circ\} \text{ and } \mathcal{V}_\delta = \{x \in \mathbb{R}^N;\ \operatorname{dist}(x, M) \le \delta\},$$

where $\delta > 0$ is small enough. It follows from assumption (V_2) that $M \neq \emptyset$. As in Section 6.3 we introduce a manifold $\Sigma = \{u \in H^1(\mathbb{R}^N);\ \int_{\mathbb{R}^N} |u|^p\, dx = 1\}$ and define a functional $J_\epsilon : \Sigma \to \mathbb{R}$ by

$$J_\epsilon(u) = \int_{\mathbb{R}^N} (\epsilon^2 |\nabla u|^2 + V(x)u^2)\, dx.$$

By $\omega_{\epsilon,\mu}$ we denote the ground state solution of the equation

$$-\epsilon^2 \Delta u + \mu u = |u|^{p-2} u \text{ in } \mathbb{R}^N \tag{6.13}$$

satisfying $||\omega_{\epsilon,\mu}||_{L^p} = 1$, where $\mu > 0$ is a parameter. It follows from Theorem 2.1 that the function $\omega_{\epsilon,\mu}$ is a minimizer of the constrained minimization problem

$$M(\epsilon, \mu) = \inf\{\int_{\mathbb{R}^N} (\epsilon^2 |\nabla u|^2 + \mu u^2)\, dx;\ u \in \Sigma\}.$$

To simplify notation, we set $\omega = \omega_{1,V_\circ}$. We now define two maps that will be used to relate the multiplicity of solutions of (P_ϵ) the category of the set $\mathcal{V}$.

Let $\delta > 0$ be fixed and let $\eta \in C^1([0, \infty))$ be such that $\eta(t) = 1$ for $0 \le t \le \frac{\delta}{2}$, $\eta(t) = 0$ for $t \ge \delta$, $0 \le \eta \le 1$ on $[0, \infty)$ and $|\eta'(t)| \le C$ for some constant $C > 0$. We define $\psi_{\epsilon,y}(x) = \eta(|x - y|)\epsilon^{-\frac{N}{p}} \omega\left(\frac{x-y}{\epsilon}\right)$ and set

$$\phi_{\epsilon,y}(x) = \frac{\psi_{\epsilon,y}(x)}{||\psi_{\epsilon,y}||_{L^p(\mathbb{R}^N)}}.$$

The mapping $\Phi_\epsilon : \mathcal{V} \to \Sigma$ is defined by $\Phi_\epsilon(y)(\cdot) = \phi_{\epsilon,y}(\cdot)$. We now check that

$$\lim_{\epsilon \to 0} \epsilon^{N\left(\frac{2}{p} - 1\right)} J_\epsilon(\Phi_\epsilon(y)) = M(1, V_\circ) \tag{6.14}$$

uniformly in $y \in \mathcal{V}$. Indeed, it follows from the exponential decay of ω that

$$J_\epsilon(\Phi_\epsilon(y)) \quad = \quad \epsilon^{N\left(1 - \frac{2}{p}\right)} \frac{\int_{\mathbb{R}^N} (|\nabla \omega|^2 + V_\circ \omega^2)\, dx + o(1)}{\left(\int_{\mathbb{R}^N} |\omega|^p\, dx + o(1)\right)^{\frac{2}{p}}}$$

$$= \epsilon^{N\left(1-\frac{2}{p}\right)}\left(M(1,V_o)+o(1)\right).$$

Letting $\epsilon \to 0$ and using the compactness of $\mathcal{V}$ relation (6.14) readily follows.

Let $\rho > 0$ be such that $\mathcal{V}_\delta \subset B(0,\rho)$. To define a barycentre we use the function χ from Section 6.2 with R replaced by ρ and set $\beta(u) = \int_{\mathbb{R}^N} |u(x)|^p \chi(x)\,dx$ for $u \in \Sigma$. It is easy to check that

$$\beta(\Phi_\epsilon(y)) = y + \int_{\mathbb{R}^N} \left(\chi(\epsilon x + y) - y\right)|\omega(x)|^p\,dx = y + o(1) \tag{6.15}$$

as $\epsilon \to 0$ uniformly in $y \in \mathcal{V}$.

Let $h : (0,\infty) \to (0,\infty)$, with $\lim_{t\to 0} h(t) = 0$ and set for $\epsilon > 0$:

$$\Sigma_\epsilon = \{u \in \Sigma;\ J_\epsilon(u) \leq M(\epsilon, V_o) + \epsilon^{N\left(1-\frac{2}{p}\right)} h(\epsilon)\}.$$

The following lemma combined with (6.15) shows that the barycentre on Σ_ϵ remains close to $\mathcal{V}$.

Lemma 6.4 *We have*

$$\lim_{\epsilon\to 0} \sup_{u\in\Sigma_\epsilon} \inf_{y\in\mathcal{V}_\delta} [\beta(u) - \beta(\phi_{\epsilon,y})] = 0.$$

Proof. Let $\epsilon_n \to 0$ as $n \to \infty$. For every n there exists $u_n \in \Sigma_{\epsilon_n}$ such that

$$\inf_{y\in M_\delta} [\beta(u_n) - \beta(\phi_{\epsilon_n,y})] = \sup_{u\in\Sigma_{\epsilon_n}} \inf_{y\in M_\delta} [\beta(u) - \beta(\phi_{\epsilon_n,y})] + o(1).$$

To complete the proof it is sufficient to find points $\{y_n\} \subset \mathcal{V}_\delta$ such that

$$\lim_{n\to\infty} [\beta(u_n) - \beta(\phi_{\epsilon_n,y_n}] = 0.$$

Rescaling u_n by $v_n(x) = \epsilon_n^{\frac{N}{p}} u_n(\epsilon x)$, we then have

$$\int_{\mathbb{R}^N} \left(|\nabla v_n|^2 + V(\epsilon_n x) v_n^2\right) dx \leq M(1, V_o) + h(\epsilon_n).$$

Since $v_n \in \Sigma$ and

$$\int_{\mathbb{R}^N} \left(|\nabla v_n|^2 + V(\epsilon_n x)\right) v_n^2\,dx \geq \int_{\mathbb{R}^N} \left(|\nabla v_n|^2 + V_o v_n^2\right) dx,$$

$\{v_n\}$ is a minimizing sequence for $M(1, V_o)$. It follows from Theorem 2.1 that there exists a sequence $\{z_n\} \subset \mathbb{R}^N$ such that $\hat{v}_n(\cdot) = v_n(\cdot + z_n) \to \hat{v}(\cdot) \not\equiv 0$ in $H^1(\mathbb{R}^N)$. Clearly, $M(1, V_o)^{\frac{1}{p-2}}\hat{v}$ is a ground state solution of the equation (6.13), with $\epsilon = 1$, $\mu = V_o$. We claim that $\{\epsilon_n z_n\}$ is a bounded sequence. Arguing indirectly, assume that $|\epsilon_n z_n| \to \infty$. We then have

$$\begin{aligned} M(1, V_o) \quad + \quad h(\epsilon_n) &\geq \int_{\mathbb{R}^N} \left(|\nabla \hat{v}_n|^2 + V(\epsilon_n(x + z_n))\hat{v}_n^2\right) dx \qquad (6.16) \\ &\geq \int_{\mathbb{R}^N} |\nabla \hat{v}_n|^2 \, dx + \int_{\mathbb{R}^N} V(\epsilon_n(x + z_n))\hat{v}_n^2 \, dx + o(1) \\ &\geq \int_{\mathbb{R}^N} |\nabla \hat{v}|^2 \, dx + \liminf_{n\to\infty} \int_{\mathbb{R}^N} V(\epsilon_n(x + z_n))\hat{v}_n^2 \, dx \\ &\geq \int_{\mathbb{R}^N} |\nabla \hat{v}|^2 \, dx + \int_{\mathbb{R}^N} \lim_{n\to\infty} V(\epsilon_n(z_n + x))\hat{v}^2 \, dx. \end{aligned}$$

Since $V_o < V(\infty) = \lim_{|x|\to\infty} V(x)$, we deduce from (6.16) that

$$M(1, V_o) \geq \int_{\mathbb{R}^N} \left(|\nabla \hat{v}|^2 + V(\infty)\hat{v}^2\right) dx \geq M(1, V(\infty)).$$

However, this is impossible because $V_o < V(\infty)$ implies that $M(1, V(\infty)) > M(1, V_o)$. Since $\{\epsilon_n z_n\}$ is bounded H, we may assume that $\epsilon_n z_n \to \hat{z}$. We now show that $\hat{z} \in \mathcal{V}$. Repeating the argument used in (6.16) we get the following inequality

$$M(1, V_o) \geq \int_{\mathbb{R}^N} \left(|\nabla \hat{v}|^2 + V(\hat{z})\hat{v}^2\right) dx \geq M(1, V(\hat{z}) \geq M(1, V_o).$$

This yields that $V_o = V(\hat{z})$, that is, $\hat{z} \in M$. Therefore, we can assume that $y_n = \epsilon_n z_n \in M_\delta$. Hence

$$\begin{aligned} |\beta(u_n) \quad - \quad \beta(\phi_{\epsilon_n, y_n})| &= \left| \int_{\mathbb{R}^N} \chi(x)|u_n(x)|^p \, dx - \int_{\mathbb{R}^N} \chi(x)|\phi_{\epsilon_n, y_n}(x)|^p \, dx \right| \\ &\leq \rho \int_{\mathbb{R}^N} \left| |u_n(x)|^p - |\phi_{\epsilon_n, y_n}(x)|^p \right| dx \\ &= \rho \int_{\mathbb{R}^N} \left| |v_n(x + z_n)|^p - \omega^p \right| dx + o(1). \end{aligned}$$

Since $v_n(\cdot + z_b) \to \omega$ in $L^p(\mathbb{R}^N)$, the result follows. □

6.5 Palais–Smale condition

The main objective of this section is to find the range of energy level of a functional

$$I_\epsilon(u) = \frac{1}{2}\int_{\mathbb{R}^N} \left(\epsilon^2|\nabla u|^2 + V(\epsilon x)u^2\right) dx - \frac{1}{p}\int_{\mathbb{R}^N} |u|^p\, dx,$$

for which the Palais–Smale condition holds.

Let $u_n \rightharpoonup u$ in $H^1(\mathbb{R}^N)$. As in Chapter 1 we define the quantity α_∞ that measures loss of mass of $\{u_n\}$ at ∞. We also set

$$\beta^1_\infty = \lim_{R\to\infty}\limsup_{n\to\infty}\int_{|x|>R} |\nabla u_n(x)|^2\, dx$$

and

$$\beta^2_\infty = \lim_{R\to\infty}\limsup_{n\to\infty}\int_{|x|>R} u_n(x)^2\, dx.$$

It follows from the proof of Proposition 1.2 that

$$M(1, V(\infty))\alpha_\infty^{\frac{2}{p}} \le \beta^1_\infty + V(\infty)\beta^2_\infty \tag{6.17}$$

and

$$M(1, V)\alpha_\infty^{\frac{2}{p}} \le \beta^1_\infty + V(\infty)\beta^2_\infty.$$

Proposition 6.1 *Let $\{u_n\} \subset H^1(\mathbb{R}^N)$ be such that*

$$I_\epsilon(u_n) \to c < \frac{p-2}{2p} M(\epsilon, V(\infty))^{\frac{p}{p-2}}$$

and $I'_\epsilon(u_n) \to 0$ in $H^{-1}(\mathbb{R}^N)$. Then $\{u_n\}$ is relatively compact in $H^1(\mathbb{R}^N)$.

Proof. For simplicity we assume that $\epsilon = 1$. It follows from Remark 2.1 that there exists a sequence $\{y_n\} \subset \mathbb{R}^N$ such that $v_n(\cdot) = u_n(\cdot + y_n) \rightharpoonup v \not\equiv 0$ in $H^1(\mathbb{R}^N)$. First, we show that $\{y_n\}$ is bounded. Arguing indirectly, assume that $|y_n| \to \infty$. We define quantities α_∞, β^1_∞ and β^2_∞ for the sequence $\{v_n\}$. Let ϕ_R be the function defined in the proof of Proposition 1.2. Since $\langle I'_1(u_n), u_n\phi_R\rangle \to 0$ as $n \to \infty$ uniformly in R we get

$$\int_{\mathbb{R}^N} |\nabla u_n|^2 \phi_R(x - y_n)\, dx + \int_{\mathbb{R}^N} \nabla u_n u_n \nabla\phi_R(x - y_n)\, dx$$
$$+ \int_{\mathbb{R}^N} V u_n^2 \phi_R(x - y_n)\, dx - \int_{\mathbb{R}^N} |u_n|^p \phi_R(x - y_n)\, dx = o(1),$$

which is equivalent to

$$\int_{\mathbb{R}^N} |\nabla v_n(x)|^2 \phi_R(x)\,dx + \int_{\mathbb{R}^N} \nabla v_n(x) v_n(x) \nabla \phi_R(x)\,dx$$
$$+ \int_{\mathbb{R}^N} V(x+y_n) v_n(x)^2 \phi_R(x)\,dx = \int_{\mathbb{R}^N} |v_n(x)|^p \phi_R(x)\,dx + o(1).$$

Letting $n \to \infty$ and then $R \to \infty$ we get

$$\beta^1_\infty + V(\infty)\beta^2_\infty = \alpha_\infty. \tag{6.18}$$

Combining (6.17) and (6.18) we then obtain

$$M(1, V(\infty))^{\frac{p}{p-2}} \le \beta^1_\infty + V(\infty)\beta^2_\infty = \alpha_\infty. \tag{6.19}$$

We now observe that

$$\begin{aligned} c &= I_1(u_n) - \frac{1}{2}\langle I_1'(u_n), u_n\rangle = \left(\frac{1}{2} - \frac{1}{p}\right) \int_{\mathbb{R}^N} |u_n|^p\,dx \\ &= \left(\frac{1}{2} - \frac{1}{p}\right) \int_{\mathbb{R}^N} |v_n(x)|^p\,dx. \end{aligned}$$

With the aid of Proposition 1.2 we deduce from this that

$$\begin{aligned} o(1) + c &= \left(\frac{1}{2} - \frac{1}{p}\right)\left(\int_{\mathbb{R}^N} |v|^p\,dx + \alpha_\infty\right) \\ &> \left(\frac{1}{2} - \frac{1}{p}\right) M(1, V(\infty))^{\frac{p}{p-2}}, \end{aligned} \tag{6.20}$$

which is impossible. Hence the sequence $\{y_n\}$ must be bounded and we can assume that $u_n \rightharpoonup u \not\equiv 0$ in $H^1(\mathbb{R}^N)$. Let α_∞, β^1_∞ and β^2_∞ correspond to the sequence $\{u_n\}$. We claim that $\alpha_\infty = 0$. In the contrary case $0 < \alpha_\infty < 1$. Repeating the first part of the proof we check that $\alpha_\infty = \beta^1_\infty + V(\infty)\beta^2_\infty$. This combined with (6.17) gives the inequality (6.20). It then follows from (6.20) that $c > \left(\frac{1}{2} - \frac{1}{p}\right) M(1, V(\infty))^{\frac{p}{p-2}}$ which is impossible. Therefore, $\int_{\mathbb{R}^N} |u_n|^p \to \int_{\mathbb{R}^N} |u|^p\,dx$ and hence $u_n \rightharpoonup u$ in $L^p(\mathbb{R}^N)$. To complete the proof we observe that

$$\langle I_1'(u_m) - I_1'(u_n), u_m - u_n\rangle = o(1),$$

which means that

$$\int_{\mathbb{R}^N} |\nabla(u_m - u_n)|^2\,dx \quad + \quad \int_{\mathbb{R}^N} V(u_m - u_n)^2\,dx$$

$$= \int_{\mathbb{R}^N} \left(|u_m|^{p-2} - |u_n|^{p-2}u_n\right)(u_m - u_n)\,dx + o(1).$$

Hence $\{u_n\}$ is the Cauchy sequence in $H^1(\mathbb{R}^N)$ and the result follows. □

Corollary 6.3 *There exists an $\epsilon_o > 0$ small enough such that J_ϵ, $0 < \epsilon < \epsilon_o$, satisfies the Palais–Smale condition on the set*

$$\{u \in \Sigma;\ J_\epsilon(u) < M(\epsilon, V(\infty))\}.$$

We only check that the set in question is nonempty. This follows if we can show that

$$\inf_{u\in\Sigma} J_\epsilon(u) < M(\epsilon, V(\infty)).$$

Indeed, if there exists a sequence $\epsilon_n \to 0$ as $n \to \infty$ such that $M(\epsilon_n, V(\infty)) < \inf_{u\in\Sigma} J_\epsilon(u)$, then by (6.14) we have

$$M(\epsilon_n, V(\infty)) \le M(\epsilon_n, V_o) + o\left(\epsilon_n^{N\left(1-\frac{2}{p}\right)}\right).$$

This in turn implies that $M(1, V(\infty)) \le M(1, V_o)$, which is impossible.

6.6 Effect of the graph topology

We are now in a position to establish the following existence result.

Theorem 6.3 *For every $\delta > 0$ small enough there exists $\epsilon_\delta = \epsilon(\delta)$ such that problem (P_ϵ) has at least $\mathrm{cat}_{\mathcal{V}_\delta}(\mathcal{V})$ solutions for every $\epsilon < \epsilon_\delta$.*

Proof. We compare the topology of the set $\mathcal{V}$ and the topology of a suitable energy level of J_ϵ. Let Φ_ϵ, β and h be mapping introduced in Section 6.4. According to Lemma 6.4 and Corollary 6.3 we can find an $\bar{\epsilon} > 0$ such that J_ϵ satisfies the Palais–Smale condition on Σ_ϵ and

$$\sup_{u\in\Sigma_\epsilon} \inf_{y\in\mathcal{V}_\delta} [\beta(u) - \beta(\phi_{\epsilon,y})] \le \frac{\delta}{2} \tag{6.21}$$

for every $0 < \epsilon < \bar{\epsilon}$. By (6.14) we can also assume that

$$J_\epsilon(\Phi_\epsilon(y)) \le M(\epsilon, V_o) + h(\epsilon)\epsilon^{N\left(1-\frac{2}{p}\right)}$$

for $0 < \epsilon < \bar{\epsilon}$. Hence, $\Phi_\epsilon(\mathcal{V}) \subset \Sigma_\epsilon$. It follows from (6.15) and (6.1) that $\mathrm{dist}(\beta(u), \mathcal{V}_\delta) < \frac{\delta}{2}$ for every $0 < \epsilon < \bar{\epsilon}$ and every $u \in \Sigma_\epsilon$. From this we deduce that $\beta(\Sigma_\epsilon) \subset \mathcal{V}_\delta$. Therefore, the map $\beta \circ \Phi_\epsilon$ is homotopic to the inclusion $j : \mathcal{V} \to \mathcal{V}_\delta$ in $\mathcal{V}_\delta$ for every $0 < \epsilon < \bar{\epsilon}$. We set $\Sigma_\epsilon^+ = \Sigma_\epsilon \cap \{u \in \Sigma;\, u \geq 0 \text{ on } \mathbb{R}^N\}$. We claim that $\operatorname{cat}_{\Sigma_\epsilon}(\Sigma_\epsilon^+) \geq \operatorname{cat}_{\mathcal{V}_\delta}(\mathcal{V})$. Indeed, assume that $\operatorname{cat}_{\Sigma_\epsilon}(\Sigma_\epsilon^+) = n$. This means that n is the least integer such that $\Sigma_\epsilon^+ = A_1 \cup \ldots \cup A_n$, where A_i are closed in Σ_ϵ^+ and contractible in Σ_ϵ, that is, there exist $\mathcal{H}_i \in C([0,1,] \times A_i, \Sigma_\epsilon)$, $i = 1, \ldots, n$, such that $\mathcal{H}_i(0, u) = u$ for every $u \in A_i$, $i = 1, \ldots, n$ and $\mathcal{H}_i(1, u) = w_i$ for every $u \in A_i$ and some $w_i \in A_i$, $i = 1, \ldots, n$. By the homotopic equivalence of $\beta \circ \Phi_\epsilon$ and $j : \mathcal{V} \to \mathcal{V}_\delta$ there exists $\mathcal{K} \in C([0,1] \times \mathcal{V}, \mathcal{V}_\delta)$ such that $\mathcal{K}(0, x) = j(x)$ and $\mathcal{K}(1, x) = \beta \circ \Phi_\epsilon(x)$ for every $x \in \mathcal{V}$. For example we can take $\mathcal{K}(t, x) = t\beta \circ \Phi_\epsilon(x) + (1-t)j(x)$. We set $K_i = \Phi_\epsilon^{-1}(A_i)$, $i = 1, \ldots, n$. Clearly, the K_i are closed and $\mathcal{V} \subset \bigcup_{i=1}^n K_i$. We now verify that the K_i are contractible in $\mathcal{V}_\delta$ using the maps $\mathcal{G}_i \in C([0,1] \times K_i, \mathcal{V}_\delta)$ defined by

$$\mathcal{G}_i(t, x) = \begin{cases} \mathcal{K}(2t, x) & \text{if } 0 \leq t \leq \frac{1}{2} \\ \beta \circ \mathcal{H}_i(2t-1, \cdot) \circ \Phi_\epsilon(x) & \text{if } \frac{1}{2} \leq t \leq 1. \end{cases}$$

This completes the proof of our claim. By a similar argument we show that $\operatorname{cat}_{\Sigma_\epsilon}(\Sigma_\epsilon^-) \geq \operatorname{cat}_{\mathcal{V}_\delta}(\mathcal{V})$, where $\Sigma_\epsilon^- = \Sigma_\epsilon \cap \{u \in \Sigma;\, u \leq 0 \text{ on } \mathbb{R}^N\}$. Since Σ_ϵ^+ and Σ_ϵ^- are disjoint in Σ_ϵ, we see that

$$\operatorname{cat}_{\Sigma_\epsilon}(\Sigma_\epsilon) = \operatorname{cat}_{\Sigma_\epsilon}(\Sigma_\epsilon^+) + \operatorname{cat}_{\Sigma_\epsilon}(\Sigma_\epsilon^-) \geq 2 \operatorname{cat}_{\mathcal{V}_\delta}(\mathcal{V}).$$

By the Lusternik - Schnirelman theory of critical points we see that J_ϵ has at least $2 \operatorname{cat}_{\mathcal{V}_\delta}(\mathcal{V})$ critical points. We now notice that these critical points belong to Σ_ϵ. As in Section 6.2 we can show that critical points in Σ_ϵ cannot change sign. Hence there exists at least $\operatorname{cat}_{\mathcal{V}_\delta}(\mathcal{V})$ critical points that are positive on $\mathbb{R}^N$. □

6.7 Concentration of solutions

According to Theorem 6.2 for each $0 < \epsilon < \epsilon_\delta$ problem (P_ϵ) admits at least one solution u_ϵ. Solutions u_ϵ can be written as $u_\epsilon = J_\epsilon(\bar{u}_\epsilon)^{\frac{1}{p-2}} \bar{u}_\epsilon$, where $\bar{u}_\epsilon$ are critical points of J_ϵ restricted (constrained) to Σ_ϵ. Let $\epsilon_n \to 0$ be an arbitrary sequence of positive numbers. It then follows from the proof of Lemma 6.4 that $\{\epsilon_n^{\frac{N}{p}} \bar{u}_\epsilon(\epsilon_n x)\}$ is a minimizing sequence for $M(1, V_\circ)$. Arguing as in the proof of Lemma 6.4, we can find a sequence $\{z_n\} \subset \mathbb{R}^N$ such that, up to a subsequence, $\bar{v}_n(\cdot) = \epsilon_n^{\frac{N}{p}} \bar{u}_{\epsilon_n}(\epsilon_n(\cdot + z_n)) \to \hat{v}$ in $H^1(\mathbb{R}^N)$,

$z_n\epsilon_n \to z$ where $V(z) = V_\circ$ and $\hat{v}$ is a minimizer for $M(1, V_\circ)$. Since

$$u_{\epsilon_n}(\epsilon_n(x+z_n)) = \left(\int_{\mathbb{R}^N} \left(|\nabla \bar{v}_n|^2 + V(\epsilon_n(x+z_n))\bar{v}_n^2\right) dx\right)^{\frac{1}{p-2}} \bar{v}_n(x),$$

we see that the sequence of solutions $\{u_{\epsilon_n}\}$ of (P_ϵ) concentrates at $z \in \mathcal{V}$ in the sense that $u_{\epsilon_n}(\epsilon_n(\cdot + z_n)) \to \omega$ in $H^1(\mathbb{R}^N)$, $\epsilon_n z_n \to z$, where ω is a ground state solution of (6.3) with $\lambda = V_\circ$ and $z_n \to z$ and $V(z) = V_\circ$. We now show that $\epsilon_n z_n$ can be taken as a unique point, where u_{ϵ_n} assumes a global maximum.

We need the following lemma.

Lemma 6.5 *Let $\phi \in C^2(\overline{B(0,a)})$ be a radial function satisfying $\phi'(0) = 0$ and $\phi''(r) < 0$ for $0 \le r \le a$. Then there exists a $\delta > 0$ such that if $\psi \in C^2(\overline{B(0,a)})$ satisfies (i) $\nabla\psi(0) = 0$ and (ii) $\|\psi - \phi\|_{C^2(\overline{B(0,a)})} < \delta$, then $\nabla\psi(x) \neq 0$ for $x \neq 0$.*

Proof. By straightforward calculations we have

$$\begin{aligned}\frac{\partial\psi}{\partial x_j}(x) &= \int_0^1 \nabla\frac{\partial\phi(tx)}{\partial x_j}\,dt\cdot x = \int_0^1 \nabla\left(\frac{\partial\psi}{\partial x_j} - \frac{\partial\phi}{\partial x_j}\right)(tx)\,dt\cdot x\\ &= \int_0^1 \phi''(tx)\,dt\cdot x_j + \int_0^1 \nabla\left(\frac{\partial\psi}{\partial x_j} - \frac{\partial\phi}{\partial x_j}\right)(tx)\,dt\cdot x.\end{aligned}$$

If $\nabla\psi(x) = 0$, then

$$-x_j\int_0^1 \phi''(tx)\,dt = \int_0^1 \nabla\left(\frac{\partial\psi}{\partial x_j} - \frac{\partial\phi}{\partial x_j}\right)(tx)\,dt\cdot x$$

for $j = 1, \ldots, N$. Let $\alpha = \min|\phi''| > 0$. Then

$$\alpha|x|^2 \le N|x|^2 \sup_{|x|\le a, i=1,\ldots,N} \left|\frac{\partial^2(\psi-\phi)}{\partial x_i\partial x_j}\right|.$$

Hence, if $\|\psi - \phi\|_{C^2\overline{B(0,a)})} < \frac{\alpha}{N}$, then $x = 0$. □

Proposition 6.2 *Let $\{u_{\epsilon'_n}\}$ be a sequence of solutions of $(P_{\epsilon'_n})$, $\epsilon'_n < \epsilon_\delta$. Then there exists a subsequence $\{\epsilon_n\}$ of $\{\epsilon'_n\}$ such that u_{ϵ_n} attains its unique maximum at x_{ϵ_n} and $u_{\epsilon_n}(\epsilon_n(\cdot + x_{\epsilon_n})) \to \omega$ in $H^1(\mathbb{R}^N)$, $x_{\epsilon_n} \to z$, where $V(z) = V_\circ$ and ω is a ground state solution of (6.3) with $\lambda = V_\circ$.*

Proof. From the discussion in the first paragraph of this section, we know that there exists a subsequence $\{\epsilon_n\}$ such that $w_n(x) = u_{\epsilon_n}(\epsilon_n x + \epsilon_n z_n) = u_{\epsilon_n}(\epsilon_n x + \bar{z}_n) \to \omega$ in $H^1(\mathbb{R}^N)$ and $\bar{z}_n \to z \in M$. Hence we may assume that $\int_{\mathbb{R}^N - B(0,R)} |w_n(x)|^{2^*}\, dx \to 0$ as $R \to 0$ uniformly in n. By the one - sided Harnack inequality [110] we have

$$\max_{y\in B(x,1)} w_n(y) \le C\left(\int_{B(x,2)} w_n(y)^{2^*}\, dy\right)^{\frac{1}{2^*}}$$

for all $x \in \mathbb{R}^N$, where C is a constant independent of x and n. Hence $w_n(x) \to 0$ as $|x| \to \infty$ uniformly in n. Consequently, $u_n(x) = u_{\epsilon_n}(x) = w_n\left(\frac{x-\bar{z}_n}{\epsilon_n}\right) \to 0$ as $n \to \infty$ uniformly in x outside any fixed ball centred at z. Let x_n be a local maximum of u_n. It follows from maximum principle that $u_n(x_n) \ge (V_\circ)^{\frac{1}{p-1}}$. Therefore $x_n \to z$ as $n \to \infty$. This argument shows also that each sequence of local maximum points converges to z. Let $\{x_n\}$ be a sequence of local maximum points of $\{u_n\}$ and set $\bar{w}_n(x) = u_n(\epsilon_n x + x_n)$. To complete the proof it is sufficient to show the uniqueness of local maximum points for $\bar{w}_n$. Since $\bar{w}_n \to 0$ as $|x| \to \infty$ uniformly with respect to n, we conclude that all local maximum points of $\bar{w}_n$ belong to a fixed ball in $\mathbb{R}^N$. Since $\bar{w}_n \to \omega$ in $H^1(\mathbb{R}^N)$ (and in $C^2_{\text{loc}}(\mathbb{R}^N)$) and 0 is the only critical point in ω, these critical points must belong to some ball $B(0,a)$, Taking a small enough, we can ensure that $w''(r) < 0$ for $0 < r \le a$. Finally, an application of Lemma 6.5 shows that for large n, $\bar{w}_n$ has only critical point at the origin. □

Solutions u_ϵ of problem (P_ϵ) exhibit an exponential decay at infinity:

Proposition 6.3 *Let $\{u_{\epsilon'_n}\}$ be a sequence of solutions of $(P_{\epsilon'_n})$ such that $u_{\epsilon'_n}(x_{\epsilon'_n}) = \max_{x\in\mathbb{R}^N} u_{\epsilon'_n}(x)$, with $x_{\epsilon'_n} \to z$ and $V(z) = V_\circ$. Then there exists a subsequence $\{u_{\epsilon_n}\}$ having the following property: for every $R > 0$ there exist an integer $n_\circ \ge 1$ and a constant $\delta > 0$ such that*

$$u_{\epsilon'_n}(x) \le C\exp\left(-\delta\sum_{i=1}^{N}\frac{|x_i - x^i_{\epsilon_n}|}{\epsilon_n}\right)$$

for all $|x - z| \ge R\epsilon_n$ and $n \ge n_\circ$, where $C > 0$ is a constant independent of n.

Proof. It follows from the proof of Proposition 6.2 that there exists a subsequence $\{u_{\epsilon_n}\}$ such that $u_{\epsilon_n}\left(\frac{x - x_{\epsilon_n}}{\epsilon_n}\right) \to 0$ as $n \to \infty$ uniformly in x outside any fixed ball centred at z. Let $H(x,\delta) = \prod_{i=1}^N \cosh\delta\left(\frac{x_i - x^i_{\epsilon_n}}{\epsilon_n}\right)$. It is easy to show that given $R > 0$, we can find $\delta_1 >$ and an integer $n_\circ \ge 1$

such that

$$-\Delta\left(\frac{1}{H}\right) + \left(V - u_{\epsilon_n}\left(\frac{x - x_{\epsilon_n}}{\epsilon_n}\right)\right)\frac{1}{H} \geq \frac{V_o}{2H}$$

for all $\delta \leq \delta_1$, $n \geq n_o$ and $|x - z| \geq \epsilon_n R$. Let $C = \sup_{n \geq n_o, |x-z| \geq \epsilon_n R} u_{\epsilon_n}\left(\frac{x - x_{\epsilon_n}}{\epsilon_n}\right)$. Then the result follows from the maximum principle. □

In the final result of the chapter we show that a point at which a sequence of positive solutions of (P_ϵ) concentrates must be a critical point of V.

Proposition 6.4 *Suppose that $V \in C^1(\mathbb{R}^N)$. Let $u_n(x) = u_{\epsilon_n}(x)$ be a sequence of positive solutions of (P_{ϵ_n}) which concentrates at point z in the following sense: for every $\epsilon > 0$ there exist constants $R > 0$ and $n_o \geq 1$ such that $u_n(x) \leq \epsilon$ for all $n \geq n_o$ and $|z - z| \geq \epsilon_n R$. Then $\nabla V(z) = 0$.*

Proof. Let $u_n(x_n) = \max_{x \in \mathbb{R}^N} u_n(x)$. We claim that $\{u_n\}$ is bounded in $L^\infty(\mathbb{R}^N)$. Arguing indirectly, assume that $\|u_n\|_{L^\infty} \to \infty$. Let $\alpha_n = \max_{x \in \mathbb{R}^N} u_n(x)$ and set $v_n(x) = \frac{1}{\alpha_n} u_n(x_n + \epsilon_n \beta_n x)$, where $\beta_n = \alpha_n^{-\frac{1}{p-2}}$. Then

$$-\Delta v_n + \beta_n^2 V(x_n + \epsilon_n \beta_n x) v_n = v_n^{p-1} \text{ in } \mathbb{R}^N,$$

$v_n(0) = 1$ and $0 \leq v_n \leq 1$ on $\mathbb{R}^N$. It follows from the Schauder estimates (see [59]) that $v_n \to v_o$ in $C^2_{\text{loc}}(\mathbb{R}^N)$. Hence

$$-\Delta v_o = v_o^{p-1} \text{ in } \mathbb{R}^N,\ v_o > 0 \text{ on } \mathbb{R}^N,$$

which is impossible [58]. Therefore, the sequence $\{u_n\}$ must be bounded in $L^\infty(\mathbb{R}^N)$. We now set $w_n(x) = u_n(z + \epsilon_n x)$. The function w_n satisfies

$$-\Delta w_n + V(z + \epsilon_n x) w_n = w_n^{p-1} \text{ in } \mathbb{R}^N, \tag{6.22}$$

It follows from Proposition 6.3 that given $R > 0$ we can find $\delta > 0$ and $n_o \geq 1$ such that

$$w_n(x) \leq C \exp\left(-\delta \sum_{i=1}^{N} |x_i|\right)$$

for $|x| \geq R$ and $n \geq n_o$. Since maximum values of w_n are greater than $(V_o)^{\frac{1}{p-2}}$, using the Schauder estimates we can show that $w_n \to w$ in

$C^2_{\text{loc}}(\mathbb{R}^N)$, where $w > 0$ on $\mathbb{R}^N$ and has exponential decay at infinity. Multiplying equation (6.22) by ∇w_n and integrating over $B(0,R)$ we get

$$\begin{aligned} 0 &= \int_{B(0,R)} (\Delta w_n \nabla w_n - \frac{1}{2}\nabla(V(z+\epsilon_n x)w_n^2)\, dx \\ &+ \frac{1}{2}\int_{B(0,R)} (\frac{\epsilon_n}{2}\nabla V(z+\epsilon_n x)w_n^2 + \frac{1}{p}\nabla w_n^p)\, dx \end{aligned}$$

and

$$\begin{aligned} &\frac{\epsilon_n}{2}\int_{B(0,R)} \nabla V(z+\epsilon_n x)w_n^2\, dx \qquad (6.23) \\ &= \int_{\partial B(0,R)} (\frac{1}{2}V(z+\epsilon_n x)w_n^2\nu - \frac{1}{p}w_n^p\nu)\, dS_x \\ &- \int_{B(0,R)} \Delta w_n \nabla w_n\, dx, \end{aligned}$$

where ν is an outward normal vector to $\partial B(0,R)$. We now observe that

$$\begin{aligned} &\int_{B(0,R)} \Delta w_n \frac{\partial w_n}{\partial x_i}\, dx \\ &= \int_{B(0,R)} (\text{div } (\nabla w_n \frac{\partial w_n}{\partial x_i}) - \nabla w_n \frac{\partial}{\partial x_i}\nabla w_n)\, dx \\ &= \int_{\partial B(0,R)} (\frac{\partial w_n}{\partial x_i}\frac{\partial w}{\partial \nu} - \frac{1}{2}\nu_i|\nabla w_n|^2)\, dS_x. \end{aligned}$$

Hence (6.24) takes the form

$$\begin{aligned} &\frac{\epsilon_n}{2}\int_{B(0,R)} \nabla V(z+\epsilon_n x)w_n^2\, dx \qquad (6.24) \\ &= \int_{B(0,R)} (\frac{1}{2}V(z+\epsilon_n x)w_n^2\nu - \frac{1}{p}w_n^p\nu - \nabla w_n\frac{\partial w_n}{\partial \nu} + \frac{1}{2}\nu|\nabla w_n|^2)\, dS_x \\ &= I_R. \end{aligned}$$

Since $w_n \in H^1(\mathbb{R}^N)$, we can find $R_m \to \infty$ such that $I_{R_m} \to 0$ as $m \to \infty$. Letting $R = R_m$ in (6.24) we get

$$\int_{\mathbb{R}^N} \nabla V(z+\epsilon_n x)w_n^2\, dx = 0.$$

Finally, letting $\epsilon_n \to$ we deduce from the Dominated Convergence Theorem that $\int_{\mathbb{R}^N} \nabla V(z)w_o^2\, dx = 0$, that is, $\nabla V(z) = 0$. □

6.8 Bibliographical notes

The results in Sections 6.1 - 6.2, which relate the multiplicity of solutions to the category of the boundary are due to Benci–Cerami and Cerami–Passaseo [15], [16], [17]. The effect of the graph topology, discussed in Sections 6.5 - 6.5, is taken from the paper of Cingolani–Lazzo [37] (see also Musina [81]). Lemma 6.5 is due to Ni–Takagi [84]. Further results on concentration phenomena for the nonlinear Schrödinger equation can be found in [57], [86], [87], [88], [93], [99], [117], [118]. In particular, the existence of standing wave solutions originates in the paper by Floer–Weinstein [57] (case $p = 3, N = 1$). This result was generalized by Oh in a series of papers [86], [87], [88], [89], who proved the existence of multi-bump solutions and his arguments were based on the Lyapunov-Schmidt reduction. The existence of a single-bump solutions was established by del Pino-Floer under the assumption that the coefficient of u is bounded from below and has strict local minimum. These results were further extended by Gui [65], who proved the existence of multi-bump solutions in the case where the coefficient of u has several strict local minima. His aproach relies on some ideas from the papers of Cotti-Zelati and Rabinowitz [43], [44].

Chapter 7

Multi-peak solutions

In this chapter we study the effect of the shape of the graph of a coefficient of u on the the number and shape of the graphs of positive solutions. We assume that $b \in C^3(\mathbb{R}_N)$ and it has $l \geq 1$ critical nondegenerate points. Under this assumpion, applying a local reduction method, we construct positive and nodal solutions with specified number of peaks.

7.1 General approach to multi-peak solutions

We consider equation

$$\Delta u + \lambda^2 u = b(x)|u|^{p-2}u \quad \text{in } \mathbb{R}_N, \tag{7.1}$$

where $2 < p < 2^*$, $N \geqq 3$, and $\lambda > 0$ is a parameter. We shall assume throughout that $b(x)$ is a bounded and nonnegative on $\mathbb{R}_N$ and of class $C^3(\mathbb{R}_N)$ function satisfying the following condition:

(B) There exists points $a^1, \ldots, a^l$ such that $b(a^j) > 0$, $Db(a^j) = 0$, and det $(D^2 b(a^j)) \neq 0$, where $D^2 b(a^j)$ is a Hessian matrix, that is,

$$D^2 b(a^j) = \left\{ \frac{\partial^2 b(a^j)}{\partial x_r \partial x_s} \right\}_{r,s=1,\ldots,N}.$$

By a change of variables $v(x) = \lambda^{-\frac{2}{p-2}} u(\frac{x}{\lambda})$ equation (7.1) takes the form

$$-\Delta v + v = b(\frac{x}{\lambda})|v|^{p-2}v \quad \text{in } \mathbb{R}_N. \tag{7.2}$$

Let U be a ground state solution of problem (see Chapter 1)

$$\begin{aligned} -\Delta U \quad &+ \quad U = U^{p-1} \quad \text{in } \mathbb{R}_N, \\ U \in H^1(\mathbb{R}_N) \qquad &\text{and } U(0) = \max_{x \in \mathbb{R}_N} U(x). \end{aligned}$$

U is a positive decreasing radially symmetric function such that

$$\lim_{|x|\to\infty} U(|x|)e^{|x|}|x|^{\frac{N-1}{2}} = c > 0.$$

For $a \in \mathbb{R}_N$ we put $U_a(y) = U(y-a)$. Motivated by Proposition 6.2 we seek a solution v_λ of (7.2) which concentrates at points $a^1, \ldots, a^l$ as $\lambda \to \infty$ and that is of the form

$$v_\lambda = \sum_{j=1}^{l} \alpha_j(\lambda) U_{\lambda a^j + x^j_\lambda} + w_\lambda,$$

where $\alpha(\lambda) > 0$, $x^j_\lambda \in \mathbb{R}_N$, $w_\lambda \in H^1(\mathbb{R}_N)$ satisfy:

$$x^j_\lambda \to 0 \quad \text{and} \quad \|w_\lambda\|_{1,2} \to 0 \text{ as } \lambda \to \infty.$$

Let $B_r = B(0,r)$. We choose $\rho > 0$ sufficiently small such that $B(a^j, 4\rho)$ are disjoint.

We set for $\rho > 0$

$$\Sigma_\rho = \{(\alpha_1, \ldots, \alpha_l, x^1, \ldots; x^l);\ \alpha_j \in \big(b(a^j)-\rho, b(a^j)+\rho\big),\ x^j \in B_\rho,\ j = 1, \ldots, l\}.$$

To simplify notation we set $\alpha = (\alpha_1, \ldots, \alpha_l)$, $z = (z^1, \ldots, z^l)$, $z^j \in \mathbb{R}_N$, $a = (a^1, \ldots, a^l)$ and

$$\phi(\alpha, z) = \sum_{j=1}^{l} \alpha_j U_{\lambda z^j}.$$

For $\eta > 0$ and $\lambda > 0$ we set

$$\begin{aligned} W(\eta, \lambda) \quad = \quad &\{u \in H^1(\mathbb{R}_N);\ \|u - \phi(\alpha, z)\|_{1,2} < \eta \text{ for some } (\alpha, z) \text{ such that} \\ &(\alpha, z - a) \in \Sigma_\rho\}. \end{aligned}$$

Lemma 7.1 *There exist constants $\eta_\circ > 0$ (sufficiently small) and $\lambda_\circ > 0$ (sufficiently large) such that if $u \in W(\eta, \lambda)$ for some $\eta \in (0, \eta_\circ)$ and $\lambda \geq \lambda_\circ$, then the minimization problem*

$$\inf\{\|u - \phi(\alpha, z)\|_{1,2};\ (\alpha, z - a) \in \Sigma_{4\rho}\} \tag{7.3}$$

has a solution and a minimizer belongs to $\Sigma_{2\rho}$.

Proof. Arguing indirectly we assume that there exist sequences $\lambda_n \to \infty$, $\eta_n \to 0$ and $u_n \in W(\eta_n, \lambda_n)$ such that a minimizer (α^n, λ^n) of (7.3) belongs to $\Sigma_{4\rho} - \Sigma_{2\rho}$ for all n. For each n we choose $(\tilde{\alpha}^n, \tilde{z}^n)$ such that $(\tilde{\alpha}^n, \tilde{z}^n - a) \in \Sigma_\rho$ and

$$||\phi(\alpha^n, z^n) - \phi(\tilde{\alpha}^n, \tilde{z}^n)||_{1,2} \to 0 \tag{7.4}$$

as $n \to \infty$. We now derive from (7.4) that

$$|\alpha_j^n - \tilde{\alpha}_j^n| \to 0 \tag{7.5}$$

and

$$\lambda_n |z^{jn} - \tilde{z}^{jn}| \to 0 \tag{7.6}$$

as $n \to \infty$ for $j = 1, \ldots, l$. This obviously contradicts the choice of (α^n, z^n). To establish (7.5) we observe that

$$\sum_{j=1}^{l} (\alpha_j^n)^2 + (\tilde{\alpha}_j^n)^2 - 2\alpha_j^n \tilde{\alpha}_j^n \langle U_{\lambda_n z^{jn}}, U_{\lambda_n \tilde{z}^{jn}} \rangle A^{-2} = o(1), \tag{7.7}$$

where $o(1) \to 0$ as $n \to \infty$ and

$$A^2 = ||U_{\lambda_n z^{jn}}||_{1,2}^2 = ||U_{\lambda_n \tilde{z}^{jn}}||_{1,2}^2.$$

Since $\langle U_{\alpha_n z^{jn}, \lambda_n \tilde{z}^{jn}} \rangle \leq A^2$, (7.7) implies that

$$\sum_{j=1}^{l} (\tilde{\alpha}_j^n - \alpha_j^n)^2 \leq \sum_{j=1}^{l} \left[(\alpha_j^n)^2 + (\tilde{\alpha}_j^n)^2 - \frac{2}{A^2} \alpha_j^n \tilde{\alpha}_j^n \langle U_{\lambda_n z^{jn}}, U_{\lambda_n \tilde{z}^{jn}} \rangle \right] = o(1)$$

and (7.5) follows. By (7.5) and (7.7) we have

$$\langle U_{\lambda_n z^{jn}}, U_{\lambda_n \tilde{z}^{jn}} \rangle = A^2 + o(1)$$

and

$$\begin{aligned} \langle U_{\lambda_n z^{jn}}, U_{\lambda_n \tilde{z}^{jn}} \rangle &= \langle U_{\lambda_n (z^{jn} - \tilde{z}^{jn})} - U, U \rangle + A^2 \\ &= \int_{\mathbb{R}_N} U^{p-1} (U_{\lambda_n (z^{jn} - \tilde{z}^{jn})}\, dx + A^2. \end{aligned}$$

Combining these two relations we get

$$\int_{\mathbb{R}_N} U^{p-1}\big(U_{\lambda_n(z^{jn}-\tilde{z}^{jn})} - U\big)\,dx = o(1),$$

and (7.6) easily follows. □

Lemma 7.2 *A minimizer of problem (7.3) is unique provided λ is sufficiently large and η sufficiently small.*

Proof. In the contrary case there exists $\eta_n \to 0$, $\lambda_n \to \infty$ and $u_n \in W(\eta_n, \lambda_n)$ such that problem (7.3) has two (α^n, z^n) and $(\tilde{\alpha}^n, \tilde{z}^n)$. Repeating the argument of Lemma 7.1 one can show that

$$\alpha_j^n - \tilde{\alpha}_j^n \to 0 \text{ and } \lambda_n|z^{jn} - \tilde{z}^{jn}| \to 0$$

$j = 1, \ldots, l$. To complete the proof it is sufficient to show that $\alpha^n = \tilde{\alpha}^n$ and $z^n = \tilde{z}^n$ for large n. Towards this end we set

$$v_n = u - \sum_{j=1}^{l} \alpha_j^n U_{\lambda_n z^{jn}}, \quad \tilde{v}_n = u - \sum_{j=1}^{l} \tilde{\alpha}_j^n U_{\lambda_n \tilde{z}^{jn}}.$$

Then for all n we have

$$\langle v_n, U_{\lambda_n z^{jn}}\rangle = \langle v_n, \frac{\partial U_{\lambda_n z^{jn}}}{\partial z_i^{jn}}\rangle = 0$$

and

$$\langle \tilde{v}_n, U_{\lambda_n \tilde{z}^{jn}}\rangle = \langle \tilde{v}_n, \frac{\partial U_{\lambda_n \tilde{z}^{jn}}}{\partial \tilde{z}_i^{jn}}\rangle = 0$$

$i = 1, \ldots, N$, $j = 1, \ldots, l$. From these two relations we deduce

$$\begin{aligned} \sum_{j=1}^{l} \langle \alpha_j^n U_{\lambda_n z^{jn}} &- \tilde{\alpha}_j^n U_{\alpha_n \tilde{z}^{jn}}, U_{\lambda_n z^{mn}}\rangle \\ &= \langle \tilde{v}_n, U_{\lambda_n z^{mn}} - U_{\lambda_n \tilde{z}^{mn}}\rangle \end{aligned} \tag{7.8}$$

$m = 1, \ldots, l$ and

$$\begin{aligned} \sum_{j=1}^{l} \langle \alpha_j^n U_{\lambda_n z^{jn}} &- \tilde{\alpha}_j^n U_{\lambda_n \tilde{z}^{jn}}, \frac{\partial U_{\lambda_n z^{mn}}}{\partial z_k^{mn}}\rangle \\ &= \langle \tilde{v}_n, \frac{\partial U_{\lambda_n z^{mn}}}{\partial z_k^{mn}} - \frac{\partial U_{\lambda_n \tilde{z}^{mn}}}{\partial \tilde{z}_k^{mn}}\rangle \end{aligned} \tag{7.9}$$

$k = 1, \ldots, N$, $m = 1, \ldots, l$. We now use (7.8) and (7.9) to examine the behaviour of $|\alpha^n - \tilde{\alpha}^n| = \sum_{j=1}^l |\alpha_j^n - \tilde{\alpha}_j^n|$ and $|z^n - \tilde{z}^n| = \sum_{j=1}^l |z^{nj} - \tilde{z}^{nj}|$. Letting $\frac{\partial U_{\lambda z}}{\partial z_i} = \frac{\partial U_y}{\partial y_i}|_{\lambda z}$ for $i = 1, \ldots, N$, we get

$$\begin{aligned}
&\sum_{j=1}^{l} \langle \alpha_j^n U_{\lambda_n z^{jn}} - \tilde{\alpha}_j^n U_{\lambda_n \tilde{z}^{jn}}, U_{\lambda_n z^{mn}} \rangle \qquad (7.10)\\
&= \sum_{j=1}^{l} \{ \tilde{\alpha}_j^n \langle U_{\lambda_n z^{jn}} - U_{\lambda_n \tilde{z}^{jn}}, U_{\lambda_n z^{mn}} \rangle \\
&+ (\alpha_j^n - \tilde{\alpha}_j^n) \langle U_{\lambda_n z^{jn}}, U_{\lambda_n z^{mn}} \rangle \} \\
&= \sum_{j=1}^{l} \Big\{ -\tilde{\alpha}_j^n \sum_{i=1}^{N} \langle \frac{\partial U_{\lambda_n z^{jn}}}{\partial z_i^{jn}}, U_{\lambda_n z^{mn}} \rangle \lambda_n (z_i^{jn} - \tilde{z}_i^{jn}) \\
&+ O(\lambda_n^2 |\tilde{z}^{jn} - z^{jn}|^2) + (\alpha_j^n - \tilde{\alpha}_j^n) \langle U_{\lambda_n z^{jn}}, U_{\lambda_n z^{mn}} \rangle \Big\}.
\end{aligned}$$

It is easy to check that

$$\langle U_{\lambda_n z^{mn}}, U_{\lambda_n z^{jn}} \rangle = o(1) \quad \text{for } j \neq m,$$

$$\langle \frac{\partial U_{\lambda_n z^{mn}}}{\partial z_i^{jn}}, U_{\lambda_n z^{jn}} \rangle = o(1) \quad \text{for } j \neq m$$

as $n \to \infty$ and

$$\langle \frac{\partial U_{\lambda_n z^{mn}}}{\partial z_i^{mn}}, U_{\lambda_n z^{mn}} \rangle = 0.$$

It then follows from (7.10) that

$$\begin{aligned}
\sum_{j=1}^{l} \langle \alpha_j^n U_{\lambda_n z^{jn}} &- \tilde{\alpha}_j^n U_{\lambda_n \tilde{z}^{jn}}, U_{\lambda_n, z^{mn}} \rangle = (\alpha_n^n - \tilde{\alpha}_m^n) A^2 \qquad (7.11)\\
&+ o(1) \Big(\sum_{j=1}^{l} \lambda_n |z^{jn} - \tilde{z}^{jn}| \Big) + o(1) |\alpha^n - \tilde{\alpha}^n|.
\end{aligned}$$

On the other hand

$$\begin{aligned}
\langle \tilde{v}_n, U_{\lambda_n z^{mn}} - U_{\lambda_n \tilde{z}^{mn}} \rangle &= \sum_{i=1}^{N} \langle \tilde{v}_n, \frac{\partial U_{\lambda_n \tilde{z}^{mn}}}{\partial \tilde{z}_i^{mn}} \rangle \lambda_n (z_i^{mn} - \tilde{z}_i^{mn}) \qquad (7.12)\\
&+ O(\|\tilde{v}_n\|_{1,2} \lambda_n^2 |z^{mn} - \tilde{z}^{mn}|^2) = O(\|\tilde{v}_n\|_{1,2} \lambda_n^2 |z^{mn} - \tilde{z}^{mn}|^2).
\end{aligned}$$

Consequently, by (7.8), (7.11) and (7.12) we obtain

$$\alpha_m^n - \tilde{\alpha}_m^n = o(1)|\alpha^n - \tilde{\alpha}^n| + o(1)\lambda_n|z^n - \tilde{z}^n|. \tag{7.13}$$

We now derive a similar identity involving $|z^{mn} - \tilde{z}^{mn}|$. We write

$$\sum_{j=1}^{l}\langle \alpha_j^n U_{\lambda_n z^{jn}} - \tilde{\alpha}_j^n U_{\lambda_n \tilde{z}^{jn}}, \frac{\partial U_{\lambda_n z^{mn}}}{\partial z_k^{mn}}\rangle \tag{7.14}$$

$$= \sum_{j=1}^{l}\Big\{\tilde{\alpha}_j^n \langle U_{\lambda_n z_j^n} - U_{\lambda_n \tilde{z}_j^n}, \frac{\partial U_{\lambda_n z^{mn}}}{\partial z_k^{mn}}\rangle + (\alpha_j^n - \tilde{\alpha}_j^n)\langle U_{\lambda_n z^{jn}}, \frac{\partial U_{\lambda_n z^{mn}}}{\partial z_k^{mn}}\rangle\Big\}$$

$$= \sum_{j=1}^{l}\Big\{-\tilde{\alpha}_j^n \sum_{i=1}^{N}\langle \frac{\partial U_{\lambda_n z^{jn}}}{\partial z_i^{jn}}, \frac{\partial U_{\lambda_n z^{mn}}}{\partial z_k^{mn}}\rangle \lambda_n\big(\tilde{z}_i^{jn} - z_i^{jn}\big)$$

$$+ O\big(\lambda_n^2|z_j^n - \tilde{z}_j^n|^2\big) + o(1)|\alpha_j^n - \alpha_j^n|\Big\}$$

$$= -\lambda_n \tilde{\alpha}_m^n \langle \frac{\partial U_{\lambda_n z^{mn}}}{\partial z_k^{mn}}, \frac{\partial U_{\lambda_n z^{mn}}}{\partial z_k^{mn}}\rangle\big(\tilde{z}_k^{mn} - z_k^{mn}\big)$$

$$+ o(1)\lambda_n|z^n - \tilde{z}^n| + o(1)|\alpha^n - \tilde{z}^n|.$$

Here we have used the following relations

$$\langle \frac{\partial U_{\lambda_n z^{jn}}}{\partial z_i^{mn}}, \frac{\partial U_{\lambda_n z^{mn}}}{\partial z_k^{mn}}\rangle = o(1) \quad \text{for } j \neq m$$

and

$$\langle \frac{\partial U_{\lambda_n z^{mn}}}{\partial z_i^{mn}}, \frac{\partial U_{\lambda_n z^{mn}}}{\partial z_k^{mn}}\rangle = 0 \quad \text{for } i \neq k.$$

Since $\|\frac{\partial U_{\lambda_n z^{mn}}}{\partial z_k^{mn}}\|_{1,2}^2 = B > 0$, where B is independent of m, n and k, we deduce from (7.14) that

$$\sum_{j=1}^{l}\langle \alpha_j^n U_{\lambda_n z^{jn}} - \tilde{\alpha}_j^n U_{\lambda_n \tilde{z}^{jn}}, \frac{\partial U_{\lambda_n z^{mn}}}{\partial z_k^{mn}}\rangle \tag{7.15}$$

$$= -\tilde{\alpha}_k^n B \lambda_m\big(\tilde{z}_k^{mn} - z_k^{mn}\big) + o(1)\lambda_n|z^n - \tilde{z}^n| + o(1)|\alpha^n - \tilde{\alpha}^n|.$$

Since we also have that

$$\langle \tilde{v}_n, \frac{\partial U_{\lambda_n z^{jn}}}{\partial z_k^{jn}} - \frac{\partial U_{\lambda_n \tilde{z}^{jn}}}{\partial \tilde{z}_k^{jn}}\rangle = o(1)\lambda_n|z^n - \tilde{z}^n|,$$

we derive from (7.9) and (7.15) the following identity

$$\lambda_n |z^{mn} - \tilde{z}^{mn}| = o(1)\lambda_n |z^n - \tilde{z}^n| + o(1)|\alpha^n - \tilde{\alpha}^n|$$

$m = 1, \ldots, l$. This combined with (7.13) yields that $\alpha^n = \tilde{\alpha}^n$ and $z^n = \tilde{z}^n$ for sufficiently large n. □

Let (α, z) be a minimizer of problem (7.3). Setting $x^j = \lambda z^j - \lambda a^j$, $j = 1, \ldots, l$, then $(\alpha, x^1, \ldots, x^l) \in \Sigma_{2\rho}$. Therefore we can deduce from Lemmas 7.1 and 7.2 the following result

Corollary 7.1 *There exist constants $\eta_0 > 0$ (sufficiently small) and λ_0 (sufficiently large) such that if $\lambda \geq \lambda_0$ and $u \in H^1(\mathbb{R}_N)$ satisfies*

$$\inf\{\|u - \sum_{j=1}^{l} \alpha_j U_{\lambda a^j + x^j}\|_{1,2};\ (\alpha_1, \ldots, \alpha_l, x^1, \ldots, x^l) \in \Sigma_\rho\} < \eta_0, \qquad (7.16)$$

then the minimization problem

$$\inf\{\|u - \sum_{j=1}^{l} \alpha_j U_{\lambda a^j + x^j}\|_{1,2};\ (\alpha_1, \ldots, \alpha_l, x^1, \ldots, x^l) \in \Sigma_{4\rho}\} \qquad (7.17)$$

has a unique solution. Moreover the minimizer of problem (7.17) belongs to $\Sigma_{2\rho}$.

Let $u \in H^1(\mathbb{R}^N)$ satisfy (7.16) and let $(\alpha, x^1, \ldots, x^l)$ be the minimizer of problem (7.17). Then u can be written in the form

$$u = \sum_{j=1}^{l} \alpha U_{\lambda a^j + x^j} + w_\lambda$$

where w_λ satisfies

$$\langle w_\lambda, U_{\lambda a^j + x^j} \rangle = 0 \qquad (7.18)$$

and

$$\langle w_\lambda, \frac{\partial U_{\lambda a^j + x^j}}{\partial x_i^j} \rangle = 0. \qquad (7.19)$$

We now define

$$E_{\lambda, x^1, \ldots, x^l} = \{w \in H^1(\mathbb{R}_N);\ w \text{ satisfies } (7.18), (7.19)\}$$

and

$$M^l_{\lambda,\eta} = \{(\alpha, x^1, \ldots, x^l, w); (\alpha, x^1, \ldots, x^l) \in \Sigma_{4\rho},\ w \in E_{\lambda,x^1,\ldots,x^l},\ \|w\|_{1,2} < \eta\},$$

where $\eta \leq \eta_o$. Let $K_\lambda : H^1(\mathbb{R}_N) - \{0\} \to \mathbb{R}$ be a functional defined by

$$K_\lambda(u) = \frac{\|u\|^2_{1,2}}{\left(\int_{\mathbb{R}_N} b(\frac{y}{\lambda})|u|^p\, dy\right)^{\frac{2}{p}}}$$

and set

$$J_\lambda(\alpha_1, \ldots, \alpha^l, x^1, \ldots, x^k, w) = K_\lambda\Big(\sum_{j=1}^{l} \alpha_j U_{\lambda a^j + x^j} + w\Big).$$

Using Corollary 7.1 we can formulate the following result

Proposition 7.1 *There exist constants $\eta_o > 0$ and $\lambda_o > 0$ such that if $\lambda \geq \lambda_o$ and $\eta \leq \eta_o$, then $(\alpha_1, \ldots, \alpha_l, x^1, \ldots, x^l, w)$ is a critical point of the functional $J_\lambda : M^l_{\lambda,\eta} \to \mathbb{R}$ if and only if $u = \sum_{j=1}^{l} \alpha_j U_{\lambda a^j + x^j} + w$ is a critical point of $K_\lambda : H^1(\mathbb{R}_N) - \{0\} \to \mathbb{R}$.*

To construct a solution of (7.2) we use the following idea. A point $(\alpha_1, \ldots, \alpha_1, x^1, \ldots, x^l, w) \in M^l_{\lambda,\eta})$ is a critical point of J_λ if and only if there exist numbers β_j, γ_{mj}, $m = 1, \ldots, N$, $j = 1, \ldots, l$ such that

$$\frac{\partial J_\lambda}{\partial \alpha_j} = 0,\ j = 1, \ldots, l, \tag{7.20}$$

$$\frac{\partial J_\lambda}{\partial x^j_i} = \sum_{m=1}^{N} \gamma_{mj} \langle w, \frac{\partial U_{\lambda a^j + x^j}}{\partial x^j_m \partial x^j_i} \rangle, \quad i = 1, \ldots, N,\ j = 1, \ldots, l \tag{7.21}$$

$$\langle \frac{\partial J_\lambda}{\partial w}, \phi \rangle = \sum_{j=1}^{l} \beta_j \langle \phi, U_{\lambda a^j + x^j} \rangle + \sum_{i=1}^{N} \sum_{j=1}^{l} \gamma_{ij} \langle \phi, D_i U_{\lambda a^j + x^j} \rangle. \tag{7.22}$$

First, following Rey [96], we show that for each $(x^1, \ldots, x^l, \lambda)$, with λ sufficiently large, there exist $w_\lambda \in E_{\lambda,x^1,\ldots,x^l}$ and constants β_j, γ_{ij}, $i = 1, \ldots, N$, $j = 1, \ldots, l$, satisfying (7.21) and (7.22). In the next step, applying Brouwer's fixed point theorem we show that there exist x^j, $j = 1, \ldots, l$, satisfying (7.20).

7.2 Preliminary estimates

Let

$$F_x = \{v \in H^1(\mathbb{R}_N);\ \langle v, U_x\rangle = \langle v, \frac{\partial U_x}{\partial x_i}\rangle = 0,\ i = 1, \ldots, N\}$$

and

$$\begin{aligned} F_{x,\lambda z} = \{v \in H^1(\mathbb{R}_N);\ &\langle v, U_x\rangle = \langle v, U_{\lambda z}\rangle = 0, \\ &\langle v, \frac{\partial U_x}{\partial x_i}\rangle = \langle v, \frac{\partial U_{\lambda z}}{\partial z_i}\rangle = 0,\ i = 1, \ldots, N\}. \end{aligned}$$

In this section we derive some integral inequalities for functions v belonging to F_x and $F_{x,\lambda z}$. An important step in the derivation of these estimates is to consider the eigenvalue problem

$$\begin{cases} -\Delta\phi \quad +\phi = \mu U_x^{p-2}\phi & \text{in } \mathbb{R}_N \\ \phi \qquad \in H^1(\mathbb{R}_N). \end{cases} \tag{7.23}$$

We also use the fact that the function

$$g(t) = \frac{\|U_x + tv\|_{1,2}^2}{\left(\int_{\mathbb{R}_N} |U_x + tv|^p\, dy\right)^{\frac{2}{p}}} \qquad t \in \mathbb{R}$$

for each fixed $v \in H^1(\mathbb{R}_N) - \{0\}$ attains its global minimum at $t = 0$. Indeed,

$$g(0) = S_p = \frac{\|U_x\|_{1,2}}{\left(\int_{\mathbb{R}_N} |U_x|^p\, dx\right)^{\frac{2}{p}}} = \inf_{u \in H^1(\mathbb{R}_N) - \{0\}} \frac{\|u\|_{1,2}}{\left(\int_{\mathbb{R}_N} |u|^p\, dx\right)^{\frac{2}{p}}}.$$

Proposition 7.2 *There exists a constant $\rho > 0$ such that each $v \in F_x$ satisfies the inequality*

$$\|v\|_{1,2}^2 - (p-1)\int_{\mathbb{R}_N} U_x^{p-2} v^2\, dy \geq \rho\|v\|_{1,2}^2$$

for $x \in B_1(0)$.

Proof. Since $g(0) = \min_{t\in\mathbb{R}} g(t)$, $g'(0) = 0$ and $g''(0) \geq 0$. Letting $W(t) = U_x + tv$, by straightforward calculations we get

$$\begin{aligned} g'(t) &= 2\int_{\mathbb{R}_N} (\nabla W(t)\nabla v + W(t)v)\, dy \left\{\int_{\mathbb{R}_N} |W(t)|^p\, dy\right\}^{-\frac{2}{p}} \\ &- 2\|W(t)\|_{1,2}^2 \left\{\int_{\mathbb{R}_N} |W(t)|^{p-2}W(t)v\, dy\right\}\left\{\int_{\mathbb{R}_N} |W(t)|^p\, dy\right\}^{-\frac{2}{p}-1} \end{aligned}$$

and

$$\begin{aligned} g''(0) &= 2\|v\|_{1,2}^2\left(\int_{\mathbb{R}_N} U_x^p\, dy\right)^{-\frac{2}{p}} - 8\langle U_x, v\rangle\left(U_x^{p-1}v\, dy\right)\left(\int_{\mathbb{R}_N} U_x^p\, du\right)^{-\frac{2}{p}-1} \\ &- 2(p-1)\|U_x\|_{1,2}^2\left(\int_{\mathbb{R}_n} U_x^{p-2}v^2\, dy\right)\left(\int_{\mathbb{R}_N} U_x^p\, dy\right)^{-\frac{2}{p}-1} \\ &+ 2p(1+\frac{2}{p})\|U_x\|_{1,2}^2\left(\int_{\mathbb{R}_N} U_x^{p-1}v\, dy\right)^2\left(\int_{\mathbb{R}_N} U_x^p\, dy\right)^{-\frac{2}{p}-2}. \end{aligned}$$

Since

$$\int_{\mathbb{R}_N} U_x^{p-1}v\, dy = \int_{\mathbb{R}_N} (-\Delta U_x + U_x)v\, dy = \langle U_x, v\rangle = 0,$$

we get

$$\begin{aligned} g''(0) &= 2\left\{\|v\|_{1,2}^2 - (p-1)\frac{\|U_x\|_{1,2}^2}{\int_{\mathbb{R}_N} U_x^p\, dy}\int_{\mathbb{R}_N} U_x^{p-2}v^2\, dy\right\}\left\{\int_{\mathbb{R}_N} U_x^p\, dy\right\}^{-\frac{2}{p}} \\ &= 2\left\{\|v\|_{1,2}^2 - (p-1)\int_{\mathbb{R}_N} U_x^{p-2}v^2\, dy\right\}\left\{\int_{\mathbb{R}_N} U_x^p\, dy\right\}^{-\frac{2}{p}} \end{aligned}$$

for all $v \in F_x$. Hence

$$\|v\|_{1,2}^2 - (p-1)\int_{\mathbb{R}_N} U_x^{p-2}v^2\, dy \geq 0 \tag{7.24}$$

for all $v \in F_x$. We now observe that for every $\phi \in H^1(\mathbb{R}_N)$ we have

$$\begin{aligned} &\int_{\mathbb{R}_N} U_x^{p-2}\phi^2\, dy \leq \left(\int_{\mathbb{R}_N} U_x^p\, dy\right)^{1-\frac{2}{p}}\left(\int_{\mathbb{R}_N} |\phi|^p\, dy\right)^{\frac{2}{p}} \\ &\quad \leq S_p^{-1}\left(\int_{\mathbb{R}_N} U_x^p\, dy\right)^{1-\frac{2}{p}}\|\phi\|_{1,2}^2 = \frac{\int_{\mathbb{R}_N} U_x^p\, dy}{\|U_x\|^2}\|\phi\|_{1,2}^2 = \|\phi\|_{1,2}^2. \end{aligned}$$

Thus the first eigenvalue μ_1 of (7.23) satisfies

$$\mu_1 = \inf\{\frac{||u||_{1,2}^2}{\int_{\mathbb{R}_N} U_x^{p-2} u^2\, dy};\ u \not\equiv 0,\ u \in H^1(\mathbb{R}_N)\} \geq 1.$$

Since U_x satifies (7.23) for $\mu = 1$, we therefore have $\mu_1 = 1$. According to (7.24) the second eigenvalue μ_2 sastisfies

$$\mu_2 = \inf\{\frac{||u||_{1,2}^2}{\int_{\mathbb{R}_N} U_x^{p-2} u^2\, dy};\ \langle U_x, u\rangle = 0,\ u \not\equiv 0\} \geq p-1.$$

Since

$$-\Delta \frac{\partial U_x}{\partial x_i} + \frac{\partial U_x}{\partial x_i} = (p-1) U_x^{p-2} \frac{\partial U_x}{\partial x_i},\quad i = 1, \ldots, N,$$

we have $\mu_2 = p-1$, and $\{\frac{\partial U}{\partial x_i};\ i = 1, \ldots, N\}$ are the corresponding eigenfunctions. By virtue of the result of Ni and Takagi [83] (see also Lemma 4.1 in [120]), the eigenspace of μ_2 is equal to span $\{\frac{\partial U_x}{\partial x_i};\ i = 1, \ldots, N\}$. Consequently the third eigenvalue of (7.23) is given by

$$\mu_3 = \inf\{\frac{||u||_{1,2}^2}{\int_{\mathbb{R}_N} U_x^{p-2} u^2\, dy};\ u \in F_x\}$$

and satisfies

$$\mu_3 > m_2 = p-1.$$

Thus

$$||v||_{1,2}^2 - (p-1)\int_{\mathbb{R}_N} U_x^{p-2} v^2\, dy \geq \left(1 - \frac{p-1}{\mu_3}\right)||v||_{1,2}^2$$

for all $v \in F_x$ and this completes the proof. □

Without loss of generality we may assume that $a^1 = 0$.

Proposition 7.3 *There exist constants $\rho > 0$ and $\delta > 0$ such that*

$$||v||_{1,2}^2 - (p-1)\int_{\mathbb{R}_N} (U_x + U_{\lambda z})^{p-2} v^2\, dy \geq \rho ||v||_{1,2}^2 \tag{7.25}$$

for all $v \in F_{x,\lambda z}$, all $x \in B(0,\delta)$ and $z \in B(a^2, \delta)$.

Proof. To prove (7.25) it is sufficient to show that

$$\Lambda_\lambda = \inf\{\frac{\|v\|_{1,2}^2}{\int_{\mathbb{R}_N}(U_x + U_{\lambda z})^{p-2}v^2\,dy};\ v \neq 0, v \in F_{x,\lambda z}\} \geq \mu_3 + o(1), \quad (7.26)$$

where $o(1) \to 0$ as $\lambda \to \infty$ uniformly for $x \in B(0,\delta)$ and $z \in B(a^2,\delta)$ for some $\delta > 0$. Arguing indirectly, we may assume that there exists $\sigma > 0$ such that $\Lambda_\lambda \leq \mu_3 - \sigma$. For each $\lambda > 0$, let $\phi_\lambda \in F_{x,\lambda z}$ be a minimizer of (7.26) such that

$$\int_{\mathbb{R}_N}(u + U_{\lambda z})^{p-2}\phi_\lambda^2\,dy = 1 \quad (7.27)$$

$$\|\phi_\lambda\|_{1,2}^2 = \Lambda_\lambda. \quad (7.28)$$

The minimizer ϕ_λ satisfies

$$\int_{\mathbb{R}_N}(\nabla\phi_\lambda\nabla\eta + \phi_\lambda\eta)\,dx - \Lambda_\lambda\int_{\mathbb{R}_N}(u_x + U_{\lambda z})^{p-2}\phi_\lambda\eta\,dx = 0 \quad (7.29)$$

for all $\eta \in F_{x,\lambda z}$. Since $\{\phi_\lambda\}$ is bounded in $H^1(\mathbb{R}_N)$, we may assume that $\phi_\lambda \rightharpoonup \phi$ in $H^1(\mathbb{R}_N)$ and $\phi_\lambda \to \phi$ in $L^2_{\rm loc}(\mathbb{R}_N)$ and moreover $\Lambda_\lambda \to \Lambda$ and $\Lambda \leq \mu_3 - \sigma$. We claim that ϕ satisfies

$$\int_{\mathbb{R}_N}(\nabla\phi\nabla\xi + \nabla\xi)\,dy - \Lambda\int_{\mathbb{R}_N}U_x^{p-2}\phi\xi\,dy = 0 \quad (7.30)$$

for all $\xi \in F_x$. Indeed, let $\xi \in F_x$. We show that there exist constants α, β_i, γ_i and δ_i, $i = 1, \ldots, N$, such that a function η given by

$$\eta = \xi - \alpha U_x - \beta U_{\lambda z} - \sum_{i=1}^{N}(\gamma_i\frac{\partial U_x}{\partial x_i} + \delta_i\frac{\partial U_{\lambda z}}{\partial z_i})$$

belongs to $F_{x,\lambda z}$. First, we observe that

$$\langle\xi, U_x\rangle = \langle\xi, \frac{\partial U_x}{\partial x_i}\rangle = 0,$$

$$\langle U_x, U_{\lambda z}\rangle = o(1),$$

$$\langle U_x, \frac{\partial U_{\lambda z}}{\partial z_i}\rangle = o(1),\ \langle U_{\lambda z}, \frac{\partial U_{\lambda z}}{\partial z_i}\rangle = 0 \text{ for } i = 1, \ldots, N,$$

$$\langle \frac{\partial U_x}{\partial x_i}, \frac{\partial U_x}{\partial x_j} \rangle = 0, \ \langle \frac{\partial U_{\lambda z}}{\partial z_i}, \frac{\partial U_{\lambda z}}{\partial z_j} \rangle = 0 \text{ for } i \neq j.$$

Therefore $\eta \in F_{x,\lambda z}$ is equivalent to solving the following system of equations

$$\alpha ||U_x||_{1,2}^2 + \beta \langle U_x, U_{\lambda z} \rangle + \sum_{i=1}^{N} \delta_i \langle U_x, \frac{\partial U_{\lambda z}}{\partial z_i} \rangle = \langle \xi, U_x \rangle = 0,$$

$$\gamma_j \left\| \frac{\partial U_x}{\partial x_i} \right\|_{1,2}^2 + \beta \langle \frac{\partial U_x}{\partial x_j}, U_{\lambda z} \rangle + \sum_{i=1}^{N} \delta_i \langle \frac{\partial U_x}{\partial x_j}, \frac{\partial U_{\lambda z}}{\partial z_i} \rangle = \langle \xi, \frac{\partial U_x}{\partial x_j} \rangle = 0,$$

$$\alpha \langle U_x, U_{\lambda z} \rangle + \beta ||U_{\lambda z}||_{1,2}^2 + \sum_{i=1}^{N} \gamma_i \langle \frac{\partial U_x}{\partial x_i}, U_{\lambda z} \rangle = \langle \xi, U_{\lambda z} \rangle = o(1),$$

$$\alpha \langle U_x, \frac{\partial U_{\lambda z}}{\partial z_j} \rangle + \sum_{i=1}^{N} \gamma_i \langle \frac{\partial U}{\partial x_i}, \frac{\partial U_{\lambda z}}{\partial z_j} \rangle + \delta_j \left\| \frac{\partial U_{\lambda z}}{\partial z_j} \right\|_{1,2}^2 = \langle \xi, \frac{\partial U_{\lambda z}}{\partial z_j} \rangle = o(1),$$

where $o(1) \to 0$ as $\lambda \to \infty$. This system is equivalent to

$$\alpha ||U||_{1,2}^2 + o(1)\beta + \sum_{i=1}^{N} o(1)\delta_i = 0,$$

$$o(1)\alpha + o(1)\beta + \gamma_j \left\| \frac{\partial U_x}{\partial x_j} \right\|_{1,2}^2 = 0,$$

$$o(1)\alpha + \beta ||U_{\lambda z}||_{1,2}^2 + \sum_{i=1}^{N} o(1)\gamma_i = o(1),$$

$$o(1)\alpha + o(1)\beta + \delta_j \left\| \frac{\partial U_{\lambda z}}{\partial z_j} \right\|_{1,2}^2 = o(1),$$

$j = 1, \ldots, N$. This system has a unique solution $\alpha = o(1)$, $\beta = o(1)$, $\gamma_i = o(1)$, $\delta_i = o(1)$, $i = 1, \ldots, N$, where $o(1) \to 0$ as $\lambda \to \infty$. We now substitute

$$\eta = \xi - \alpha U_x - \beta U_{\lambda z} - \sum_{i=1}^{N} \left(\gamma_i \frac{\partial U_x}{\partial x_i} + \delta_i \frac{\partial U_{\lambda z}}{\partial z_i}\right)$$

into (7.29) and use the fact that

$$(U_x + U_{\lambda z})^{p-2} \to U_x^{p-2} \text{ as } \lambda \to \infty$$

uniformly for $x \in B(0, R)$ for each $R > 0$, to get (7.30). Substituting $\xi = \phi$ into (7.30) we obtain that

$$||\phi||_{1,2}^2 - \Lambda \int_{\mathbb{R}_N} U_x^{p-2} \phi^2 \, dy = 0. \tag{7.31}$$

On the other hand $||\phi||_{1,2}^2 \geq \mu \int_{\mathbb{R}_N} U_x^{p-2} \phi^2 \, dy$, since $\phi \in F_x$. This contradicts (7.31) and consequently $\phi \equiv 0$. This means that $\phi_\lambda \to 0$ in $L^2_{\text{loc}}(\mathbb{R}_N)$. We set

$$\tilde{\phi}_\lambda(x) = \phi_\lambda(x + \lambda z).$$

It is easy to check that $\tilde{\phi}_\lambda$ is a minimizer of the problem

$$\tilde{\Lambda}_\lambda = \inf \left\{ \frac{||v||_{1,2}^2}{\int_{\mathbb{R}_N} (U_{x-\lambda z} + U_o)^{p-2} v^2 \, dz}; \; v \in F_o \right\}.$$

Obviously, we have $\tilde{\Lambda}_\lambda = \Lambda_\lambda$. Finally, we can easily prove using the above argument that $\tilde{\phi}_\lambda \to 0$ in $L^2_{\text{loc}}(\mathbb{R}_N)$. This means that

$$\int_{\mathbb{R}_N} (U_x + U_{\lambda z})^{p-2} \phi_\lambda^2 \, dy \to 0$$

as $\lambda \to \infty$ which contradicts (7.27) and the result follows. □

7.3 One peak solution

The objective of this section is to construct a one peak solution. Let a^j be a critical point of b. For simplicity we assume that $a^j = 0$ and $b(0) = 1$. The idea is to construct a a critical point of K_λ of the form

$$u = U_\lambda + w_\lambda, \quad w_\lambda \in E_{\lambda,x}.$$

By virtue of Corollary 7.1 and Proposition 7.1 this is equivalent to findining a critical point $(1, x, w) \in M^1_{\lambda,\eta}$ of the functional J_λ. To find a critical point of J_λ it is sufficient to show that for $\eta > 0$ small enough and $\lambda > 0$ large enough there exist constants γ, β_j, $j = 1, \ldots, N$, such that

$$\frac{\partial J_\lambda}{\partial x_i} = \sum_{j=1}^{N} \beta_j \langle w, \frac{\partial^2 U_x}{\partial x_j \partial x_i} \rangle, \quad i = 1, \ldots, N, \tag{7.32}$$

$$\langle \frac{\partial J_\lambda}{\partial w}, \phi \rangle = \gamma \langle U_x, \phi \rangle + \sum_{j=1}^{N} \beta_j \langle \frac{\partial U_x}{\partial x_i}, \phi \rangle \tag{7.33}$$

for every $\phi \in H^1(\mathbb{R}_N)$.

Proposition 7.4 *There exist a constant $\lambda_o > 0$ and a C^1-mapping*

$$\Psi : [\lambda_o, \infty) \times \overline{B(0,1)} \to E_{\lambda,x}$$

such that $w_{\lambda,x} = \Psi(\lambda, x)$ satisfies (7.33) for some $(\gamma, \beta_1, \ldots, \beta_N) \in \mathbb{R}_{N+1}$. Furthermore, there exists a constant $C > 0$ such that

$$||w_{\lambda,x}||_{1,2} \leq \frac{C}{\lambda^2}. \tag{7.34}$$

Proof. Following Bahri [10] (see also [27]) we expand J_λ in a neighbourhood of $w = 0$. Let $w \in E_{\lambda,x}$, then

$$\begin{aligned} J_\lambda(t, x, w) &= K_\lambda(U_x + w) = J_\lambda(t, x, 0) + f_{\lambda,x}(w) + Q_{\lambda,x}(w) \\ &+ R_{\lambda,x}(w), \end{aligned}$$

where $f_{\lambda,x}(w)$ is a continuous linear functional on $E_{\lambda,x}$, $Q_{\lambda,x}(w)$ is a continuous quadratic form on $E_{\lambda,x}$ and $R_{\lambda,x}$ is a higher order term. Letting $\bar{p} = \min(p, 3)$ we obtain by straightforward calculations that

$$J_\lambda(1, x, w) = \frac{||U_x + w||^2_{1,2}}{\left(\int_{\mathbb{R}_N} b(\frac{y}{\lambda}) |U_x + w|^p \, dy\right)^{\frac{2}{p}}}$$

$$= ||U_x + w||^2_{1,2} \Big[\int_{\mathbb{R}_N} b\big(\frac{y}{\lambda}\big) \big(U_x^p + pU_x^{p-1} w + \frac{p(p-1)}{2} U_x^{p-2} w^2$$

$$+ O\big(|w|^{\bar{p}} U_x^{p-\bar{p}}\big)\big) \, dy\Big]^{-\frac{2}{p}}$$

$$
\begin{aligned}
&= \|U_x + w\|_{1,2}^2 \Bigg[\Big(\int_{\mathbb{R}_N} b(\frac{y}{\lambda}) U_x^p \, dy \Big)^{-\frac{2}{p}} \\
&-2 \Big(\int_{\mathbb{R}_N} b(\frac{y}{\lambda}) U_x^p \, dy \Big)^{-\frac{2}{p}-1} \int_{\mathbb{R}_N} b(\frac{y}{\lambda}) U_x^{p-1} w \, dy \\
&+(2+p) \Big(\int_{\mathbb{R}_N} b(\frac{y}{\lambda}) U_x^p \, dy \Big)^{-\frac{2}{p}-2} \Big(\int_{\mathbb{R}_N} b(\frac{y}{\lambda}) U_x^{p-1} w \, dy \Big)^2 \\
&+O\big(\int_{\mathbb{R}_N} |w|^{\bar{p}} \, dy \big) \Bigg]
\end{aligned}
$$

and consequently

$$
f_{\lambda,x}(w) = - \frac{-2\|U_x\|_{1,2}^2}{\Big(\int_{\mathbb{R}_N} b(\frac{y}{\lambda}) U_x^p \, dy \Big)^{\frac{2}{p}+1}} \int_{\mathbb{R}_N} b(\frac{y}{\lambda}) U_x^{p-1} w \, dy,
$$

$$
\begin{aligned}
Q_{\lambda,x}(w) = \frac{1}{\Big(\int_{\mathbb{R}_N} b(\frac{y}{\lambda}) U_x^p \, dy \Big)^{\frac{2}{p}}} \Bigg[& \|w\|_{1,2}^2 \\
& - \frac{(p-1)\|U\|_{1,2}^2}{\Big(\int_{\mathbb{R}_N} b(\frac{y}{\lambda}) U_x^p \, dy \Big)} \Big(\int_{\mathbb{R}_N} b(\frac{y}{\lambda}) U_x^{p-2} w^2 \, dy \Big) \\
& + \frac{(p+2)}{\Big(\int_{\mathbb{R}_N} b(\frac{y}{\lambda}) U_x^p \, dy \Big)^2} \Big(\int_{\mathbb{R}_N} b(\frac{y}{\lambda}) U_x^{p-1} w \, dy \Big)^2 \Bigg],
\end{aligned}
$$

$$
R_{\lambda,x}(w) = O\Big(\int_{\mathbb{R}_N} |w|^{\bar{p}} U_x^{p-\bar{p}} \, dy \Big) = O\big(\|w\|_{1,2}^{\bar{p}} \big),
$$

$$
R'_{\lambda,x}(w) = O\big(\|w\|_{1,2}^{\bar{p}-1} \big) \quad \text{and} \quad R'_{\lambda,x}(w) = O\big(\|w\|_{1,2}^{\bar{p}-2} \big).
$$

We now show that there exist $\rho' > 0$ and $\lambda_o > 0$ such that

$$
Q_{\lambda,x}(w) \geq \rho' \|w\|_{\lambda,2}^2 \tag{7.35}
$$

for all $w \in E_{\lambda,x}$ and $\lambda \geq \lambda_o$. Indeed, we have

$$
\int_{\mathbb{R}_N} b(\frac{y}{\lambda}) U_x^p \, dy \;\; = \;\; \int_{\mathbb{R}_N} U_x^p \, dy + \int_{\mathbb{R}_N} \big(b(\frac{y}{\lambda}) - b(0) \big) U_x^p \, dy
$$

$$= \int_{\mathbb{R}_N} U_x^p\, dy + O\Big(\frac{1}{\lambda^2}\int_{\mathbb{R}_N} |y|^2 U_x^p\, dy\Big)$$
$$= \int_{\mathbb{R}_N} U_x^p\, dy + O\Big(\frac{1}{\lambda^2}\Big)$$

and

$$\int_{\mathbb{R}_N} b\Big(\frac{y}{\lambda}\Big) U_x^{p-2} w^2\, dy = \int_{\mathbb{R}_N} U_x^{p-2} w^2\, dy + O\Big(\frac{1}{\lambda^2}\|w\|_{1,2}^2\Big).$$

Since $\|U_x\|_{1,2}^2 = \int_{\mathbb{R}_N} U_x^p\, dy$, we see that

$$Q_{\lambda,x}(w) \geq \frac{1}{\Big(\int_{\mathbb{R}_N} U_x^p\, dy + O(\frac{1}{\lambda^2})\Big)^{\frac{2}{p}}}\Big[\|w\|_{1,2}^2 - (p-1)\int_{\mathbb{R}_N} U_x^{p-2} w^2\, dy + O\Big(\frac{1}{\lambda}\Big)\|w\|_{1,2}^2\Big].$$

We now apply Proposition 7.2 to find $\lambda_\circ > 0$ and $\rho' > 0$ such that (7.35) holds. From this we deduce the existence of a unique symmetric continuous coercive operator $A_{\lambda,x} : E_{\lambda,x} \to E_{\lambda,x}$ such that

$$Q_{\lambda,x}(w) = \langle A_{\lambda,x} w, w\rangle.$$

Since $f_{\lambda,x}$ is a continuous linear functional on $E_{\lambda,x}$ equipped with a scalar product from $H^1(\mathbb{R}_N)$, there exists a unique element of $E_{\lambda,x}$, denoted also by $f_{\lambda,x}$, such that $f_{\lambda,x}(w) = \langle f_{\lambda,x}, w\rangle$. We claim that

$$\|f_{\lambda,x}\| = O\Big(\frac{1}{\lambda^2}\Big). \tag{7.36}$$

Indeed, applying the Hölder and Sobolev inequalities and the fact that

$$\int_{\mathbb{R}_N} U_x^{p-1} w\, dy = \int_{\mathbb{R}_N} (-\Delta U_x + U_x) w\, dy = \langle U_x, w\rangle = 0$$

we get

$$\Big|\int_{\mathbb{R}_N} b\Big(\frac{y}{\lambda}\Big) U_x^{p-1} w\, dy\Big| = \Big|\int_{\mathbb{R}_N} \Big(b\Big(\frac{y}{\lambda}\Big) - b(0)\Big) U_x^{p-1} w\, dy\Big|$$
$$\leq \frac{C}{\lambda^2}\int_{\mathbb{R}_N} |y|^2 U_x^{p-1}|w|\, dy \leq \frac{C_1}{\lambda^2}\,\|w\|_{1,2}$$

for some constants $C > 0$ and $C_1 > 0$. Using $f_{x,\lambda}$, $A_{x,\lambda}$ we can write

$$\frac{\partial J_\lambda}{\partial v} |_{E_{x,\lambda}} (x, \lambda, w) = f_{x,\lambda} + 2A_{x,\lambda} w + R_{x,\lambda}{}'(w). \tag{7.37}$$

Consequently there exists $(\gamma, \beta_1, \ldots, \beta_N) \in \mathbb{R}_{N+1}$ such that (7.33) holds if and only if

$$f_{x,\lambda} + 2A_{x,\lambda} w + R'_{x,\lambda}(w) = 0.$$

We now define a mapping $F_{x,\lambda} : E_{x,\lambda} \times E_{x,\lambda} \to \mathbb{R}$ by

$$F_{x,\lambda}(f, w) = f + 2A_{x,\lambda} w + R'_{x,\lambda}(w).$$

It is easy to see that $F_{x,\lambda}(0, 0) = 0$ and

$$\frac{\partial F_{x,\lambda}(f, w)}{\partial w} = 2A_{x,\lambda} + R'_{x,\lambda}(w).$$

It follows from (7.35) and the fact that $R'_{x,\lambda}(w) = O\big(\|w\|_{1,2}^{\bar{p}-2}\big)$ that there exists $\eta > 0$ such that $\frac{\partial F_{x,\lambda}(f,w)}{\partial w}$ is invertible for $\lambda \geq \lambda_\circ$ and $\|w\|_{1,2} < \eta$. By the implicit function theorem there exists a C^1-branch $(f, \Psi_{x,\lambda}(f))$ emanating from $(0, 0)$ and satisfying $F_{x,\lambda}(f, \Psi_{x,\lambda}(f)) = 0$ and which can be continued provided $\|\Psi_{x,\lambda}(f)\|_{1,2} < \eta$. We now notice that if $F_{x,\lambda}(f, w) = 0$, then

$$\langle f, w\rangle + 2\langle A_{x,\lambda} w, w\rangle + \langle R'_{x,\lambda}(w), w\rangle = 0.$$

Hence (7.35) and the fact that $R'_{x,\lambda}(f) = O\big(\|w\|_{1,2}^{\bar{p}-1}\big)$ imply that there exists $\eta' > 0$ and $k > 0$ such that $\|w\|_{1,2} \leq k\|f\|_{1,2}$ provided $\|w\|_{1,2} \leq \eta'$. This combined with (7.36) shows the existence of η such that the branch $(f, \Psi_{x,\lambda}(f))$ may be continued provided $\|f\|_{1,2} < \eta$. The estimate (7.34) follows from (7.36). □

Proposition 7.5 *Let $w_{x,\lambda}$ and $(\gamma, \beta_1, \ldots, \beta_N)$ be solutions of (7.33) obtained in Proposition 7.4. Then the system (7.32) is equivalent to*

$$x = W(x, \lambda), \tag{7.38}$$

where $W : \bar{B}(0, 1) \to \mathbb{R}_N$ is continuous mapping satisfying $W(x, \lambda) = o(1)$, where $o(1) \to 0$ uniformly in $x \in \bar{B}(0, 1)$ as $\lambda \to \infty$.

Proof. Let $w = w_{x,\lambda}$. A direct computation shows that

$$\begin{aligned}\frac{\partial J_\lambda}{\partial x_i} &= \frac{2\langle U_x + w, \frac{\partial U_x}{\partial x_i}\rangle}{\left(\int_{\mathbb{R}^N} b(\frac{y}{\lambda})|U_x + w|^p\, dy\right)^{\frac{2}{p}}} \\ &- 2\|U_x + w\|_{1,2}^2 \frac{\int_{\mathbb{R}^N} b(\frac{y}{\lambda})|U_x + w|^{p-2}(U_x + w)\frac{\partial U_x}{\partial x_i}\, dy}{\left(\int_{\mathbb{R}^N} b(\frac{y}{\lambda})|U_x + w|^p\, dy\right)^{\frac{2}{p}+1}}.\end{aligned}$$

Since $w \in E_{x,\lambda}$ and

$$\langle U_x, \frac{\partial U_x}{\partial x_i}\rangle = \frac{1}{2}\frac{\partial}{\partial x_i}\langle U_x, U_x\rangle = \frac{1}{2}\frac{\partial}{\partial x_i}\langle U, U\rangle = 0,$$

we see that

$$\frac{\partial J_\lambda}{\partial x_i} = -2\|U_x + w\|_{1,2}^2 \frac{\int_{\mathbb{R}^N} b(\frac{y}{\lambda})|U_x + w|^{p-2}(U_x + w)\frac{\partial U_x}{\partial x_i}\, dy}{\left(\int_{\mathbb{R}^N} b(\frac{y}{\lambda})|U_x + w|^p\, dy\right)^{\frac{2}{p}+1}}. \tag{7.39}$$

It follows from (7.34) that

$$\frac{\|U_x + w\|_{1,2}^2}{\left(\int_{\mathbb{R}^N} b(\frac{y}{\lambda})|U_x + w|^p\, dy\right)^{\frac{2}{p}+1}} = A^{-\frac{2}{p}} + O(\frac{1}{\lambda^2}), \tag{7.40}$$

where $A = \|U_x\|_{1,2}^2 = \|U\|_{1,2}^2 = \int_{\mathbb{R}^N} U_x^p\, dy$. To estimate the remaining integral in (7.39) we write

$$\begin{aligned}&\int_{\mathbb{R}^N} b(\frac{y}{\lambda})|U_x + w|^{p-2}(U_x + w)\frac{\partial U_x}{\partial x_i}\, dy \\ &\quad = \int_{\mathbb{R}^N} b(\frac{y}{\lambda})\Big(U_x^{p-1}\frac{\partial U_x}{\partial x_i} + (p-1)U^{p-2}w\frac{\partial U_x}{\partial x_i} \\ &\quad + \frac{\partial U_x}{\partial x_i}O(|w|^{\bar{p}-1})\Big)\, dy = I_1 + I_2 + I_3,\end{aligned}$$

where $\bar{p} = \min\{p, 3\}$ and by (7.34)

$$I_3 = O(\|w\|^{\bar{p}-1}) = o(\frac{1}{\lambda^2}). \tag{7.41}$$

Since $w \in E_{x,\lambda}$, we have

$$\begin{aligned}\int_{\mathbb{R}_N} U_x^{p-2} w \frac{\partial U_x}{\partial x_i}\,dy &= \frac{1}{p-1}\int_{\mathbb{R}_N} \frac{\partial}{\partial x_i}(U_x^{p-1})\, w\, dy \\ &= \frac{1}{p-1}\int_{\mathbb{R}_N} \frac{\partial}{\partial x_i}(-\Delta U_x + U_x)\, w\, dy \\ &= \frac{1}{p-1}\langle \frac{\partial U_x}{\partial x_i}, w\rangle = 0.\end{aligned}$$

It therefore follows from (7.34) that

$$\begin{aligned}I_2 &= \int_{\mathbb{R}_N} (b(\frac{y}{\lambda}) - 1)(p-1) U_x^{p-2} w \frac{\partial U_x}{\partial x_i}\,dy \\ &+ (p-1)\int_{\mathbb{R}_N} U_x^{p-2} w \frac{\partial U_x}{\partial x_i}\,dy \\ &= (p-1)\int_{\mathbb{R}_N} (b(\frac{y}{\lambda}) - 1) U_x^{p-2} w \frac{\partial U_x}{\partial x_i}\,dy = O(\frac{1}{\lambda^2})\|w\|_{L^p} = O(\frac{1}{\lambda^2}).\end{aligned} \tag{7.42}$$

To estimate I_1 we observe that

$$\begin{aligned}\int_{\mathbb{R}_N} U_x^{p-1} \frac{\partial U_x}{\partial x_i}\,dy &= \frac{1}{p}\int_{\mathbb{R}_N} \frac{\partial}{\partial x_i} U_x^p\,dy = \frac{1}{p}\frac{\partial}{\partial x_i}\int_{\mathbb{R}_N} U_x^p\,dy \\ &= \frac{1}{p}\frac{\partial}{\partial x_i}\int_{\mathbb{R}_N} U^p\,dy = 0.\end{aligned}$$

Then we obtain

$$\begin{aligned}I_1 &= \int_{\mathbb{R}_N} (b(\frac{y}{\lambda}) - 1) U_x^{p-1} \frac{\partial U_x}{\partial x_i}\,dy + \int_{\mathbb{R}_N} U_x^{p-1}\frac{\partial U_x}{\partial x_i}\,dy \\ &= \frac{1}{\lambda^2}\sum_{j,m=1}^N \int_{\mathbb{R}_N} D_{jm} b(0) y_j y_m U_x^{p-1} \frac{\partial U_x}{\partial x_i}\,dy + O(\frac{1}{\lambda^2}) \\ &= \frac{1}{\lambda^2}\sum_{j,m=1}^N \int_{\mathbb{R}_N} D_{jm} b(0) y_j y_m U_x^{p-1} \frac{\partial U_x}{\partial y_i}\,dy + O(\frac{1}{\lambda^2}) \\ &= \frac{1}{\lambda^2}\sum_{j,m=1}^N D_{jm} b(0)\int_{\mathbb{R}_N} (y_j + x_j)(y_m + x_m) U^{p-1} \frac{\partial U}{\partial y_i}\,dy + O(\frac{1}{\lambda^2}).\end{aligned}$$

Using the radial symmetry of U we get

$$\int_{\mathbb{R}_N} y_j y_m U^{p-1}\frac{\partial U}{\partial y_i}\,dy = \int_{\mathbb{R}_N} y_j y_m U^{p-1}(|y|)\frac{d}{d|y|}U(|y|)\frac{y_i}{|y|}\,dy = 0,$$

$$\int_{\mathbb{R}_N} U^{p-1} \frac{\partial U}{\partial y_i}\, dy = \int_{\mathbb{R}_N} U^{p-1} \frac{dU(|y|)}{d|y|} \frac{y_i}{|y|}\, dy = 0$$

and

$$\int_{\mathbb{R}_N} x_m y_j U^{p-1} \frac{\partial U}{\partial y_i}\, dy = \int_{\mathbb{R}_N} x_m y_j U^{p-1} \frac{\partial U(|y|)}{\partial |y|} \frac{y_i}{|y|}\, dy = 0$$

for $i \neq j$. Hence

$$I_1 = \frac{2}{\lambda^2} \sum_{m=1}^{N} D_{mi} b(0) \int_{\mathbb{R}_N} x_m y_i U^{p-1} \frac{\partial U}{\partial y_i}\, dy + O\big(\frac{1}{\lambda^3}\big). \tag{7.43}$$

Since

$$B = -\int_{\mathbb{R}_N} y_i \frac{\partial U}{\partial y_i} = -\int_{\mathbb{R}_N} \frac{y_i^2}{|y|^2} \frac{dU(|y|)}{d|y|}\, dy = -\frac{1}{N} \int_{\mathbb{R}_N} |y| \frac{dU(|y|)}{d|y|}\, dy > 0,$$

we derive from (7.41), (7.42) and (7.43) that

$$\int_{\mathbb{R}_N} b\big(\frac{y}{\lambda}\big) |U_x + w|^{p-2} (U_x + w) \frac{\partial U_x}{\partial x_i}\, dy = -\frac{2B}{\lambda^2} \sum_{m=1}^{N} D_{mi} b(0) x_m + o\big(\big(\frac{1}{\lambda^2}\big).$$

Inserting this and (7.40) into (7.39) we get

$$\begin{aligned} \frac{\partial J_\lambda}{\partial x_i} &= \Big(A^{-\frac{2}{p}} + O\big(\frac{1}{\lambda^2}\big)\Big) \Big(\frac{2B}{\lambda^2} \sum_{m=1}^{N} D_{im} b(0) x_m + o\big(\frac{1}{\lambda^2}\big)\Big) \\ &= \frac{2BA^{-\frac{2}{p}}}{\lambda^2} \sum_{m=1}^{N} D_{im} b(0) x_m + o\big(\frac{1}{\lambda^2}\big). \end{aligned}$$

The above formula for $\frac{\partial J_\lambda}{\partial x_i}$ shows that the system of equations (7.32) is equivalent to

$$\frac{2BA^{-\frac{2}{p}}}{\lambda^2} \sum_{m=1}^{N} D_{im} b(0) x_m = \sum_{j=1}^{N} \beta_j \langle w, \frac{\partial^2 U_x}{\partial x_j \partial x_i} \rangle + o\big(\frac{1}{\lambda^2}\big), \tag{7.44}$$

$i = 1, \ldots, N$. It follows from (7.34) that

$$|\langle w, \frac{\partial^2 U_x}{\partial x_j \partial x_i} \rangle| \leq C \|w\|_{1,2} = O\big(\frac{1}{\lambda^2}\big). \tag{7.45}$$

We claim that

$$\beta_j = O\big(\frac{1}{\lambda^2}\big). \tag{7.46}$$

Indeed, substituting $\phi = \frac{\partial U_x}{\partial x_i}$ into (7.33) we obtain

$$\langle \frac{\partial J_\lambda}{\partial w}, \frac{\partial U_x}{\partial x_i} \rangle = \gamma \langle U_x, \frac{\partial U_x}{\partial x_i} \rangle + \sum_{j=1}^{N} \beta_j \langle \frac{\partial}{\partial x_j} U_x, \frac{\partial}{\partial x_i} U_x \rangle.$$

Since

$$\langle U_x, \frac{\partial}{\partial x_i} U_x \rangle = \frac{\partial}{\partial x_i} \langle U_x, U_x \rangle = \frac{\partial}{\partial x_i} \langle U, U \rangle = 0$$

and for $i \neq j$

$$\begin{aligned}
\langle \frac{\partial}{\partial x_j} U_x, \frac{\partial}{\partial x_i} U_x \rangle &= \int_{\mathbb{R}_N} \left[\nabla (\frac{\partial U_x}{\partial x_j}) \nabla (\frac{\partial U_x}{\partial x_i})\, dy + \frac{\partial U_x}{\partial x_j} \frac{\partial U_x}{\partial x_i} \right] dy \\
&= \int_{\mathbb{R}_N} \frac{\partial U_x}{\partial x_j} (p-1) U_x^{p-2} \frac{\partial U_x}{\partial x_i}\, dy = (p-1) \int_{\mathbb{R}_N} \frac{\partial U_x}{\partial y_j} \frac{\partial U_x}{\partial y_i} U_x^{p-2}\, dy \\
&= (p-1) \int_{\mathbb{R}_N} \frac{\partial U}{\partial y_j} \frac{\partial U}{\partial y_i} U^{p-2}\, dy \\
&= (p-1) \int_{\mathbb{R}_N} U^{p-2} (\frac{U(|y|)}{d|y|})^2 \frac{y_i y_j}{|y|^2}\, dy = 0,
\end{aligned}$$

we get that

$$\beta_j = \frac{\langle \frac{\partial J_\lambda}{\partial w}, \frac{\partial U_x}{\partial x_i} \rangle}{\left\| \frac{\partial U_x}{\partial x_i} \right\|_{1,2}^2}.$$

We now observe that

$$\left| \langle \frac{\partial J_\lambda}{\partial w}, \frac{\partial U_x}{\partial x_i} \rangle \right| = \left| \frac{\partial J_\lambda}{\partial x_i} \right| = O(\frac{1}{\lambda^2})$$

and $\left\| \frac{\partial U_x}{\partial x_i} \right\|_{1,2}^2 = C_\circ > 0$. Consequently the claim (7.46) easily follows. Finally, by virtue of (7.44), (7.45) and (7.46) we have

$$\frac{2BA^{-\frac{2}{p}}}{\lambda^2} D^2 b(0) x = o(\frac{1}{\lambda^2}).$$

Since det $D^2 b(0) \neq 0$, we have for $x \in \overline{B(0,1)}$, $x = W(x, \lambda)$, with $W(x, \lambda) = o(1))$. □

We are now in a position to establish the existence of one peak solution.

Theorem 7.1 *Let $x = 0$ be a nondegenerate critical point of b. Then there exists a $\lambda_\circ > 0$ such that for $\lambda \geq \lambda_\circ$ equation (7.2) has a positive solution of the form*

$$u_\lambda(y) = \alpha_\lambda U(\lambda y - x_\lambda) + w_\lambda(y),$$

where $\alpha_\lambda \to b(0)^{-\frac{1}{p-2}}$, $x_\lambda \to 0$, $w_\lambda \to 0$ in $H^1(\mathbb{R}_N)$ as $\lambda \to \infty$ and $w_\lambda \in E_{x_\lambda,\lambda}$.

Proof. According to Proposition 7.5, $W(x,\lambda)$ maps $\overline{B(0,1)}$ into itself provided $\lambda \geq \lambda_\circ$ ($\lambda_\circ > 0$ sufficiently lage). By the Brouwer fixed point theorem equation (7.38) has a solution. It follows from our construction that

$$u_\lambda = \alpha_\lambda \big(U_x + w_\lambda\big),$$

with

$$\alpha_\lambda = \left(\frac{\|U_x + w_\lambda\|_{1,2}^2}{\int_{\mathbb{R}_N} b(\frac{y}{\lambda})|U_x + w_\lambda|^p\, dy}\right)^{\frac{1}{p-2}}.$$

Since

$$\alpha_\lambda = \left(\frac{\|U_x\|_{1,2}^2 + o(1)}{\int_{\mathbb{R}_N} U_x^p\, dy + o(1)}\right)^{\frac{1}{p-2}} \to b(0) = 1, \text{ as } \lambda \to \infty,$$

it follows from Proposition 7.4 that $\alpha_\lambda w_\lambda \to 0$ in $H^1(\mathbb{R}_N)$. It remains to show that $u_\lambda > 0$ on $\mathbb{R}_N$. Let $u_\lambda^- = \max(0, -u_\lambda)$. Then by the Sobolev inequality

$$S_p \left(\int_{\mathbb{R}_N} (u_\lambda^-)^p\, dx\right)^{\frac{2}{p}} \leq \|u_\lambda^-\|_{1,2}^2 = \int_{\mathbb{R}_N} b\big(\frac{y}{\lambda}\big)\,(u_\lambda^-)^p\, dx \leq C \int_{\mathbb{R}_N} (u_\lambda^-)^p\, dx$$

for some constant $C > 0$. If $u_\lambda^- \not\equiv 0$, then this inequality shows that $\int_{\mathbb{R}_N} (u_\lambda^-)^p\, dx \geq C_1$ for some constant $C_1 > 0$ independent of λ. On the other hand

$$u_\lambda^- = \big(\alpha_\lambda U_x + \alpha_\lambda w_\lambda\big)^- \leq \alpha_\lambda |w_\lambda|$$

and consequently $u_\lambda^- \to 0$ in $L^p(\mathbb{R}_N)$ which is impossible. □

7.4 Multi-peak solutions

To establish the existence of multi-peak solutions we follow the ideas described in Section 7.1. We restrict ourselves to the construction of two peak solution. We assume that $a^1 = 0$ and a^2 are two nondegenerate critical points of b, that is $Db(a^i) = 0$, $\det D^2b(a^i) \neq 0$, $i = 1, 2$. For simplicity we assume that $b(0) = 1$.

Let $M^2_{\lambda,\eta}$, E_{λ,x^1,x^2}, K_λ and J_λ be defined as in Section 7.1. We construct a two peak solution of (7.2) in the form

$$u_\lambda(y) = \alpha_1(\lambda)U(y - x^1(\lambda)) + \alpha_2(\lambda)U(y - \lambda a^2 - x^2(\lambda)) + w_\lambda(y). \quad (7.47)$$

According to Proposition 7.1 we need to show the existence of a critical point $(1, \gamma, x^1, x^2, w) \in M^2_{\lambda,\eta}$ of K_λ. This will follow if we can prove that for small $\eta > 0$ and large $\lambda > 0$ there exist constants $\alpha_1, \alpha_2, \beta^1_1, \ldots, \beta^1_N, \beta^2_N, \ldots, \beta^2_N$ such that

$$\frac{\partial J_\lambda}{\partial \gamma} = 0 \quad (7.48)$$

$$\frac{\partial J_\lambda}{\partial x^1_i} = \sum_{i=1}^{N} \beta^1_j \langle \frac{\partial^2 U_{x^1}}{\partial x_j \partial x_i}, w\rangle \quad (7.49)$$

$$\frac{\partial J_\lambda}{\partial x^2_i} = \sum_{j=1}^{N} \beta^2_j \langle \frac{\partial^2 U_{\lambda a^2 + x^2}}{\partial x_j \partial x_i}, w\rangle \quad (7.50)$$

$$\begin{aligned} \langle \frac{\partial J_\lambda}{\partial w}, \phi\rangle &= \alpha_1\langle U_{x^1}, \phi\rangle + \sum_{j=1}^{N} \beta^1_j \langle \frac{\partial U^1_x}{\partial x_j}, \phi\rangle + \alpha_2 \langle U_{\lambda a^2 + x^2}, \phi\rangle \quad (7.51) \\ &+ \sum_{j=1}^{N} \beta^2_j \langle \frac{\partial U_{\lambda a^2 + x^2}}{\partial x_j}, \phi\rangle \end{aligned}$$

for every $\phi \in H^1(\mathbb{R}_N)$.

We commence by extending Proposition 7.4 to E_{λ,x^1,x^2}.

Proposition 7.6 *There exist $\lambda_\circ > 0$ and a C^1-map $\Psi : [\lambda_\circ, \infty) \times \overline{B(0,1)} \times \overline{B(0,1)} \to E_{\lambda,x^1,x^2}$ such that $w_{\lambda,x^1,x^2} = \Psi(\lambda, x^1, x^2)$ satisfies*

(7.51) for some $(\alpha_1, \alpha_2, \beta_1^1, \dots, \beta_N^1, \beta_1^2, \dots, \beta_N^2) \in \mathbb{R}_{2N+2}$. *Moreover, there exists a constant* $C > 0$ *such that*

$$\|w_{\lambda,x^1,x^2}\|_{1,2} \leq \frac{C}{\lambda^2}.$$

Proof. The proof is similar to that of Proposition 7.4. First we expand J_λ in a neighbourhood of $w = 0$. Let $w \in E_{\lambda,x^1,x^2}$. Then

$$J_\lambda(1, \lambda, x^1, x^2, w) = J_\lambda(1, \lambda, 0, x^1, x^2, 0) + f(w) + Q(w) + R(w),$$

where f is a continuous linear functional on E_{λ,x^1,x^2}, Q is a continuous quadratic form on E_{λ,x^1,x^2} and R is a higher order term. Let $Z(x^1, x^2) = U_{x^1} + \gamma U_{\lambda a^2 + x^2}$. Terms f and Q have the following form

$$f(w) = \frac{-2\|Z(x^1,x^2)\|_{1,2}^2 \int_{\mathbb{R}_N} b(\frac{y}{\lambda}) Z(x^1,x^2)^{p-1} w\, dy}{\left(\int_{\mathbb{R}_N} b(\frac{y}{\lambda}) Z(x^1,x^2)^p\, dy\right)^{\frac{2}{p}}},$$

$$\begin{aligned} Q(w) &= \left(\int_{\mathbb{R}_N} b(\frac{y}{\lambda}) Z(x^1,x^2)^p\, dy\right)^{-\frac{2}{p}} \Bigg\{ \|w\|_{1,2}^2 \\ &- (p-1) \frac{\|Z(x^1,x^2)\|_{1,2}^2}{\int_{\mathbb{R}_N} b(\frac{y}{\lambda}) Z(x^1,x^2)^p\, dy} \int_{\mathbb{R}_N} b(\frac{y}{\lambda}) Z(x^1,x^2)^{p-2} w^2\, dy \\ &+ (p+2) \frac{\|Z(x^1,x^2)\|_{1,2}^2}{\left(\int_{\mathbb{R}_N} b(\frac{y}{\lambda}) Z(x^1,x^2)^p\, dy\right)^2} \left(\int_{R_N} b(\frac{y}{\lambda}) Z(x^1,x^2)^{p-1}\, dy\right)^2 \Bigg\} \end{aligned}$$

Step I. We commence by showing that

$$\|f\| = O(\frac{1}{\lambda^2}). \tag{7.52}$$

First we observe that for every $l > 0$ and $h > 0$ there exists a constant $\tau > 0$ such that

$$\int_{\mathbb{R}_N} e^{-l|x|} e^{-h|x-\lambda a^2|}\, dx = O(e^{-\tau\lambda}). \tag{7.53}$$

Indeed, without loss of generality we may assume that $a^2 = (a_1^2, 0, \dots, 0)$ and we have

$$\int_{\mathbb{R}_N} e^{-l|x|} e^{-h|x-\lambda a^2|}\, dx = \int_{x_1 < \frac{\lambda}{2} a_1^2} e^{-l|x|} e^{-h|x-\lambda a^2|}\, dx$$

$$+ \int_{x_1 \geq \frac{\lambda}{2} a_1^2} e^{-l|x|} e^{-h|x-\lambda a^2|}\, dx.$$

Since

$$I_1 \leq e^{-h\frac{\lambda}{2}a_1^2} \int_{\mathbb{R}_N} e^{-l|x|}\, dx = Ce^{-h\frac{\lambda}{2}a_1^2}$$

and

$$I_2 \leq e^{-\frac{l}{2}\lambda a_1^2} \int_{\mathbb{R}_N} e^{-h|x|}\, dx = Ce^{-l\frac{\lambda}{2}a_1^2}$$

and (7.52) follows. To prove (7.52) we write using (7.53)

$$\int_{\mathbb{R}_N} b(\frac{y}{\lambda}) Z(x^1, x^2)^{p-1} w\, dx = \int_{\mathbb{R}_N} b(\frac{y}{\lambda}) U_{x_1}^{p-1} w\, dy \tag{7.54}$$
$$+\gamma^{p-1} \int_{\mathbb{R}_N} b(\frac{y}{\lambda}) U_{\lambda a^2 + x^2}^{p-1} w\, dy + O\big(e^{-\tau\lambda}\big)\|w\|_{1,2}.$$

As in the proof of Proposition 7.4 we show that

$$|\int_{\mathbb{R}_N} b(\frac{y}{\lambda}) U_{x^1}^{p-1} w\, dy| \leq \frac{C}{\lambda^2}\|w\|_{1,2}$$

and

$$|\int_{\mathbb{R}_N} b(\frac{y}{\lambda}) U_{\lambda a^2+x^2}^{p-1} w\, dy| \leq \frac{C}{\lambda^2}\|w\|_{1,2}$$

for some constant $C > 0$. This combined with (7.54) implies (7.52).
Step II. There exists a constant $\rho' > 0$ such that

$$Q(w) \geq \rho' \|w\|_{1,2}^2 \tag{7.55}$$

for all $w \in E_{\lambda, x^1, x^2}$. It follows from Step I that

$$\|Z(x^1, x^2)\|_{1,2}^2 = \|U\|_{1,2}^2 (1 + \gamma^2) + O\big(e^{-\tau\lambda}\big).$$

Hence

$$\int_{\mathbb{R}_N} b(\frac{y}{\lambda}) Z(x^1, x^2)^p\, dy = \int_{\mathbb{R}_N} b(\frac{y}{\lambda}) U_{x^1}^p\, dy + \gamma^p \int_{\mathbb{R}_N} b(\frac{y}{\lambda}) U_{\lambda a^2 + x^2}^p\, dy$$
$$+O\big(e^{-\tau\lambda}\big) = \int_{\mathbb{R}_N} b(0) U_{x^1}^p\, dy + \int_{\mathbb{R}_N} \big(b(\frac{y}{\lambda}) - b(0)\big) U_{x^1}^p\, dy$$
$$+\gamma^p \int_{\mathbb{R}_N} b(a^2) U_{\lambda a^2 + x^2}\, dy$$

$$+\gamma^p \int_{\mathbb{R}_N} \left(b(\frac{y}{\lambda}) - b(a^2)\right) U_{\lambda a^2+x^2}\, dy + O\left(e^{-\tau\lambda}\right)$$
$$= \left(1 + \gamma^p b(a^2)\right) \|U\|_{1,2}^2 + O\left(\frac{1}{\lambda^2}\right)$$

and

$$\int_{\mathbb{R}_N} b(\frac{y}{\lambda}) Z(x^1, x^2)^{p-2} w^2\, dy = \int_{\mathbb{R}_N} b(\frac{y}{\lambda}) U_{x^1}^{p-2} w^2\, dy$$
$$+\gamma^{p-2} \int_{\mathbb{R}_N} b(\frac{y}{\lambda}) U_{\lambda a^2+x^2} w^2\, dy + O\left(e^{-\tau\lambda}\right) \|w\|_{1,2}^2$$
$$= \int_{\mathbb{R}_N} U_{x^1}^{p-2} w^2\, dy + \gamma^{p-2} b(a^2) \int_{\mathbb{R}_N} U_{\lambda a^2+x^2}^{p-2} w^2\, dy + O\left(\frac{1}{\lambda^2}\right) \|w\|_{1,2}^2$$
$$= \int_{\mathbb{R}_N} \left(U_{x^1} + \gamma b(a^2)^{\frac{1}{p-2}} U_{\lambda a^2+x^2}\right)^{p-2} w^2\, dy + O\left(\frac{1}{\lambda^2}\right) \|w\|_{1,2}^2.$$

It therefore follows that

$$\frac{\|Z(x^1, x^2)\|_{1,2}^2}{\int_{\mathbb{R}_N} b(\frac{y}{\lambda}) Z(x^1, x^2, a^2)^p\, dy} \int_{\mathbb{R}_N} b(\frac{y}{\lambda}) Z(x^1, x^2)^{p-2} w^2\, dy$$
$$= \frac{1+\gamma^2}{1+\gamma^p b(a^2)} \int_{\mathbb{R}_N} \left(U_{x^1} + \gamma b(a^2)\right)^{\frac{1}{p-2}} \left(U_{\lambda a^2+x^2}\right)^{p-2} w^2\, dy$$
$$+ \quad O\left(\frac{1}{\lambda^2}\right) \|w\|_{1,2}^2.$$

We now observe that if ρ is sufficiently small then for $\gamma \in \left(b(a^2)^{-\frac{1}{p-2}} - \rho, b(a^2)^{-\frac{1}{p-2}} + \rho\right)$, both terms $\frac{1+\gamma^2}{1+\gamma^p b(a^2)}$ and $\gamma b(a^2)^{\frac{1}{p-2}}$ are close to 1. Thus (7.55) follows from Proposition 7.3. To establish the existence of a mapping $\Psi : [\lambda_\circ, \infty) \times \overline{B(0,1)} \times \overline{B(0,1)} \to R_{\lambda,x^1,x^2}$ such that $\Psi(\lambda, x^1, x^2) = w_{\lambda,x^1,x^2}$ satisfies (7.49) and (7.50) we repeat the second part of the proof of Proposition 7.4. □

Proposition 7.7 *For $w = w_{\lambda,x^1,x^2}$ and $(1, \gamma, x^1, x^2) \in \Sigma_\rho$, the equation*

$$\frac{\partial J_\lambda}{\partial \gamma} = 0$$

is equivalent to

$$\gamma - b(a^2)^{-\frac{1}{p-2}} = V\left(\gamma, x^1, x^2, \lambda\right),$$

where V is a continuous function of $(\gamma, x^1, x^2, \lambda)$ and

$$V(\gamma, x^1, x^2, \lambda) = O(\frac{1}{\lambda^2}),$$

Proof. It is easy to check that

$$\begin{aligned}\frac{\partial J_\lambda}{\partial \gamma} &= \frac{2\langle Z(x^1,x^2), U_{\lambda a^2+x^2}\rangle}{\left(\int_{\mathbb{R}_N} b(\frac{y}{\lambda}) Z(x^1,x^2)+w|^p\,dy\right)^{\frac{2}{p}}} \\ &- 2\|Z(x^1,x^2)\|_{1,2}^2 \frac{\int_{\mathbb{R}_N} b(\frac{y}{\lambda})|Z(x^1,x^2)|^{p-2}(Z(x^1,x^2)+w)U_{\lambda a^2+x^2}\,dy}{\left(\int_{\mathbb{R}_N} b(\frac{y}{\lambda})|Z(x^1,x^2)+w|^p\,dy\right)^{\frac{2}{p}+1}} \\ &= \left(\int_{\mathbb{R}_N} b(\frac{y}{\lambda})|Z(x^1,x^2)+w|^p\,dy\right)^{-\frac{2}{p}} \Big\{2\gamma\|U\|_{1,2}^2 + O(e^{-\tau\lambda}) \\ &- 2\Big((1+\gamma^2)\|U\|_{1,2}^2 + O(e^{-\lambda\tau})\Big)\left(\int_{\mathbb{R}_N} b(\frac{y}{\lambda}) Z(x^1,x^2)+w|^{p-2}\right. \\ &\times \left.(Z(x^1,x^2)+w)U_{\lambda a^2+x^2}\,dy\right)\left(\int_{\mathbb{R}_N} b(\frac{y}{\lambda})|Z(x^1,x^2)+w|^p\,dy\right)^{-1}\Big\}\end{aligned} \tag{7.56}$$

and

$$\begin{aligned}\int_{\mathbb{R}_N} b(\frac{y}{\lambda})|Z(x^1,x^2)+w|^p\,dy &= \int_{\mathbb{R}_N} b(\frac{y}{\lambda}) Z(x^1,x^2)^p\,dy \\ &+p\int_{\mathbb{R}_N} b(\frac{y}{\lambda}) Z(x^1,x^2)^{p-1} w\,dy + O(\|w\|_{1,2}^2).\end{aligned} \tag{7.57}$$

Since

$$\begin{aligned}&|\int_{\mathbb{R}_N} b(\frac{y}{\lambda})\left[Z(x_1,x_2)^{p-1} - U_{x^1}^{p-1} - \gamma^{p-1}U_{\lambda a^2+x^2}\right] w\,dy| \\ &\le C\int_{\mathbb{R}_N} (U_{x^1}U_{\lambda a^2+x^2}^{p-2} + U_{x^1}^{p-2}U_{\lambda a^2+x^2})|w|\,dy \\ &\le Ce^{-\tau\lambda}\|w\|_p = O(e^{-\tau\lambda})\end{aligned}$$

for some constant $C > 0$, we have that

$$\begin{aligned}\int_{\mathbb{R}_N} b(\frac{y}{\lambda}) Z(x^1,x^2)^{p-1} w\,dy &= \int_{\mathbb{R}_N} b(\frac{y}{\lambda}) U_{x^1}^{p-1} w\,dy \\ &+\gamma^{p-1}\int_{\mathbb{R}_n} b(\frac{y}{\lambda}) U_{\lambda a^2+x^2}^{p-1} w\,dy + O(e^{-\tau\lambda})\end{aligned} \tag{7.58}$$

$$= O\big(\frac{1}{\lambda^2}\big)\|w\|_{1,2}^2 + O\big(e^{-\tau\lambda}\big) = O\big(\frac{1}{\lambda^2}\big).$$

Similarly

$$\int_{\mathbb{R}_N} b\big(\frac{y}{\lambda}\big) Z(x^1, x^2)^p \, dy = \int_{\mathbb{R}_N} b\big(\frac{y}{\lambda}\big) U_{x^1}^p \, dy \tag{7.59}$$
$$+\gamma^p \int_{\mathbb{R}_N} b\big(\frac{y}{\lambda}\big) U_{\lambda a^2 + x^2}^p \, dy$$
$$+O\big(e^{-\lambda\tau}\big) = \big(1 + \gamma^2 b(a^2)\big)\|U\|_{1,2}^2 + O\big(\frac{1}{\lambda^2}\big) + O\big(e^{-\tau\lambda}\big).$$

Inserting (7.58) and (7.59) into (7.57) we obtain

$$\int_{\mathbb{R}_N} b\big(\frac{y}{\lambda}\big)|Z(x^1, x^2) + w|^p \, dy = \big(1 + \gamma^p b(a^2)\|U\|_{1,2}^2\big) + O\big(\frac{1}{\lambda^2}\big). \tag{7.60}$$

Finally, we observe that

$$\int_{\mathbb{R}_N} b\big(\frac{y}{\lambda}\big)|Z(x^1, x^2) + w|^{p-2}\big(Z(x^1, x^2) + w\big) U_{\lambda a^2 + x^2} \, dy \tag{7.61}$$
$$= \int_{\mathbb{R}_N} b\big(\frac{y}{\lambda}\big)\Big[\big(Z(x^1, x^2)\big)^{p-1} + \big(Z(x^1, x^2)\big)^{p-2} w$$
$$+O\big(|w|^{\tilde{p}}\big)\Big] U_{\lambda a^2 + x^2} \, dy$$
$$= b(a^2)\gamma^{p-1}\|U\|_{1,2}^2 + \gamma^{p-1}\int_{\mathbb{R}_N} b\big(\frac{y}{\lambda}\big) U_{\lambda a^2 + x^2}^{p-1} w \, dy$$
$$+O\big(\|w\|_{1,2}^{\tilde{p}}\big) + O\big(\frac{1}{\lambda^2}\big)$$
$$= b(a^2)\gamma^{p-2}\|U\|_{1,2}^2 + O\big(\frac{1}{\lambda^2}\big),$$

where $\tilde{p} = \min(2, p-1)$. Consequently, combining (7.60), (7.61) with (7.56) and letting $D(\gamma, \lambda, a^2) = (1 + \gamma^p b(a^2)\|U\|_{1,2}^2 + O\big(\frac{1}{\lambda^2}\big)$ we get

$$\frac{1}{2}\frac{\partial J_\lambda}{\partial \gamma} = \gamma \frac{\|U\|_{1,2}^2}{D(\gamma, \lambda, a^2)} - \frac{(1 + \gamma^2)\|U\|_{1,2}^2 + O\big(e^{-\tau\lambda}\big)\,\big[b(a^2)\gamma^{p-1}\|U\|_{1,2}^2 + O\big(\frac{1}{\lambda^2}\big)\big]}{D(\gamma, \lambda, a^2)^{1+\frac{2}{p}}}$$
$$+ \frac{O\big(e^{-\tau\lambda}\big)}{D(\gamma, \lambda, a^2)^{\frac{2}{p}}}.$$

From this we deduce that $\frac{\partial J_\lambda}{\partial \gamma} = 0$ is equivalent to

$$\gamma - \frac{(1+\gamma^2)b(a^2)\gamma^{p-1}}{1+\gamma^p b(a^2)} = O\left(\frac{1}{\lambda^2}\right)$$

which completes the proof. □

Proposition 7.8 *Let $w = w_{\lambda,x^1,x^2}$, $(1,\gamma,x^1,x^2) \in \Sigma_p$. Then*

$$\frac{\partial J_\lambda}{\partial x_k^i} = \frac{A_i}{\lambda^2}\sum_{j=1}^{N} D_{jk} b(a^i) x_j^i + o\left(\frac{1}{\lambda^2}\right),\ i = 1,2,$$

for some constants $A_i > 0$.

Combining Propositions 7.7 and 7.8, we thus have the following equivalent form of equations (7.48), (7.49) and (7.50).

Proposition 7.9 *Equations (7.48), (7.49) and (7.50) are equivalent to*

$$\begin{aligned} \gamma &- b(a^2)^{-\frac{1}{p-2}} = V(\gamma, x^1, x^2, \lambda) \\ x^i &= W_i(\gamma, x^1, x^2, \lambda), \quad i = 1,2, \end{aligned}$$

where V and W_i, $i = 1,2$ are continuous functions of $(\gamma, x^1, x^2, \lambda)$ satisfying $|W_i| \to 0$ as $\lambda \to \infty$ uniformly in $(\gamma, x^1, x^2), (1,\gamma,x^1,x^2) \in \Sigma_p$ and $V = O(\frac{1}{\lambda^2})$.

Repeating the argument from the proof of Theorem 7.1 we can establish, with the aid of Proposition 7.9, the existence of two peak solutions of (7.2).

Theorem 7.2 *Suppose that (B) holds with $a^1 = 0$. Then there exists a constant $\lambda_o > 0$ such that for $\lambda \geq \lambda_o$ equation (7.2) admits a positive solution of the form*

$$u_\lambda(y) = \alpha_1(\lambda)U(y - x^1(\lambda)) + \alpha_2(\lambda)U(y - \lambda a^2 - x^2(\lambda)) + w_\lambda(y),$$

where

$$\alpha_i(\lambda) \to b(a^i)^{-\frac{1}{p-2}}, \quad i = 1,2,$$

$$x^i(\lambda) \to 0, \quad i = 1,2,$$

$$\|w_\lambda\|_{1,2} \to 0 \text{ in } H^1(\mathbb{R}_N), \quad w_\lambda \in E_{\lambda,x^1(\lambda),x^2(\lambda)}.$$

If condition (B) holds with $k \geq 2$, then we can construct a nodal solution (solution varying in sign). Indeed, it is sufficient to notice that constants α^i in Corollary 7.1 can be taken negative and in Proposition 7.3 $U_{\lambda z}$ can be replaced by $-U_{\lambda z}$. If we choose $\alpha_i \in \left(b(a^2)^{-\frac{1}{p-2}} - \rho, b(a^2)^{-\frac{1}{p-2}} + \rho\right)$ we can establish the existence of a solution of (7.2) of the form

$$u_\lambda(y) = \lambda^{\frac{2}{\lambda-2}}\left[\alpha_1(\lambda)U(\lambda(y-a^1-x^1(\lambda))-\alpha_2(\lambda)U(\lambda(y-a^2-x^2(\lambda))\right]+w_\lambda(y),$$

with

$$\alpha_i(\lambda) \to b(a^i)^{\frac{1}{p-2}},$$

$$x^i(\lambda) \to 0,\ i = 1, 2,$$

$$||w_\lambda||_{1,2} \to 0$$

as $\lambda \to \infty$.

The above remark and Theorem 7.2 show that we can construct at least $2^k - 1$ positive solutions with a number of peaks from 1 to k and nodal solutions with any specific number m of peaks, with $m = 2, 3, \ldots, k$.

We only formulate the existence resullt concerning positive solutions.

Theorem 7.3 *There exists $\lambda_\circ > 0$ such that for $\lambda \geq \lambda_\circ$ equation (7.1) has at least $2^k - 1$ positive solutions such that for any $l = 1, \ldots, k$, there exists $\frac{k(k-1)\ldots(k-l+1)}{l!}$ solutions of the form*

$$u_{J,\lambda}(y) = \lambda^{\frac{2}{p-2}}\left[\sum_{i=1}^{l} \alpha_{j_i} U\big(\lambda(y - a^{j_i} - x^{j_i}\lambda))\big) + w_{J,\lambda}(\lambda y)\right],$$

where $J = (j_1, \ldots, j_l)$, $j_i \in \{1, \ldots, k\}$, $j_i \neq j_m$, for $i \neq m$, $x^{j_i}(\lambda) \in \mathbb{R}_N$, $w_{J,\lambda} \in H^1(\mathbb{R}_N)$ and $x^{j_i}(\lambda) \to 0$, $\alpha_{j_i}(\lambda) \to b(a^{j_i})^{-\frac{1}{p-2}}$, $w_{J,\lambda} \to 0$ in $H^1(\mathbb{R}_N)$ as $\lambda \to \infty$.

7.5 Multi-peak solutions generated by local minima

In Section 7.4 we proved that k different critical points of the coefficient b allow us to construct a multi-peak solution, with k, peaks concentrating at the corresponding critical points of b. Here we study the question whether we can associate with a given critical point a multi-peak solution with peaks concentrating at that critical point. In general, this is not always possible in the case where $b(x)$ is a radial function. It is known that the equation

$$-\lambda^2 \Delta u + u = b(|x|)|u|^{p-2}u \quad \text{in} \, \mathbb{R}_N,$$

where $b(|x|)$ is a nonincreasing continuous function, has a radially symmetric and nonincreasing solution u. Therefore u has exactly one peak. However, a strict local minimum of b leads to the existence of a multi-peak solution. In this section we modify the construction from Section 7.1 - 7.4 to obtain a multi-peak solution for the equation

$$-\lambda^2 \Delta u + u = b(x)|u|^{p-2}u \quad \text{in } \mathbb{R}_N. \tag{7.62}$$

We assume that

(b_1) The coefficient $b(x)$ has a strict local minimum at a point $x_o \in \mathbb{R}_N$, that is $b(x) > b(x_o)$ for all $0 < |x - x_o| < \delta$ for some $\delta > 0$.

(b_2) There exist constants $C > 0$ and $\theta > 0$ such that

$$|b(x) - b(y)| < C|x - y|^\theta$$

for all $|x - x_o| \leq \delta$ and $|y - x_o| \leq \delta$.

As in Section 7.1 we assume that $x_o = 0$ and $b(0) = 1$. In the sequel we use the following notations

$$U_{\lambda,y}(\cdot) = U\left(\frac{\cdot - y}{\lambda}\right), \quad y \in \mathbb{R}_N,$$

$$\langle u, v\rangle_\lambda = \lambda^2 \int_{\mathbb{R}_N} (\nabla u \nabla v + uv)\, dx,$$

$$\|u\|_\lambda^2 = \langle u, v\rangle_\lambda \quad \text{for } u, v \in H^1(\mathbb{R}_N).$$

If $\lambda = 1$, we use notation $\|\cdot\|_{1,2}$. Let $I_\lambda : H^1(\mathbb{R}_N) \to \mathbb{R}$ be a functional given by

$$I_\lambda(u) = \frac{1}{2}\|u\|_\lambda^2 - \frac{1}{p}\int_{\mathbb{R}_N} b(x)|u|^p\, dx.$$

To use a local reduction method we introduce the following sets:

$$D^k_{\lambda,R,\delta} = \{(x^1, \ldots, x^k) \in \mathbb{R}_{kN},\, x^j \in \overline{B_\delta},\, j = 1, \ldots, k,\, \frac{|x^i - x^j|}{\lambda} \geq R \text{ for } i \neq j\},$$

$$E_{\lambda,x} = \{v \in H^1(\mathbb{R}_N); \langle v, U_{x^j,\lambda}\rangle_\lambda = \langle v, \frac{\partial U_{x^j,\lambda}}{\partial x_i^j}\rangle_\lambda = 0,\ j =, \ldots, k, i = 1, \ldots, N\}$$

$$M_{\lambda,R,\delta} = \{(\alpha^1, \ldots, \alpha^k, x^1, \ldots, x^k, v);\ (x^1, \ldots, x^k) \in D^k_{\lambda,R,\delta},$$
$$|\alpha^i - 1| \leq \delta,\ i = 1, \ldots, k,\ v \in E_{\lambda,k},\ \|v\|^2_\lambda \leq \delta\lambda^N\}.$$

The following two results are obvious modifications of Propositions 7.1 and 7.2.

Proposition 7.10 *There exist constants $R_\circ > 0$, $\delta_\circ > 0$ and $\lambda_\circ > 0$ such that for $\lambda \in (0, \lambda_\circ]$, $R \geq R_\circ$ and $\delta \in (0, \delta_\circ]$, a point $(\alpha^1, \ldots, \alpha^k, x^1, \ldots, x^k, v) \in M_{\lambda,R,\delta}$ is a critical point of J_λ if and only if $u = \sum_{j=1}^k \alpha^j U_{x^j,\lambda} + v$ is a crtical point of I_λ in $H^1(\mathbb{R}_N)$.*

Proposition 7.11 *There exist constants $\delta_\circ > 0$, $\lambda_\circ > 0$ and $R_\circ > 0$ such that for $\lambda \in (0, \lambda_\circ]$, $\delta \in (0, \delta_\circ]$ and $R \geq R_\circ$*

$$\|v\|^2_\lambda - (p-1)\int_{\mathbb{R}_N}\left(\sum_{j=1}^k U_{x^j,\lambda}\right)^{p-2} v^2\,dy \geq \rho\|v\|^2_\lambda$$

for all $(x^1, \ldots, x^k) \in D^k_{\lambda,R,\delta}$, $v \in E_{\lambda,k}$ and some constant $\rho > 0$.

As in Section 7.4 we observe that $(\alpha^1, \ldots, \alpha^k, x^1, \ldots x^k, v) \in M_{\lambda,R,\delta}$ is a critical point of J_λ if and only if there exist constants A_l, $B_{l,h}$, $l = 1, \ldots, k$, $h = 1, \ldots, N$ such that

$$\frac{\partial J_\lambda}{\partial x_i^j} = \sum_{h=1}^N B_{j,h}\langle \frac{\partial^2 U_{x^j,\lambda}}{\partial x_i^j \partial x_h^j}, v\rangle_\lambda, \tag{7.63}$$

$$\frac{\partial J_\lambda}{\partial \alpha^i} = 0, \tag{7.64}$$

$$\langle \frac{\partial J_\lambda}{\partial v}, \phi\rangle = \sum_{l=1}^k A_l\langle U_{x^l,\lambda}, \phi\rangle_\lambda + \sum_{l=1}^k\sum_{h=1}^N B_{l,h}\langle \frac{\partial U_{x^l,\lambda}}{\partial x_h^l}, \phi\rangle_\lambda. \tag{7.65}$$

The construction of a multi-peak solution will be carried out in two steps. First we show that for a given sequence of points $(x^1, \ldots, x^k)$ and $\lambda > 0$ sufficiently small there exist α^i, $i = 1, \ldots, k$, $v_\lambda \in E_{\lambda,x}$ and constants A_l, $A_{l,k}$, $l = 1, \ldots, k$, $h = 1, \ldots, N$ such that (7.64) and (7.65) are satisfied and that the mappings $(x^1, \ldots, x^k) \to \alpha^i(x^1, \ldots, x^k)$ and $(x^1, \ldots, x^k) \to$

$v(x^1,\dots,x^k)$ are of class C^1. In the second step we show that for $\lambda > 0$ sufficiently small there exists a sequence of points $(x^1,\dots,x^k) \in D^k_{\lambda,R,\delta}$ such that $(\alpha^1,\dots,\alpha^k,x^1,\dots,x^k) \in M_{\lambda,R,\delta}$ and satisfies (7.63).

7.6 Multi-peak solutions concentrating at one point

For the ease of notations we set $\alpha = (\alpha^1,\dots,\alpha^k) \in \mathbb{R}^k$, $x = (x^1,\dots,x^k) \in \mathbb{R}_{kN}$. We introduce inner products in $\mathbb{R}_k$ and $\mathbb{R}_k \times E_{\lambda,x}$ by

$$(\lambda,\gamma)_\lambda = \lambda^2 \sum_{j=1}^k \alpha^j \gamma^j$$

for $\alpha, \gamma \in \mathbb{R}_k$,

$$\big((\alpha,v),(\gamma,u)\big)\big)_\lambda = (\alpha,\gamma)_\lambda + (u,v)_\lambda$$

for (α,v) and $(\gamma,u) \in \mathbb{R}_k \times E_{\lambda,x}$. We also set

$$\|\alpha\|^2_\lambda = (\alpha,\alpha)_\lambda \text{ and } \|(\alpha,v)\|^2_\lambda = \big((\alpha,v),(\alpha,v)\big)_\lambda .$$

In order to construct C^1-mappings $(x^1,\dots,x^k) \to \alpha(x^1,\dots,x^k)$ and $(x^1,\dots,x^k) \to v(x^1,\dots,x^k)$ we need to expand J_λ around $(x,1,0)$. We shall use the following algebraic inequalities
(i) for $2 < p \leq 3$

$$\big||a+b|^p - a^p - b^p - pa^{p-1}b - pab^{p-1}\big| \leq \begin{cases} C|b|^{p-1}|a| & \text{if } |b| \leq |a|, \\ C|a|^{p-1}|b| & \text{if } |b| \leq |a| \end{cases} \leq |a|^{\frac{p}{2}}|b|^{\frac{p}{2}}$$

for some constant $C > 0$,
(ii) for $p > 3$

$$\big|a+b|^p - a^p - b^p - pa^{p-1}b - pab^{p-1}\big| \leq C\big(a^{p-2}b^2 + a^2b^{p-2}\big)$$

for some constant $C > 0$.

Lemma 7.3 *For every $x \in D_{\delta,R,\lambda}$ we have*

$$\begin{aligned} J_\lambda(x,1,0) &= \left(\frac{k}{2} - \frac{1}{p}\sum_{i=1}^k b(x^i)\right)\|U\|^2_{1,2}\lambda^N \\ &- \int_{\mathbb{R}_N} b(y) \sum_{i=1}^{k-1} U_{x^i,\lambda} \left(\sum_{j=i+1}^k U_{x^j,\lambda}\right)^{p-1} dy \end{aligned}$$

$$+ \quad O\Big(\lambda^N(\lambda^\theta + \sum_{i=1}^k |1-b(x^i)|^2 + \sum_{i\neq j} e^{-(1+\sigma)\frac{|x^i-x^j|}{\lambda}})\Big)$$

for some constant $\sigma > 0$.

Proof. Let $H_{\lambda,x} = \sum_{j=1}^k U_{x^j,\lambda}$. It is clear that

$$\begin{aligned} \|H_{\lambda,x}\|_\lambda^2 &= k\lambda^N \|U\|_{1,2}^2 + \sum_{i\neq j}(U_{x^i,\lambda}, U_{x^j,\lambda}) \\ &= k\lambda^N \|U\|_{1,2}^2 + 2\sum_{i<j}\int_{\mathbb{R}_N} U_{x^i,\lambda}^{p-1} U_{x^j,\lambda}\, dy. \end{aligned} \tag{7.66}$$

Applying (i) and (ii) we obtain

$$\begin{aligned} &\int_{\mathbb{R}_N} b(y) H_{\lambda,x}^p\, dy - \int_{\mathbb{R}_N} b(y)\sum_{j=1}^k U_{x^j,\lambda}^p\, dy \\ &= \int_{\mathbb{R}_N} b(y)\Big(\sum_{j=2}^k U_{x^j,\lambda}\Big)^p dy \\ &- \int_{\mathbb{R}_N} b(y)\sum_{j=1}^k U_{x^j,\lambda}^p\, dy + p\int_{\mathbb{R}_N} b(y) U_{x^1}^{p-1}\Big(\sum_{j=2}^k U_{x^j,\lambda}\Big) dy \\ &+ p\int_{\mathbb{R}_N} b(y) U_{x^1}\Big(\sum_{j=2}^k U_{x^j,\lambda}\Big)^{p-1} dy + O\Big(\lambda^N \sum_{i\neq j} e^{-(1+\sigma)\frac{|x^i-x^j|}{\lambda}}\Big). \end{aligned}$$

Repeated application of (i) and (ii) leads to the following estimate

$$\begin{aligned} \int_{\mathbb{R}_N} b(y) H_{\lambda,x}^p\, dy &= \int_{\mathbb{R}_N} b(y)\sum_{j=1}^k U_{x^j,\lambda}^p\, dy \\ &+ p\int_{\mathbb{R}_N} b(y)\sum_{i<j} U_{x^i,\lambda}^{p-1} U_{x^j,\lambda}\, dy \\ &+ p\int_{\mathbb{R}_N} b(y)\sum_{i=1}^{k-1} U_{x^i,\lambda}\Big(\sum_{j=i+1}^k U_{x^j,\lambda}\Big)^{p-1} dy \\ &+ O\Big(\lambda^N \sum_{i\neq j} e^{-(1+\sigma)\frac{|x^i-x^j|}{\lambda}}\Big). \end{aligned} \tag{7.67}$$

On the other hand using the assumption (b_2) we get

$$\int_{\mathbb{R}_N} b(y)U^p_{x^j,\lambda}\,dy = \int_{B(x^j,\delta)} b(y)U_{x^j,\lambda}\,dy \tag{7.68}$$
$$+O\big(\lambda^N e^{-\frac{\delta}{\lambda}}\big)$$
$$= b(x^j)\int_{B(x^i,\delta)} U^p_{x^j,\lambda}\,dy + O\big(\lambda^{N+\theta}\big)$$
$$= b(x^j)\|U\|^2_{1,2}\lambda^N + O\big(\lambda^{N+\theta}\big),$$

and

$$\int_{\mathbb{R}_N} b(y)U^{p-1}_{x^i,\lambda}U_{x^j,\lambda}\,dy = \int_{\mathbb{R}_N} b(y)U^{p-1}_{x^i,\lambda}U_{x^j,\lambda}\,dy \tag{7.69}$$
$$+O\big(\lambda^N e^{-\frac{(p-1)\delta}{\lambda}}\big) = b(x^i)\int_{B(x^i,\delta)} U^{p-1}_{x^i,\lambda}U_{x^j,\lambda}\,dy + O\big(\lambda^{N+\theta}\big)$$
$$= b(x^i)\int_{\mathbb{R}_N} U^{p-1}_{x^i,\lambda}U_{x^j,\lambda}\,dy + O\big(\lambda^{N+\theta}\big).$$

Inserting (7.66),(7.67),(7.68) and (7.69) into the formula for $J_\lambda(x,1,0)$ the result readily follows. □

Proposition 7.12 *For every integer $k \geq 1$ there exist constants $\lambda_o > 0$, $\delta_o > 0$ and $R_o > 0$ such that for $\lambda \in (0,\lambda_o]$, $R \geq R_o$ and $\delta \in (0,\delta_o]$ there exists a C^1-mapping such that*

$$(\alpha_\lambda, v_\lambda) : D_{\lambda,R,\delta} \to R_k \times E_{\lambda,x}$$

such that (7.64) and (7.65) hold. Moreover $(\alpha_\lambda, v_\lambda)$ satisfies the estimate

$$\lambda^{\frac{N}{2}}\sum_{j=1}^{k}|\alpha^i_2 - 1| \quad + \quad \|v_\lambda\|_\lambda \tag{7.70}$$
$$= \quad \lambda^{\frac{N}{2}}O\Big(\lambda^\theta + \sum_{i\neq j} e^{-\min\left(\frac{q-1}{2},1\right)\frac{|x^i-x^j|}{\lambda}} + \sum_{j=1}^{k}|b(x^j)-1|\Big).$$

Proof. Following Bahri [10] and Rey [96] (see also Yan - Noussair [85]) we expand J_λ in a neighbourhood of $(\alpha,v)=(1,0)$. We set $\beta^i = \alpha^i - 1$, $\beta = (\beta^1,\ldots,\beta^k)$ and $u = (\beta, v) \in \mathbb{R}_k \times E_{\lambda,x}$ and define

$$J^*_\lambda(x,u) = J_\lambda(\alpha,x,v).$$

Then

$$J_\lambda^*(x,u) = J_\lambda^*(x,0) + f_{\lambda,x}(u) + Q_{\lambda,x}(u) + R_{\lambda,x}(u), \tag{7.71}$$

where $u = (\beta, v)$ and

$$\begin{aligned} f_{\lambda,x}(u) &= -\int_{\mathbb{R}_N} b(y) H_{\lambda,x}^{p-1}(y) v(y)\, dy \qquad (7.72) \\ &+ \sum_{j=1}^{k} \left[\langle H_{\lambda,x}, U_{x^j,\lambda}\rangle_\lambda - \int_{\mathbb{R}_N} b(y) H_{\lambda,x}^{p-1} U_{x^j,\lambda} \right] \beta_j\, dy, \end{aligned}$$

$$\begin{aligned} Q_{\lambda,x}(u) &= Q_{\lambda,x}^1(\beta) + Q_{\lambda,x}^2(v) + Q_{\lambda,x}^3(\beta,v), \\ Q_{\lambda,x}^1(\beta) &= \sum_{i,j=1}^{k} \Big[\langle U_{x^i,\lambda}, U_{x^j,\lambda}\rangle_\lambda \\ &- (p-1) \int_{\mathbb{R}_N} b(y) \big(H_{\lambda,x}\big)^{p-2} U_{x^i,\lambda} U_{x^j,\lambda} \Big] \beta^i \beta^j\, dy, \\ Q_{\lambda,x}^2(v) &= \|v\|_\lambda^2 - (p-1) \int_{\mathbb{R}_N} b(y) \big(H_{\lambda,x}\big)^{p-2} v^2\, dy, \\ Q_{\lambda,x}^3(\beta,v) &= -\sum_{j=1}^{k} \int_{\mathbb{R}_N} b(y) \big(H_{\lambda,x}\big)^{p-2} U_{x^j,\lambda} v \beta^j\, dy. \end{aligned}$$

The higher order term $R_{\lambda,x}$ satisfies

$$R_{\lambda,x} = O\big(\|u_\lambda\|_\lambda^{\min(3,p)}\big), \quad R'_{\lambda,x}(u) = O\big(\|u\|_\lambda^{\min(2,p-1)}\big),$$

$$R''_{\lambda,x}(u) = O\big(\|u\|_\lambda^{\min(1,p-2)}\big).$$

As in the proof of Proposition 7.4 we see that there exist $f_{\lambda,x} \in \mathbb{R}_k \times E_{\lambda,x}$ and a continuous linear operator $A_{\lambda,x} : \mathbb{R}_k \times E_{\lambda,x} \to \mathbb{R}_k \times E_{\lambda,x}$ such that $f_{\lambda,x}(u) = (f_{\lambda,x}, u)_\lambda$ and

$$Q_{\lambda,x}(u) = (A_{\lambda,x} u, u) = Q_{\lambda,x}^1(\beta) + Q_{\lambda,x}^2(v) + Q_{\lambda,x}^3(\beta,v).$$

We now show that for $\lambda > 0$, $\delta > 0$ sufficiently small and $R > 0$ sufficiently large $A_{\lambda,x}$ has a bounded inverse. For $R_\circ > 0$ sufficiently large and $\delta_\circ > 0$

sufficiently small we have

$$
\begin{aligned}
Q^1_{\lambda,x}(\beta) &= \sum_{i=1}^{k}|\beta^i|^2\left[\|U_{x^i,\lambda}\|^2_\lambda-(p-1)\int_{\mathbb{R}_N} b(y)U^p_{x^i,\lambda}\,dy\right] \qquad (7.73)\\
&+ O(\lambda^N e^{-R})|\beta|^2-\lambda^N\Big((2-p)\|U\|^2_{1,0}+o_\delta(1)+O(e^{-R})\Big)|\beta|^2\\
&\le -C\lambda^N|\beta|^2=-C\|\beta\|^2_\lambda,
\end{aligned}
$$

where $C>0$ is a constant depending on N, $o_\delta(1)\to 0$ as $\delta\to 0$, and

$$
\begin{aligned}
|Q^3_{\lambda,}(\beta,v)| &\le \sum_{i=1}^{k}\Big|\int_{\mathbb{R}_N} b(y)H^{p-2}_{\lambda,x}U_{x^i,\lambda}v\,dy\Big||\beta_i| \qquad (7.74)\\
&= \sum_{i=1}^{k}\left[\int_{\mathbb{R}_N} b(y)U^{p-1}_{\lambda,x^i}v\,dy+O(\lambda^{\frac{N}{2}}e^{-R})\|v\|_\lambda\right]|\beta^i|\\
&= \sum_{i=1}^{k}\left[\int_{\mathbb{R}_N} U^{p-1}_{\lambda,x^i}v+o_\delta(1)\lambda^{\frac{N}{2}}\|v\|_\lambda+O(\lambda^{\frac{N}{2}}e^{-R})\|v\|_\lambda\right]|\beta^i|\\
&= \Big(O(e^{-R})+o(\delta)\|\beta\|_\lambda\Big)\|v\|_\lambda.
\end{aligned}
$$

It follows from these estimates that for $\lambda\in(0,\lambda_o]$, $R\ge R_o$ and $\delta\in(0,\delta_o]$ there exists a unique linear operator $B_{\lambda,x}:\mathbb{R}_k\times E_{\lambda,x}\to\mathbb{R}_k\times E_{\lambda,x}$ such that

$$(B_{\lambda,x}u,u)=Q^1_{\lambda,x}(\beta)+Q^2_{\lambda,x}(v).$$

According to Proposition 7.11 and estimate (7.73) $B_{\lambda,x}$ is invertible and $\|B^{-1}_{\lambda,x}\|\le C$ for some constant $C>0$ independent of λ and x. It now follows from (7.74) that

$$\|A_{\lambda,x}-B_{\lambda,x}\|=\|Q^3_{\lambda,x}\|\le O(e^{-R})+o_\delta(1).$$

Thus, we can choose constants $R_o>0$, δ_o and $\lambda_o>0$ such that for $\lambda\in(0,\lambda_o]$, $R\ge R_o$ and $\delta\in(0,\delta_o]$, $A_{\lambda,x}$ has an inverse and $\|A^{-1}_{\lambda,x}\|\le C$ for some constant $C>0$. Since

$$\frac{\partial J^*_\lambda}{\partial u}|_{\mathbb{R}_k\times E_{\lambda,x}}(x,u)=f_{\lambda,x}+2A_{\lambda,x}u+R'_{\lambda,x}(u),$$

we see that equations (7.64) and (7.65) are equivalent to the equation

$$f_{\lambda,x}+2A_{\lambda,x}u+R'_{\lambda,x}(u)=0. \qquad (7.75)$$

As in Proposition 7.4 we use the implicit function theorem to obtain the existence of C^1-mapping

$$u_\lambda = (\alpha_\lambda, v_\lambda) : D_{\lambda,R,\delta} \to R_k \times E_{\lambda,x}$$

for $\lambda \in (0, \lambda_\circ]$, $R \geq R_\circ$, $\delta \in (0, \delta_\circ]$ for some $\lambda_\circ > 0$, $R_\circ > 0$ and $\delta_\circ > 0$, satisfying (7.75) and

$$\|u_\lambda\| \leq C\|f_{\lambda,x}\| \tag{7.76}$$

for some $C > 0$. We now estimate $\|f_{\lambda,x}\|$. Towards this end we show that

$$\begin{aligned}\int_{\mathbb{R}_N} b(y) H_{\lambda,x}^{p-1} v\, dy &= \sum_{i=1}^k \int_{\mathbb{R}_N} b(y) U_{x^i,\lambda}^{p-1} v\, dy \\ &+ O\big(\lambda^{\frac{N}{2}} \sum_{i\neq j} e^{-\min(\frac{p-1}{2},1)\frac{|x^i-x^j|}{\lambda}}\big)\|v\|_\lambda.\end{aligned} \tag{7.77}$$

Indeed, it follows from (i), if $2 < p < 3$, that

$$\begin{aligned}\left|\int_{\mathbb{R}_N} b(y)\left(H_{\lambda,x}^{p-1} - \sum_{i=1}^k U_{x^i,\lambda}^{p-1}\right) v\, dy\right| &\leq C\int_{\mathbb{R}_N} \sum_{i\neq j} U_{x^i,\lambda}^{\frac{p-1}{2}} U_{x^i,\lambda}^{\frac{p-1}{2}} |v|\, dy \\ &\leq C\lambda^{\frac{N}{2}}\|v\|_\lambda \left(\int_{\mathbb{R}_N} \sum_{i\neq j} U^{\frac{p-1}{2}} U_{1,\frac{|x^i-x^j|}{\lambda}}^{\frac{p-1}{2}}\, dy\right)^{\frac{p-1}{p}} \\ &= O\big(\lambda^{\frac{N}{2}} \sum_{i\neq j} e^{-(p-1)\frac{|x^i-x^j|}{2\lambda}}\big)\|v\|_\lambda\end{aligned}$$

and, if $p > 3$, that

$$\begin{aligned}&\left|\int_{\mathbb{R}_N} \left(H_{\lambda,x}^{p-1} - \sum_{i=1}^k U_{x^i,\lambda}\right) v\, dy\right| \\ &\leq C\sum_{i\neq j}\int_{\mathbb{R}_N} U_{x^i,\lambda}^{p-2} U_{x^j,\lambda}|v|\, dy = O\left(\lambda^{\frac{N}{2}} \sum_{i\neq j} e^{-\frac{|x^i-x^j|}{\lambda}}\right)\|v\|_\lambda.\end{aligned}$$

The relation (7.77) follows from the last two estimates. We now estimate the integral on the right hand side of (7.77). Applying (b_2) we obtain

$$\int_{\mathbb{R}_N} b(y) U_{x^i,\lambda}^{p-1} v\, dy = \int_{B(x^i,\delta)} U_{x^i,\lambda}^{p-1} v\, dy + O\big(\int_{|y-x^i|\geq\delta} U_{x^i,\lambda}^{p-1}|v|\, dy\big)$$

$$
\begin{aligned}
&= b(x^i)\int_{B(x^i,\delta)} U_{x^i,\lambda}^{p-1} v\,dy + O\Big(\int_{B(x^i,\delta)} |y-x^i|^\theta U_{x^i,\lambda}^{p-1}|v|\,dy\Big)\\
&+ O\Big(\int_{|y-x^i|\ge\delta} U_{x^i,\lambda}^{p-1}|v|\,dy\Big)\\
&= O\Big(\int_{B(x^i,\delta)} |y-x^i|^\theta U_{x^i,\lambda}^{p-1}|v|\,dy\Big) + O\Big(\int_{|y-x^i|\ge\delta} U_{x^i,\lambda}^{p-1}|v|\,dy\Big)\\
&= O\Big(\lambda^{N+\theta}\int_{\mathbb{R}_N} |y|^\theta U^{p-1}|v(\lambda y + x^i)|\,dy\Big)\\
&+ O\Big(\lambda^{N}\int_{|y|>\frac{\delta}{\epsilon}} U^{p-1}|v(\lambda y + x^i)|\,dy\Big)\\
&= O\big(\lambda^{\frac{N}{2}+\theta} + \lambda^{\frac{N}{2}}e^{-\frac{\delta}{\epsilon}}\big)\|v\|_\lambda = O\big(\lambda^{\frac{N}{2}+\theta}\big)\|v\|_\lambda.
\end{aligned}
$$

This combined with (7.77) gives

$$
\int_{\mathbb{R}_N} b(y)H_{\lambda,x}^{p-1}\,dy = O\big(\lambda^{\frac{N}{2}+\theta} + \lambda^{\frac{N}{2}}\sum_{i\neq j} e^{-\min(\frac{p-1}{2},1)\frac{|x^i-x^j|}{\lambda}}\big)\|v\|_\lambda. \tag{7.78}
$$

We also have using (b_2)

$$
\begin{aligned}
\int_{\mathbb{R}_N} b(y)H_{\lambda,x}^{p-1}U_{x^i,\lambda}\,dy &= \int_{B(x^i,\delta)} b(y)H_{\lambda,x}^{p-1}U_{x^i,\lambda}\,dy\\
&+O\Big(\lambda^N\big(\int_{|y|\ge\frac{\delta}{\epsilon}} U^p\,dy\big)^{\frac{1}{p}}\Big)\\
&= b(x^i)\int_{B(x^i,\delta)} H_{\lambda,x}^{p-1}U_{x^i,\lambda}\,dy\\
&+O\big(\int_{B(x^i,\delta)} |y-x^i|^\theta H_{\lambda,x}^{p-1}U_{x^i,\lambda}\,dy\big) + O\big(\lambda^N e^{-\frac{\delta}{\lambda}}\big)\\
&= b(x^i)\int_{B(x^i,\delta)} U_{x^i,\lambda}^{p}\,dy + O\big(\lambda^{N+\theta} + \lambda^N\sum_{i\neq j} e^{-\frac{x^i-x^j|}{\epsilon}}\big)\\
&= b(x^i)\lambda^N\|U\|_{1,2}^2 + O\big(\lambda^{N+\theta} + \lambda^N\sum_{i\neq j} e^{-\frac{|x^i-x^j|}{\lambda}}\big).
\end{aligned}
$$

This estimate leads

$$
\langle H_{\lambda,x}, U_{x^i,\lambda}\rangle_\lambda \quad - \quad \int_{\mathbb{R}_N} b(y)H_{\lambda,x}^{p-1}U_{x^i,\lambda}\,dy \tag{7.79}
$$

$$\begin{aligned} &= \lambda^N\Big(\|U\|_{1,2}^2 + O(\sum_{i\neq j} e^{-\frac{-|x^i-x^j|}{\lambda}})\Big) \\ &- b(x^i)\lambda^N\|U\|_{1,2}^2 + O(\lambda^{N+\theta} + \lambda^N\sum_{i\neq j} e^{-\frac{|x^i-x^j|}{\lambda}}) \\ &= O(\lambda^\theta + \sum_{i=1}^k |1-b(x^i)| + \sum_{i\neq j} e^{-\frac{|x^i-x^j|}{\lambda}})\lambda^N. \end{aligned}$$

Inserting (7.78) and (7.79) into (7.72) we obtain

$$|(f_{\lambda,x},u)_\lambda| = \lambda^{\frac{N}{2}} O\Big(\lambda^\theta + \sum_{i=1}^k |1-b(x^i)| + \sum_{i\neq j} e^{-\min(\frac{p-1}{2},1)\frac{|x^i-x^j|}{\lambda}}\Big)\|v\|_\lambda$$

which implies that

$$\|f_{\lambda,x}\|_\lambda = \lambda^{\frac{N}{2}} O\Big(\lambda^\theta + \sum_{i\neq j} e^{-\max(\frac{p-1}{2},1)\frac{|x^i-x^j|}{\lambda}} + \sum_{i=1}^k |b(x^i)-1|\Big).$$

This combined with (7.76) gives the second part of the assertion of our Proposition. □

Let $\big(\alpha_\lambda(x), v_\lambda(x)\big)$ be a C^1-mapping from Proposition 7.12. We define

$$\tilde{J}_\lambda(x) = J_\lambda(\alpha_\lambda(x), x, v_\lambda(x))$$

for $x \in D_{\lambda,R,\delta}$, $\lambda \in (0,\lambda_o]$, $\delta \in (0,\delta_o]$ and $R \geq R_o$.

Let $x_\lambda = (x_\lambda^1, \ldots, x_\lambda^k) \in D_{\lambda,R,\delta}$ be such that

$$\tilde{J}_\lambda(x_\lambda) = \max\{\tilde{J}_\lambda(x);\ x \in D_{\lambda,R,\delta}\}. \tag{7.80}$$

We show below that x_λ for λ sufficiently small is an interior point of $D_{\lambda,R,\delta}$ and it is therefore a critical point of $\tilde{J}_\lambda$.

Proposition 7.13 *Let x_λ satisfy (7.80). Then*

$$x_\lambda^i \to 0, \quad i = 1, \ldots, k,$$

$$\frac{|x_\lambda^i - x_\lambda^j|}{\lambda} \to \infty, \quad i \neq j,$$

as $\lambda \to \infty$.

Proof. Expanding J_λ^* as in Proposition 7.11 and observing that $J_\lambda^*(x,0) = J_\lambda(x,1,0)$ we derive from Lemma 7.3 that

$$\begin{aligned}
J_\lambda(\alpha_\lambda(x_\lambda), x_\lambda, v_\lambda(x_\lambda)) &= J_\lambda^*(x_\lambda, u_\lambda) \qquad (7.81)\\
&= J_\lambda^*(x_\lambda, 0) + O(\|f_{\lambda,x_\lambda}\|_\lambda^2 + \|u\|_\lambda^2)\\
&= J_\lambda^*(x_\lambda, 0) + O\Big(\lambda^N(\lambda^{2\theta} + \sum_{i\neq j} e^{-\min(p-1,2)\frac{|x^i-x^j|}{\lambda}} + \sum_{i=1}^{k}|b(x^i)-1|^2)\Big)\\
&= \Big(\frac{k}{2} - \frac{1}{p}\sum_{j=1}^{k} b(x^j)\Big)\|U\|_{1,2}^2\lambda^N - \int_{\mathbb{R}^N} b(y)\sum_{j=1}^{k-1} U_{x^i,\lambda}\Big(\sum_{j=i+1}^{k} U_{x^j,\lambda}\Big)^{p-1} dy\\
&\quad +O\Big(\lambda^N(\lambda^{2\theta} + \sum_{j=1}^{k}|1-b(x^j)|^2 + \sum_{i\neq j} e^{-\min(p-1,2)\frac{|x^i-x^j|}{\lambda}})\Big).
\end{aligned}$$

Let $z_\lambda^i = \lambda^\tau e_i$, $i = 1,\ldots,k$, for some $\tau \in (\frac{1}{2}, 1)$ and some vectors $e_1, \ldots, e_k$ with $e_i \neq e_j$ for $i \neq j$. Then

$$\frac{|z_\lambda^i - z_\lambda^j|}{\lambda} = \frac{|e_i - e_j|}{\lambda^{1-\tau}} \to \infty \quad \text{as } \lambda \to 0.$$

Therefore for sufficiently small λ, $z_\lambda = (z_\lambda^1, \ldots, z_\lambda^k) \in D_{\lambda,R,\delta}$. It then follows from (7.80) and (7.81) that

$$\begin{aligned}
J_\lambda(\alpha_\lambda(x_\lambda), x_\lambda, v_\lambda(x_\lambda)) &\geq J_\lambda(\alpha_\lambda(z_\lambda), z_\lambda, v_\lambda(z_\lambda)) \qquad (7.82)\\
&\geq (\frac{k}{2} - \frac{1}{p}\sum_{i=1}^{k} b(z_\lambda^i))\|U\|_{1,2}^2\lambda^N\\
&\quad +O\Big(\lambda^N(\lambda^{2\theta} + \sum_{i=1}^{k} b(z_\lambda^i))\Big)\|U\|_{1,2}^2\lambda^N\\
&\quad +O\Big(\lambda^N(\lambda^{2\theta} + \sum_{j=1}^{k}|1-b(z_\lambda^i)|^2 + \sum_{i\neq j} e^{-(1+\sigma)\frac{|e_i-e_j|}{\lambda^{1-\tau}}})\Big)\\
&= (\frac{1}{2} - \frac{1}{p})k\|U\|_{1,2}^2\lambda^{2N} + O(\lambda^{N+\theta_1}),
\end{aligned}$$

where $\theta_1 = \min(\tau, 2\theta)$ and $1+\sigma = \min(p-1, 2)$. It then follows from (7.81) and (7.82) that

$$(\frac{k}{2} - \frac{1}{p}\sum_{i=1}^{k} b(x_\lambda^i))\|U\|_{1,2}^2\lambda^N \qquad (7.83)$$

$$
\begin{aligned}
&-\int_{\mathbb{R}_N} b(y)\sum_{i=1}^{k-1} U_{x_\lambda^i,\lambda}\Big(\sum_{j=i+1}^{k} U_{x_\lambda^j,\lambda}\Big)^{p-1}dy\\
&+O\Big(\lambda^N\big(\lambda^{2\theta}+\sum_{i=1}^{k}|1-b(x_\lambda^i)|^2+\sum_{i\neq j}e^{-(1+\sigma)\frac{|x_\lambda^i-x_\lambda^j|}{\lambda}}\big)\Big)\\
&\geq (1-\frac{1}{p})k\|U\|_{1,2}^2\lambda^N+O(\lambda^{N+\theta_1}),
\end{aligned}
$$

which yields

$$
\begin{aligned}
&\frac{1}{p}\sum_{i=1}^{k}\big(b(x_\lambda^i)-1\big)\|U\|_{1,2}^2\lambda^N+\int_{\mathbb{R}_N} b(y)\sum_{i=1}^{k} U_{x_\lambda^i,\lambda}\Big(\sum_{j=i+1}^{k} U_{x_\lambda^j,\lambda}\Big)^{p-1}dy\\
&\quad\leq O\Big(\lambda^N\big(\lambda^{2\theta}+\sum_{i=1}^{k}|1-b(x_\lambda^i)|^2+\sum_{i\neq j}e^{-(1+\lambda)\frac{|x_\lambda^i-x_\lambda^j|}{\lambda}}\big)\Big).
\end{aligned}
$$

On the other hand we have the following estimates

$$
\begin{aligned}
&\int_{\mathbb{R}_N} b(y)\sum_{i=1}^{k-1} U_{x_\lambda^i,\lambda}\Big(\sum_{j=i+1}^{k} U_{x_\lambda^i,\lambda}\Big)^{p-1}dy\\
&\quad\geq \int_{\mathbb{R}_N}\sum_{i=1}^{k-1} U_{x_\lambda^i,\lambda}\Big(\sum_{j=i+1}^{k} U_{x_\lambda^j,\lambda}\Big)^{p-1}dy\\
&\quad+\int_{|y|\leq\delta}\big(b(y)-1\big)\sum_{i=1}^{k-1} U_{x_\lambda^i,\lambda}\Big(\sum_{j=i+1}^{k} U_{x_\lambda^j,\lambda}\Big)^{p-1}dy\\
&\quad+\int_{|y|\geq\delta}\big(b(y)-1\big)\sum_{i=1}^{k-1} U_{x_\lambda^j,\lambda}\Big)^{p-1}\Big(\sum_{j=i+1}^{k} U_{x_\lambda^j,\lambda}\Big)^{p-1}dy
\end{aligned}
$$

and

$$
\begin{aligned}
&\left|\int_{|y|\geq\delta}\big(b(y)-1\big)\sum_{i=1}^{k-1} U_{x_\lambda^i,\lambda}\Big(\sum_{j=i+1}^{k} U_{x_\lambda^j,\lambda}\Big)^{p-1}dy\right|\\
&\quad\leq C\lambda^N\sum_{j\neq i}\int_{|y|\geq\frac{\delta}{\lambda}} U(y)U\big(y-\frac{(x_\lambda^j-x_\lambda^i)}{\lambda}\big)^{p-1}dy\\
&\quad\leq C\lambda^N\sum_{i\neq j}\int_{|y|\geq\frac{\delta}{\lambda}} e^{-|y|}e^{-(p-1)\left|y-\frac{(x_\lambda^i-x_\lambda^j)}{\lambda}\right|}dy=O(\lambda^N e^{-\frac{\delta}{\lambda}}).
\end{aligned}
$$

The last two estimates and assumption (b_1) imply that

$$\int_{\mathbb{R}_N} b(y)\sum_{i=1}^{k-1} U_{x^i_\lambda,\lambda}\Big(\sum_{j=i+1}^{k} U_{x^i_\lambda,\lambda}\Big)^{p-1} dy \tag{7.84}$$
$$\geq C\sum_{i\neq j} e^{-(1+\frac{\sigma}{2})\frac{|x^i_\lambda - x^j_\lambda|}{\lambda}}\lambda^N + O\big(\lambda^N e^{-\frac{\epsilon}{\delta}}\big),$$

where $C > 0$ is a constant. Combining (7.83) and (7.84) we deduce that

$$\lambda^N\sum_{i=1}^{k}\big(b(x^i_\lambda) - 1\big) + C\sum_{i\neq j} e^{-(1+\frac{\sigma}{2})\frac{|x^i_\lambda - x^j_\lambda|}{\lambda}}\lambda^N \leq O\big(\lambda^{N+2\theta}\big)$$

and consequently

$$b(x^i_\lambda) \to 1 = b(0), \quad x^i_\lambda \to 0$$

and

$$\frac{|x^i_\lambda - x^j_\lambda|}{\lambda} \to \infty, \quad i \neq j,$$

as $\lambda \to 0$. □

We are now in a position to establish the existence of a multi-peak solution concentrating at a local minimum of b.

Theorem 7.4 *Suppose that (b_1) and (b_2) hold. Then, for each $k = 1, 2, \ldots$, there exists $\lambda_\circ = \lambda_\circ(k)$ such that for each $\lambda \in (0, \lambda_\circ)$ problem (7.62) has a multi-peak solution u_ϵ of the form*

$$u_\lambda = \sum_{j=1}^{k} \alpha^j_\lambda U_{x^j_\lambda,\lambda} + v_\lambda$$

for some constants $\alpha^j_\lambda > 0$, points $x^j_\lambda \in \mathbb{R}_N$, $j = 1, \ldots, k$ and $H^1(\mathbb{R}_N)$, satisfying

$$x^j_\lambda \to x_\circ, \quad \frac{|x^i_\lambda - x^j_\lambda|}{\lambda} \to \infty, \quad i \neq j,$$

$$\alpha^j_\lambda \to b(x_\circ)^{-\frac{1}{p-2}}, \quad \|v_\lambda\|_\lambda = o\big(\lambda^{\frac{N}{2}}\big)$$

as $\lambda \to 0$.

Proof. By Proposition 7.12, for each k, there exist λ_o, R_o, δ_o and a C^1-mapping

$$(\alpha_\lambda(x_\lambda), v_\lambda(x_\lambda)) : D_{\lambda,R,\delta} \to \mathbb{R}_k \times E_{\lambda,x}$$

for each $\lambda \in (0, \lambda_o]$, $R \geq R_o$, $\delta \in (0, \delta_o]$ such that (7.64) and (7.65) hold. According to Proposition 7.13 we can choose x_λ so that $x_\lambda^j \to 0$, $\frac{|x_\lambda^i - x_\lambda^j|}{\lambda} \to \infty$ as $\lambda \to 0$, $i, j = 1, \ldots, N$ and $i \neq j$ and $\frac{\partial \tilde{J}_\lambda(x_\lambda)}{\partial x_i^j} = 0$. Since (7.64) and (7.65) hold we have

$$\begin{aligned} 0 &= \sum_{l=1}^{k} \frac{\partial J_\lambda}{\partial \alpha^l} \frac{\partial \alpha^l}{\partial x_l^j} + \frac{\partial J_\lambda}{\partial x_i^j} + \langle \frac{\partial J_\lambda}{\partial v}, \frac{\partial v_\lambda}{\partial x_i^j} \rangle_\lambda \qquad (7.85) \\ &= \frac{\partial J_\lambda}{\partial x_i^j} + \sum_{l=1}^{k} A_l \langle U_{x_\lambda^l}, \frac{\partial v_\lambda}{\partial x_i^j} \rangle_\lambda \\ &+ \sum_{l=1}^{k} \sum_{h=1}^{N} B_{l,h} \langle \frac{\partial U_{x_\lambda^l,\lambda}}{\partial x_k^l}, \frac{\partial v_\lambda}{\partial x_i^j} \rangle_\lambda . \end{aligned}$$

Since $v_\lambda \in E_{\lambda,x}$, integrating by parts we get $\langle U_{x_\lambda^l,\lambda}, \frac{\partial v_\lambda}{\partial x_i^j} \rangle_\lambda = 0$. It therefore follows from (7.85) that

$$\frac{\partial J_\lambda}{\partial x_i^j} = \sum_{h=1}^{N} B_{jh} \langle \frac{\partial^2 U_{x_\lambda^l,\lambda}}{\partial x_i^j, \partial x_h^j}, v_\lambda \rangle_\lambda ,$$

that is, (7.63) holds. □

7.7 Bibliographical notes

The results of this chapter are based on the works of Cao-Noussair-Yan [27] and Noussair-Yan [85] which use some ideas from the articles of Rey [96] and Bahri-Coron [6].

Chapter 8

Multiple positive and nodal solutions

In this chapter we investigate the effect of the shape of the graph of a coefficient b of the equation (2.3) in $\mathbb{R}_N$ on the number of positive and nodal solutions. We only assume that b is continuous, nonnegative and bounded on $\mathbb{R}_N$. The main objective of this chapter is to show that each absolute maximum of the coefficient b generates at least one nodal and one positive solution. These solutions are obtained using Ekeland's variational principle.

8.1 Localized minima

We consider the equation (2.6) in $\mathbb{R}_N$, that is,

$$-\Delta u + \lambda u = b(x)|u|^{p-2}u \text{ in } \mathbb{R}_N. \tag{8.1}$$

By the change of variables $\mu = \frac{1}{\sqrt{\lambda}}$, $v(x) = \mu^{\frac{2}{p-2}}u(\mu x)$, equation (8.1) is transformed to

$$-\Delta v + v = b_\mu(x)|v|^{p-2}v \text{ in } \mathbb{R}_N, \tag{8.2}$$

where $b_\mu(x) = b(\mu x)$. Throughout this chapter it is assumed that b is a continuous, nonnegative and bounded function on $\mathbb{R}_N$ and has k strict maxima attained at points $a^1, \ldots, a^k$ such that

$$b_* = b(a^j) = \max\{b(x);\ x \in \mathbb{R}_N\}. \tag{8.3}$$

For $l > 0$ let

$$Q_l(a^j) = \prod_{i=1}^{N}(a_i^j - l, a_i^j + l).$$

We choose numbers $K > 0$ and $l > 0$ so that $b(x) < b_*$ for $x \in \partial Q_l(a^j)$ and $\bar{Q}_l(a^j)$ are disjoint and $\bigcup_{j=1}^k Q_l(a^j) \subset (-K, K)^N$.

Let a be a continuous and bounded function and let $c \in \mathbb{R}$. We set

$$G_a(u) = \frac{1}{2}\int_{\mathbb{R}_N}(|\nabla u|^2 + u^2)\,dx - \frac{1}{p}\int_{\mathbb{R}_N} a(x)|u|^p\,dx,$$

$$\mathcal{N}_a^c = \{u \in H^1(\mathbb{R}_N) - \{0\};\ \langle G_a{}'(u), u\rangle = c\}$$

and

$$M_a^c = \inf\{G_a(u);\ u \in \mathcal{N}_a^c\}.$$

If $c = 0$, we abbreviate $M_a = M_a^0$ and $\mathcal{N}_a = \mathcal{N}_a^0$.

Let $\phi_\mu : \mathbb{R} \to \mathbb{R}$ be given by

$$\phi_\mu(t) = \begin{cases} \frac{2K}{\mu} & \text{for } t > \frac{2K}{\mu} \\ t & \text{for } -\frac{2K}{\mu} \le t \le \frac{2K}{\mu} \\ -\frac{2K}{\mu} & \text{for } t < -\frac{2K}{\mu}. \end{cases}$$

We define continuous functionals $g_\mu^i : H^1(\mathbb{R}_N) - \{0\} \to \mathbb{R}$, $i = 1, \dots, N$, by

$$g_\mu^i(u) = \frac{\int_{\mathbb{R}_N}\phi_\mu(x^i)|u|^p\,dx}{\int_{\mathbb{R}_N}|u|^p\,dx}$$

and set $g_\mu(u) = \big(g_\mu^1(u), \dots, g_\mu^N(u)\big)$. We now define the following subsets of $\mathcal{N}_{b_\mu}$:

$$N_\mu^j = \{u \in H^1(\mathbb{R}_N);\ u \in \mathcal{N}_{b_\mu},\ g_\mu(u) \in Q_{\frac{l}{\mu}}^j\},$$

$$O_\mu^j = \{u \in H^1(\mathbb{R}_N);\ u \in \mathcal{N}_{b_\mu},\ g_\mu(u) \in \partial Q_{\frac{l}{\mu}}^j\},$$

$$R_\mu^j = \{u \in H^1(\mathbb{R}_N);\ u^\pm \in N_\mu^j\}$$

and

$$\partial R_\mu^j = \{u \in H^1(\mathbb{R}_N);\ u^\pm \in N_\mu^j \cup O_\mu^j \text{ and } u^+ \text{ or } u^- \in O_\mu^j\}.$$

We use these sets to define the following constrained minimization problems

$$J^j_\mu = \inf\{G_{b_\mu}(u);\ u \in R^j_\mu\},$$

$$\bar{J}^j_\mu = \inf\{G_{b_\mu}(u);\ u \in \partial R^j_\mu\}$$

$$I^j_\mu = \inf\{G_{b_\mu}(u);\ u \in N^j_\mu\}$$

and

$$\bar{I}^j_\mu = \inf\{G_{b_\mu}(u);\ u \in O^j_\mu\}.$$

We first derive basic properties of these quantities.

Lemma 8.1 *Suppose that a is a continuous bounded and nonnegative function on $\mathbb{R}_N$. Then*

$$M^c_a = \frac{c}{2} \text{ for } c > 0$$

and

$$M_a \leq M^c_a + M^{-c}_a - \frac{p-2}{2p}|c| \text{ for every } c \in \mathbb{R}.$$

Proof. If $u \in M^c_a$ and $c > 0$, then

$$\int_{\mathbb{R}_N} (|\nabla u|^2 + u^2)\, dx = \int_{\mathbb{R}_N} a|u|^p\, dx + c \geq c$$

and

$$\begin{aligned} G_a(u) &= \frac{p-2}{2p}\int_{\mathbb{R}_N}(|\nabla u|^2 + u^2)\, dx + \frac{c}{p} \\ &\geq \frac{p-2}{2p}c + \frac{c}{p} = \frac{c}{2}. \end{aligned}$$

To establish the equality let $v \in H^1(\mathbb{R}_N)$ with $\int_{\mathbb{R}_N} |\nabla v|^2\, dx = c$ and set $u_\sigma(x) = \sigma^{\frac{N-2}{2}} v(\sigma x)$, $w_\sigma(x) = (1+\delta)u_\sigma(x)$. We now observe that

$$\int_{\mathbb{R}_N} |u_\sigma|^q\, dx = \sigma^{\frac{(N-2)q}{2} - N} \int_{\mathbb{R}_N} |v|^q\, dx \to 0$$

as $\sigma \to \infty$ for every $q \in (2, 2^*)$. It then follows from the intermediate value theorem that there exists $\delta = \delta(\sigma)$, with $\delta(\sigma) \to 0$ as $\sigma \to \infty$, such that

$w_\sigma \in \mathcal{N}_a^c$. Hence

$$G_a(w_\sigma) = \frac{1}{2}\int_{\mathbb{R}_N} (|\nabla w_\sigma|^2 + w_\sigma^2)\,dx - \frac{1}{p}\int_{\mathbb{R}_N} a|w_\sigma|^p\,dx \to \frac{c}{2}$$

as $\sigma \to \infty$ and the assertion follows. To show the second part let $c > 0$ and $u \in \mathcal{N}_a^{-c}$. Then

$$\int_{\mathbb{R}_N} (|\nabla u|^2 + u^2)\,dx = \int_{\mathbb{R}_N} a|u|^p\,dx - c < \int_{\mathbb{R}_N} a|u|^p\,dx.$$

It is easy to check that there exists $t \in (0,1)$ such that $v = tu \in \mathcal{N}_a$. Therefore

$$\begin{aligned} G_a(v) &= (\frac{1}{2} - \frac{1}{p})\int_{\mathbb{R}_N} (|\nabla v|^2 + v^2)\,dx \\ &= (\frac{1}{2} - \frac{1}{p})t^2\int_{\mathbb{R}_N} (|\nabla u|^2 + u^2)\,dx \\ &< (\frac{1}{2} - \frac{1}{p})\int_{\mathbb{R}_N} (|\nabla u|^2 + u^2)\,dx \\ &= G_a(u) + \frac{c}{2} + (\frac{1}{p} - \frac{1}{2})c = G_a(u) + M_a^c - \frac{p-2}{2p}c. \end{aligned}$$

The assertion follows by taking the infimum over $\mathcal{N}_a^{-c}$. □

Lemma 8.2 *For every $\epsilon > 0$ there exists $\mu_\epsilon > 0$ such that*

$$J_\mu^j < 2M_{b_*} + \epsilon \tag{8.4}$$

and

$$I_\mu^j < M_{b_*} + \epsilon \tag{8.5}$$

for $j = 1, \ldots, k$ and $\mu \in (0, \mu_\epsilon)$.

Proof. To prove (8.4) it is sufficient to construct functions $w_\mu \in R_\mu^j$, $\mu > 0$, such that

$$G_{b_\mu}(w_\mu) \to 2M_{b_*} \text{ as } \mu \to 0.$$

Let $u \in H^1(\mathbb{R}_N)$ be a positive (radially symmetric) solution of the equation

$$-\Delta u + u = b_*|u|^{p-2}u \text{ in } \mathbb{R}_N. \tag{8.6}$$

For $\mu > 0$ satisfying $2\sqrt{\mu} < 1$ we define a C^1 - function $\psi_\mu : \mathbb{R}_N \to [0,1]$ such that $\psi_\mu(x) = 1$ for $|x| < \frac{1}{2\sqrt{\mu}} - 1$, $\psi_\mu(x) = 0$ for $|x| > \frac{1}{2\sqrt{\mu}}$ and $|\nabla\psi_\mu| \le 2$ on $\mathbb{R}_N$. Let $x^\mu = \frac{1}{2\sqrt{\mu}}(1,1,\ldots,1)$ and set

$$w_\mu(x) = t_\mu^+ u\big(x - \frac{a^j}{\mu} + x^\mu\big)\psi_\mu\big(x - \frac{a^j}{\mu} + x^\mu\big) - t_\mu^- u\big(x - \frac{a^j}{\mu} - x^\mu\big)\psi\big(x - \frac{a^j}{\mu} - x^\mu\big),$$

where $t_\mu^\pm > 0$ are such that $w_\mu^\pm \in \mathcal{N}_{b_\mu}$. This requires that

$$(t_\mu^\pm)^{p-2} = \frac{\|u(\cdot - \frac{a^j}{\mu} \pm x^\mu)\psi(\cdot - \frac{a^j}{\mu} \pm x^\mu)\|_{1,2}^2}{\int_{\mathbb{R}_N} b(\mu x)|u(x - \frac{a^j}{\mu} \pm x^\mu)\psi_\mu(x - \frac{a^j}{\mu} \pm x^\mu)|^p\,dx}.$$

It is easy to see that $t_\mu^\pm \to 1$ as $\mu \to 0$. We now check that $g_\mu(w_\mu^\pm) \in Q^j_{\frac{l}{\mu}}$. Since $\psi_\mu\big(x - \frac{a^j}{\mu} \pm x^\mu\big) = 0$ for $|x_i - \frac{a_i^j}{\mu}| > \frac{1}{\sqrt{\mu}}$, it follows from the definition of ψ_μ that

$$g_\mu^i(w_\lambda^\pm) = \frac{\int_{Q^j_{\frac{l}{\mu}}} \phi_\mu(x_i)u^p\big(x - \frac{a^j}{\mu} \pm x^\mu\big)\psi_\mu^p\big(x - \frac{a^j}{\mu} \pm x^\mu\big)\,dx}{\int_{Q^j_{\frac{l}{\mu}}} u^p\big(x - \frac{a^j}{\mu} \pm x^\mu\big)\psi_\mu^p\big(x - \frac{a^j}{\mu} \pm x^\mu\big)\,dx}$$

provided $\frac{1}{\sqrt{\mu}} < \frac{l}{\mu}$. By the definition of ϕ_μ, $g(w_\mu^\pm) \in Q^j_{\frac{l}{\mu}}$, that is, $w_\mu^\pm \in N_\mu^j$. We now evaluate $G_{b_\mu}(w_\mu)$:

$$\begin{aligned}
G_{b_\mu}(w_\mu) &= \frac{(t_\mu^+)^2}{2}\|u(\cdot - \frac{a^j}{\mu} + x^\mu)\psi_\mu(\cdot - \frac{a^j}{\mu} + x^\mu)\|_{1,2}^2 \\
&\quad - \frac{(t_\mu^+)^p}{p}\int_{\mathbb{R}_N} b(\mu x)|u(x - \frac{a^j}{\mu} + x^\mu)\psi_\mu(x - \frac{a^j}{\mu} + x^\mu)|^p\,dx \\
&\quad + \frac{(t_\mu^-)^2}{2}\|u(\cdot - \frac{a^j}{\mu} - x^\mu)\psi_\mu(\cdot - \frac{a^j}{\mu} - x^\mu)\|_{1,2}^2 \\
&\quad - \frac{(t_\mu^-)^p}{p}\int_{\mathbb{R}_N} b(\mu x)|u(x - \frac{a^j}{\mu} - x^\mu)\psi(x - \frac{a^j}{\mu} - x^\mu)|^p\,dx \\
&= \frac{(t_\mu^+)^2}{2}\|u\|_{1,2}^2 - \frac{(t_\mu^+)^p}{p}\int_{\mathbb{R}_N} b(\mu x + a^j - \mu x^\mu)|u(x)|^p\,dx \\
&\quad + \frac{(t_\mu^-)^2}{2}\|u\|_{1,2}^2 - \frac{(t_\mu^-)^p}{p}\int_{\mathbb{R}_N} b(\mu x + a^j + \mu x^\mu)|u(x)|^p\,dx + o(1),
\end{aligned}$$

where $o(1) \to 0$ as $\mu \to 0$. It is easy to deduce from the last equalities that $G_{b_\mu}(w_\mu) \to 2I_{b_*}$ as $\mu \to 0$. The proof of (8.5) is similar and is omitted. $\square$

Lemma 8.3 *There exist numbers $\epsilon > 0$ and $\mu_\epsilon > 0$ such that*

$$\bar{J}^j_\mu > 2M_{b_*} + \epsilon \tag{8.7}$$

and

$$\bar{I}^j_\mu > M_{b_*} + \epsilon \tag{8.8}$$

for $j = 1, \ldots, k$ and $\mu \in (0, \mu_\epsilon)$.

Proof. We only provide the proof of (8.7). Arguing indirectly we assume that there exists a sequence $\mu_n \to 0$ such that 0 $\bar{J}^j_{\mu_n} \to c \le 2M_{b_*}$. Then there exists a sequence $\{u_n\} \subset \partial R^j_{\mu_n}$ such that

$$\langle G_{b_{\mu_n}}{}'(u_n^\pm), u_n^\pm \rangle = 0 \tag{8.9}$$

and either $g_{\mu_n}(u_n^+) \in \partial Q^j_{\frac{1}{\mu_n}}$ or $g_{\mu_n}(u_n^-) \in \partial Q^j_{\frac{1}{\mu_n}}$. Without loss generality we may assume that $g_{\mu_n}(u_n^+) \in \partial Q^j_{\frac{1}{\mu_n}}$. Since $u_n^\pm \in \mathcal{N}_{b_{\mu_n}}$, we deduce from the Sobolev imbedding theorem that $\|u_n^\pm\|_{1,2} \ge \nu > 0$ for some constant ν and all n. Applying Lemma 1.1 we deduce the existence of sequences $\{y_n^\pm\} \subset \mathbb{R}_N$ such that

$$\int_{B(y_n^\pm,1)} |u_n^\pm|^2 \, dx \ge \alpha > 0$$

for some constant α and all n. Let $\tilde{u}_n(x) = u_n^+(x + y_n)$. Since $\{\tilde{u}_n\}$ is bounded in $H^1(\mathbb{R}_N)$ we may assume that $\tilde{u}_n \rightharpoonup u_o \not\equiv 0$, $\tilde{u}_n \to u_o$ in $L^p_{\text{loc}}(\mathbb{R}_N)$ and a. e. on $\mathbb{R}_N$. We set $v_n = \tilde{u}_n - u_o$ and write using Lemma 1.3

$$\int_{\mathbb{R}_N} b(\mu_n x + \mu_n y_n)|\tilde{u}_n|^p \, dx = \int_{\mathbb{R}_N} b(\mu_n x + \mu_n y_n)|u_o|^p \, dx + \int_{\mathbb{R}_N} b(\mu_n x + \mu_n y_n)|v_n|^p \, dx + o(1)$$

and by the weak convergence of $\tilde{u}_n$ to u_o we get

$$\int_{\mathbb{R}_N} (|\nabla \tilde{u}_n|^2 + \tilde{u}_n^2) \, dx = \int_{\mathbb{R}_N} (|\nabla u_o|^2 + u_o^2) \, dx + \int_{\mathbb{R}_N} (|\nabla v_n|^2 + v_n^2) \, dx + o(1).$$

We now observe that (8.9) can be written in the following way

$$\int_{\mathbb{R}_N} (|\nabla \tilde{u}_n|^2 + \tilde{u}_n^2)\, dx = \int_{\mathbb{R}_N} b(\mu_n x + \mu_n y_n)|\tilde{u}_n|^p\, dx.$$

Hence combining the last three relations we obtain

$$\begin{aligned}\int_{\mathbb{R}_N} (\nabla v_n|^2 + v_n^2)\, dx - \int_{\mathbb{R}_N} b(\mu_n x + \mu_n y_n)|v_n|^p\, dx \\ = \int_{\mathbb{R}_N} b(\mu_n x + \mu_n y_n)|u_\circ|^p\, dx - \int_{\mathbb{R}_N} (|\nabla u_\circ|^2 + u_\circ^2)\, dx + o(1).\end{aligned} \tag{8.10}$$

We distinguish the cases : (I) $||v_n||_{1,2} \to 0$ and (II) $||v_n||1, 2 \to d > 0$.
Case (I). It follows from assumtpion (8.3) that there exists a number $\delta > 0$ such that

$$b(x) < b_* \text{ for } x \in \bar{Q}^j_{l+\delta} - Q^j_{l-\delta}. \tag{8.11}$$

By selecting a subsequence if necessary we may assume that $\{\mu_n y_n\}$ satisfies one of the following conditions

(i) $\{\mu_n y_n\} \subset Q^j_{l+\delta} - \bar{Q}^j_{l-\delta}$,
(ii) $\{\mu_n y_n\} \subset \bar{Q}^j_{l-\delta}$,
(iii) $\{\mu_n y_n\} \subset \mathbb{R}_N - Q^j_{l+\delta}$ and $\{\mu_n y_n\}$ is bounded,
(iv) $\{\mu_n y_n\}$ is unbounded.

Let $\epsilon > 0$. We choose $R_\epsilon > 0$ so that

$$\int_{|x| \geq R_\epsilon} |\tilde{u}_n|^p\, dx \leq \epsilon \int_{\mathbb{R}_N} |\tilde{u}_n|^p\, dx. \tag{8.12}$$

(i) We may assume that $\mu_n y_n \to \bar{y} \in \bar{Q}^j_{l+\delta} - Q^j_{l-\delta}$ and $b(\bar{y}) < b_*$. Since by (8.9), $u_\circ \in \mathcal{N}_{b(\bar{y})}$, we see that

$$\begin{aligned}&\lim_{n\to\infty} G_{b_{\mu_n}}(u_n^+) \\ &= \lim_{n\to\infty} \left(\frac{1}{2} \int_{\mathbb{R}_N} (|\nabla u_n^+|^2 + |u_n^+|^2)\, dx - \frac{1}{p} \int_{\mathbb{R}_N} b(\mu_n x)|u_n^+|^p\, dx \right) \\ &= \lim_{n\to\infty} \left(\frac{1}{2} \int_{\mathbb{R}_N} (|\nabla \tilde{u}_n|^2 + \tilde{u}_n^2)\, dx - \frac{1}{p} \int_{\mathbb{R}_N} b(\mu_n x + \mu_n y_n)|\tilde{u}_n|^p\, dx \right) \\ &= \frac{1}{2} \int_{\mathbb{R}_N} (|\nabla u_\circ|^2 + u_\circ^2)\, dx - \frac{1}{p} \int_{\mathbb{R}_N} b(\bar{y})|u_\circ|^p\, dx > M_{b_*}.\end{aligned} \tag{8.13}$$

On the other hand since

$$\int_{\mathbb{R}_N} (|\nabla u_n^-|^2 + |u_n^-|^2) = \int_{\mathbb{R}_N} b_{\mu_n}(x)|u_n^-|^p\,dx$$
$$\leq b_* \int_{\mathbb{R}_N} |u_n^-|^p\,dx$$

we see that there exists $t_n \in (0,1]$ such that $t_n u_n^- \in \mathcal{N}_{b_*}$. Therefore

$$\begin{aligned} G_{b_{\mu_n}}(u_n^-) &= (\frac{1}{2} - \frac{1}{p}) \int_{\mathbb{R}_N} (|\nabla u_n^-|^2 + |u_n^-|^2)\,dx \\ &\geq (\frac{1}{2} - \frac{1}{p}) t_n^2 \int_{\mathbb{R}_N} (|\nabla u_n^-|^2 + |u_n^-|^2)\,dx \geq M_{b_*}. \end{aligned}$$

This, combined with (8.12), yields

$$\lim_{n\to\infty} G_{b_{\mu_n}}(u_n) \geq \lim_{n\to\infty} G_{b_{\mu_n}}(u_n^+) + \lim_{n\to\infty} G_{b_{\mu_n}}(u_n^-) \geq 2M_{b_*},$$

which is impossible.
(ii) First we note that for $|x_i| < R_\epsilon$ we have

$$x_i + (y_n)_i \in \left(\frac{a_i^j - (l-\delta)}{\mu_n} - R_\epsilon, \frac{a_i^j + (l-\delta)}{\mu_n} + R_\epsilon\right) \subset (-\frac{2K}{\mu_n}, \frac{2K}{\mu_n})$$

provided n is sufficiently large. We then deduce from (8.11) and the definition of ϕ_{μ_n} that

$$g_{\mu_n}^i(u_n^+) > \left(\frac{a_i^j - (l-\delta)}{\mu_n} - R_\epsilon\right)(l-\epsilon) - \frac{2K\epsilon}{\mu_n}$$

and

$$g_{\mu_n}^i(u_n^+) < \left(\frac{a_i^j + (l-\delta)}{\mu_n} + R_\epsilon\right)(1-\epsilon) + \frac{2K\epsilon}{\mu_n}.$$

Consequently if ϵ and $\delta > \epsilon$ are chosen sufficiently small, we get

$$g_{\mu_n}^i(u_n^+) \in \left(\frac{a_i^j - l}{\mu_n}, \frac{a_i^j + l}{\mu_n}\right) \text{ for large } n,$$

which is impossible since $g_{\mu_n}(u_n^+) \in \partial Q^j_{\frac{l}{\mu_n}}$.
(iii) We may assume that $\mu_n y_n \to \bar{y} \in \bar{Q}^j_{l+\delta}$, with $\bar{y}_i \geq a_i^j + l + \delta$ for some

i, and $(y_n)_i > \frac{a_i^j+l+\frac{\delta}{2}}{\mu_n}$ for all n. If $|x_i| < R_\epsilon$ then

$$x_i + (y_n)_i > \frac{a_i^j + l + \frac{\delta}{2}}{\mu_n} - R_\epsilon$$

and

$$g_{\mu_n}^i(u_n^+) > \left(\frac{a_i^j + l + \frac{\delta}{2}}{\mu_n} - R_\epsilon\right)(1-\epsilon) - \frac{2K\epsilon}{\mu_n} > \frac{a_i^j + l}{\mu_n},$$

provided $\epsilon > 0$ and $\delta > \epsilon$ are sufficiently small and n is large. We again obtain a contradiction with $g_{\mu_n}(u_n^+) \in \partial Q^j_{\frac{l}{\mu_n}}$.

(iv) This case can be ruled out by assuming $\mu_n(y_n)_i > 2K$ (or $\mu_n(y_n)_i < -2K$) for some i and modifying the argument from the case (iii).

Case (II). $\|v_n\|_{1,2} \to L > 0$ as $n \to \infty$. Letting

$$\int_{\mathbb{R}_N} (|\nabla u_\circ|^2 + u_\circ^2)\, dx - \int_{\mathbb{R}_N} b(\mu_n x + \mu_n y_n)|u_\circ|^p\, dx = A + o(1)$$

we therefore have by (8.9) that

$$\int_{\mathbb{R}_N} (|\nabla v_n|^2 + v_n^2)\, dx - \int_{\mathbb{R}_N} b(\mu_n x + \mu_n y_n)|v_n|^p\, dx = -A + o(1). \quad (8.14)$$

Without loss of generality we may assume that $A > 0$. It then follows from (8.13) that there exists $t_n \to 1$ such that $w_n = t_n v_n$ belongs to $\mathcal{N}^{A+o(1)}_{b(\mu_n x+\mu_n y_n)}$. Using this and the fact that $u_\circ \in \mathcal{N}^{A+o(1)}_{b(\mu_n x+\mu_n y_n)}$ we get that

$$\begin{aligned}
G_{b_{\mu_n}}(u_n^+) &= \frac{1}{2}\int_{\mathbb{R}_N} (|\nabla u_\circ|^2 + u_\circ^2)\, dx - \frac{1}{p}\int_{\mathbb{R}_N} b(\mu_n x + \mu_n y_n)|u_\circ|^p\, dx \\
&\quad + \frac{1}{2}\int_{\mathbb{R}_N} (|\nabla v_n|^2 + v_n^2)\, dx - \frac{1}{p}\int_{\mathbb{R}_N} b(\mu_n x + \mu_n y_n)|v_n|^p\, dx + o(1) \\
&\geq \frac{A}{2} + \frac{1}{2}\int_{\mathbb{R}_N} (|\nabla w_n|^2 + w_n^2)\, dx - \frac{1}{p}\int_{\mathbb{R}_N} b(\mu_n x + \mu_n y_n)|w_n|^p\, dx + o(1) \\
&= M^A_{b(\mu_n x+\mu_n y_n)} + M^{-A}_{b(\mu_n x+\mu_n y_n)} + o(1) \\
&> M_{b(\mu_n x+\mu_n y_n)} + \left(\frac{1}{2} - \frac{1}{p}\right)A + o(1) \\
&\geq M_{b_*} + \left(\frac{1}{2} - \frac{1}{p}\right)A + o(1).
\end{aligned}$$

This implies that

$$\begin{aligned}\lim_{n\to\infty} G_{b_{\mu_n}}(u_n) &= \lim_{n\to\infty}\left(G_{b_{\mu_n}}(u_n^+) + G_{b_{\mu_n}}(u_n^-)\right)\\ &\geq 2M_{b_*} + \left(\frac{1}{2}-\frac{1}{p}\right)|A|,\end{aligned}$$

which is impossible. Finally, if $A = 0$ there exist $s_n > 0$ and $t_n > 0$ with $t_n \to 1$ and $s_n \to 1$ such that $\bar{v}_n = t_n v_n$ and $\bar{w}_n = s_n u_o$ belong to $\mathcal{N}_{b(\mu_n x+\mu_n y_n)}$. Thus

$$\begin{aligned}&\lim_{n\to\infty} G_{b_{\mu_n}}(u_n) = \lim_{n\to\infty} G_{b_{\mu_n}}(u_n^+) + \lim_{n\to\infty} G_{b_{\mu_n}}(u_n^-)\\ &= \lim_{n\to\infty}\left[\frac{1}{2}\int_{\mathbb{R}_N}(\nabla\bar{w}_n|^2+\bar{w}_n^2)\,dx - \frac{1}{p}\int_{\mathbb{R}_N} b(\mu_n x+\mu_n y_n)|\bar{w}_n|^p\,dx\right.\\ &\left.+\frac{1}{2}\int_{\mathbb{R}_N}(|\nabla\bar{v}_n|^2+\bar{v}_n^2)\,dx - \frac{1}{p}\int_{\mathbb{R}_N} b(\mu_n x+\mu_n y_n)|\bar{v}_n|^p\,dx\right]\\ &+\lim_{n\to\infty} G_{b_{\mu_n}}(\bar{u}_n) > 3M_{b_*}\end{aligned}$$

and we have arrived at a contradiction. □

8.2 Palais–Smale sequences

Palais–Smale sequences will be obtained by applying Ekeland's variational principle [52]. The following two lemmas provide an important step in the construction of Palais–Smale sequences.

Lemma 8.4 *For every $u \in R_\mu^j$, there exist $\epsilon > 0$ and differentiable functions $t_\pm = t_\pm(w) > 0$ defined for $w \in B(0,\epsilon) \subset H^1(\mathbb{R}_N)$ such that $t_\pm(0) = 1$ and the function $z = t_+(w)(u-w)^+ - t_-(w)(u-w)^-$ maps $B(0,\epsilon)$ into R_μ^j and*

$$\langle t_\pm'(0), v\rangle = \frac{2\int_{\mathbb{R}_N}(\nabla u^\pm\nabla v + u^\pm v)\,dx - p\int_{\mathbb{R}_N} b_\mu|u^\pm|^{p-2}u^\pm v\,dx}{\int_{\mathbb{R}_N}(|\nabla u^\pm|^2+|u^\pm|^2)\,dx - (p-1)\int_{\mathbb{R}_N} b_\mu|u^\pm|^p\,dx}$$

for all $v \in H^1(\mathbb{R}_N)$.

Proof. We apply the implicit function theorem to a function $F^+ : \mathbb{R}\times H^1(\mathbb{R}_N)\to\mathbb{R}$ given by

$$F^+(t,w) = t\int_{\mathbb{R}_N}(|\nabla(u-w)^+|^2+|(u-w)^+|^2)\,dx - t^{p-1}\int_{\mathbb{R}_N} b_\mu|(u-w)^+|^p\,dx.$$

Since $u \in R^j_\mu$ we see that $F^+(1,0) = 0$ and

$$\frac{d}{dt}F^+(1,0) = \int_{\mathbb{R}_N} \left(|\nabla u^+|^2 + |u^+|^2\right) dx - (p-1)\int_{\mathbb{R}_N} b_\mu |u^+|^p \, dx \neq 0.$$

According to the implicit function theorem there exists a function $t_+ : B(0,\epsilon) \to \mathbb{R}$ such that $t_+(0) = 1$ and $F^+(t_+(w), w) = 0$ for $w \in B(0,\epsilon)$. This is equivalent to $\langle \nabla G_{b_\mu}{}'(t_+(w)(u-w)^+, t_+(w)(u-w)^+\rangle = 0$. Taking $\epsilon > 0$ sufficiently small we check that $g_\mu(t_+(w)(u-w)^+) \in Q^j_{\frac{l}{\mu}}$. The second function t_- is obtained by applying the implicit function theorem to the functional

$$F^-(t,w) = t\int_{\mathbb{R}_N} \left(|\nabla(u-w)^-|^2 + |(u-w)^-|^2\right) dx - t^{p-1}\int_{\mathbb{R}_N} b_\mu |(u-w)^-|^p \, dx.$$

□

In a similar manner we can prove the following:

Lemma 8.5 *Let $u \in N^j_\mu$. Then there exists $\epsilon > 0$ and a differentiable function $t : B(0,\epsilon) \subset H^1(\mathbb{R}_N) \to \mathbb{R}$ such that $t(0) = 1$ and the function $z = t(w)(u-w)$ maps $B(0,\epsilon)$ into N^j_μ and moreover*

$$\langle t'(0), v\rangle = \frac{2\int_{\mathbb{R}_N} (\nabla u \nabla v + uv)\, dx - p\int_{\mathbb{R}_N} b_\mu |u|^{p-2} uv \, dx}{\int_{\mathbb{R}_N} (|\nabla u|^2 + u^2)\, dx - (p-1)\int_{\mathbb{R}_N} b_\mu |u|^p \, dx}$$

for all $v \in H^1(\mathbb{R}_N)$.

We are now in a position to generate Palais–Smale sequences for the minimization problems J^j_μ and I^j_μ.

Proposition 8.1 *There exists $\mu_o > 0$ such that $\mu \in (0, \mu_o)$, $j = 1, \ldots, k$*

(1) $J^j_\mu < 3M_{b_*}$ and J^j_μ admits a minimizing sequence which is a Palais–Smale sequence for G_{b_μ},

(2) $I^j_\mu < 2M_{b_*}$ and I^j_μ admits a a minimizing sequence which is a Palais–Smale sequence for G_{b_μ}.

Proof. First we notice that $\bar{R}^j_\mu = R^j_\mu \cup \partial R^j_\mu$. It then follows from Lemmas 8.2 and 8.3 that

$$J^j_\mu < \min\left(3M_{b_*}, \bar{J}^J_\mu\right),\ j = 1, \ldots, k, \tag{8.15}$$

for $\mu \in (0, \mu_o)$. Hence

$$J^j_\mu = \inf\left(G_{b_\mu}(u);\ u \in \bar{R}^j_\mu\right),\ j = 1, \ldots, k. \tag{8.16}$$

We now fix j. By virtue of Ekeland's variational principle (see [52]) there exists a sequence $\{u_n\} \subset \bar{R}^j_\mu$ such that

$$G_{b_\mu}(u_n) < J^j_\mu + \frac{1}{n} \tag{8.17}$$

and

$$G_{b_\mu}(u_n) < G_{b_\mu}(w) + \frac{1}{n}||w - u_n||_{1,2} \tag{8.18}$$

for $w \in R^j_\mu$. By (8.15) we may assume that $u_n \in R^j_\mu$ for n sufficiently large. We now apply Lemma 8.4 to u_n to get the functions $t^n_\pm : B(0, \epsilon_n) \to \mathbb{R}$, for some $\epsilon_n > 0$, such that $t^n_+(w)(u_n - w)^+ - t^n_-(w)(u_n - w)^- \in R^j_\mu$. Let $0 < \delta < \epsilon_n$ and $u \in H^1(\mathbb{R}_N)$ with $u \not\equiv 0$. We set $w_\delta = \frac{\delta u}{||u||_{1,2}}$ and

$$z_\delta = t^n_-(w_\delta)(u_n - w_\delta)^+ - t^n_-(w_\delta)(u_n - w_\delta)^-.$$

Since $z_\delta \in R^j_\mu$, we deduce from (8.18) that

$$G_{b_\mu}(z_\delta) - G_{b_\mu}(u_n) \geq -\frac{1}{n}||z_\delta - u_n||_{1,2}.$$

It then follows from the mean value theorem that

$$\langle G_{b_\mu}{}'(u_n), z_\delta - u_n\rangle + o(||z_\delta - u_n||_{1,2}) \geq -\frac{1}{n}||z_\delta - u_n||_{1,2}.$$

Since

$$(u_n - w_\delta)^+ - (u_n - w_\delta)^- = u_n - w_\delta$$

we get that

$$\begin{aligned} &\langle G_{b_\mu}{}'(u_n), -w_\delta\rangle + \big(t^n_+(w_\delta) - 1\big)\langle G_{b_\mu}{}'(u_n), (u_n - w_\delta)^+\rangle \\ &- \quad \big(t^n_-(w_\delta) - 1\big)\langle G_{b_\mu}{}'(u_n), (u_n - w_\delta)^-\rangle \\ &\geq \quad -\frac{1}{n}||z_\delta - u_n||_{1,2} + o(||u_n - z_\delta||_{1,2}). \end{aligned} \tag{8.19}$$

We now observe that $t^n_\pm(w_\delta)(u_n - w_\delta)^\pm \in \mathcal{N}_{b_\mu}$ and consequently we get from (8.19) that

$$\begin{aligned} &-\delta\langle G_{b_\mu}{}'(u_n), \frac{u}{||u||_{1,2}}\rangle + \frac{\big(t^n_+(w_\delta) - 1\big)}{t^n_+(w_\delta)}\langle G_{b_\mu}{}'(z_\delta), t^n_+(w_\delta)(u_n - w_\delta)^+\rangle \\ &-\frac{t^n_-(w_\delta) - 1\big)}{t^n_-(w_\delta)}\langle G_{b_\mu}{}'(z_\delta), t^n_-(w_\delta)(u_n - w_\delta)^-\rangle \end{aligned}$$

$$+(t_+^n(w_\delta)-1)\langle G_{n_\mu}{}'(u_n)-G_{b_\mu}{}'(z_\delta),(u_n-w_\delta)^+\rangle$$
$$-(t_-^n(w_\delta)-1)\langle G_{b_\mu}{}'(u_n)-G_{b_\mu}{}'(z_\delta),(u_n-w_\delta)^-\rangle$$
$$\geq -\frac{1}{n}\|z_\delta-u_n\|_{1,2}+o(\|z_\delta-u_n\|_{1,2}).$$

We now write this inequality in the following form

$$\langle G_{b_\mu}{}'(u_n),\frac{u}{\|u\|_{1,2}}\rangle \leq \frac{\|z_\delta-u_n\|_{1,2}}{n\delta}+\frac{o(\|z_\delta-u_n\|_{1,2})}{\delta} \tag{8.20}$$
$$+\frac{t_+^n(w_\delta)-1}{\delta}\langle G_{b_\mu}{}'(u_n)-G_{b_\mu}{}'(z_\delta),(u_n-w_\delta)^+\rangle$$
$$-\frac{t_-^n(w_\delta)-1}{\delta}\langle G_{b_\mu}{}'(u_n)-G_{b_\mu}{}'(z_\delta),(u_n-w_\delta)^-\rangle.$$

On the other hand we can find a constant $C>0$, independent of δ, such that

$$\|z_\delta-u_n\|_{1,2}\leq \delta + C\big(|t_-^n(w_\delta)-1|+|t_+^n(w_\delta)-1|\big)$$

and

$$\lim_{\delta\to 0}\frac{|t_\pm^n(w_\delta)-1|}{\delta}\leq \|t_\pm^n{}'(0)\|_{H^{-1}}\leq C.$$

If we let $\delta\to 0$ in (8.18) for a fixed n and use the fact that $z_\delta\to u_n$ in $H^1(\mathbb{R}_N)$ as $\delta\to 0$, we get

$$\langle G_{b_\mu}{}'(u_n),\frac{u}{\|u\|_{1,2}}\rangle < \frac{C}{n}.$$

This shows that $\{u_n\}$ is a Palais–Smale sequence for G_{b_μ}. The proof of (ii) is analogous and therefore is omitted. □

8.3 Existence theorems

The main tool of our approach to the existence results is the following compactness property of Palais–Smale sequences constructed in Proposition 8.1.

Proposition 8.2 *Let $\mu_\circ>0$ be constants from Proposition 8.1 and let* $0<\mu<\mu_\circ$.

(1) Let $\{u_n\} \subset R^j_\mu$ *be a sequence satisfying*

$$G_{b_\mu}(u_n) \to J^j_\mu \tag{8.21}$$

and

$$G_{b_\mu}{}'(u_n) \to 0 \text{ in } H^{-1}(\mathbb{R}_N). \tag{8.22}$$

Then $\{u_n\}$ *is relatively compact in* $H^1(\mathbb{R}_N)$.

(2) Let $\{u_n\} \subset N^j_\mu$ *be a sequence satisfying* $G_{b_\mu} \to I^j_\mu$ *and* $G_{b_\mu}{}'(u_n) \to 0$ *in* $H^{-1}(\mathbb{R}_N)$. *Then* $\{u_n\}$ *is relatively compact in* $H^1(\mathbb{R}_N)$.

Proof. We only prove part (i). Part (ii) of the proposition is proved in the same way. Since $\{u_n\}$ is bounded in $H^1(\mathbb{R}_N)$ we may assume that $u_n \rightharpoonup u_\circ$ in $H^1(\mathbb{R}_N)$ and $u_n \to u_\circ$ a.e. on $\mathbb{R}_N$. Our objective is to show that

$$u_\circ^+ \not\equiv 0 \text{ and } u_\circ^- \not\equiv 0 \tag{8.23}$$

and

$$u_n \to u_\circ \text{ in } H^1(\mathbb{R}_N). \tag{8.24}$$

To prove (8.23) suppose that

$$u_\circ^+ \equiv 0 \text{ or } u_\circ^- \equiv 0. \tag{8.25}$$

As in the proof of Lemma 8.3 we show that there exist a sequence $\{y_n\} \subset \mathbb{R}_N$ and a constant $\alpha > 0$ such that

$$\int_{B(0,1)} |u_n(x+y_n)|^2\, dx \geq \alpha \text{ for all.} \tag{8.26}$$

Let $\tilde{u}_n(x) = u_n(x+y_n)$. Since $\{\tilde{u}_n\}$ is bounded in $H^1(\mathbb{R}_N)$ we may assume that

$$\tilde{u}_n \rightharpoonup \bar{u} \not\equiv 0 \text{ in } H^1(\mathbb{R}_N). \tag{8.27}$$

We now distinguish two cases: 1) $\bar{u}^+ \not\equiv 0$ and $\bar{u}_- \not\equiv 0$ and 2) either $\bar{u}_+ \not\equiv 0$ or $\bar{u}^- \not\equiv 0$.

Case 1) We set $v_n = \tilde{u}_n - \bar{u}$. If $\|v_n\|_{1,2} \to 0$ and $\{y_n\}$ is unbounded we show as in Lemma 8.3 that $g^i_\mu(u_n^+) > a^j_i + \frac{3l}{2\mu}$ or $g^i_\mu(u_n^+) < a^j_i - \frac{3l}{\mu}$ for large n and some i. This contradicts the fact that $\{u_n\} \subset R^j_\mu$. If $\|v_n\|_{1,2} \to 0$ and $\{y_n\}$ is bounded, then we may assume that $y_n \to y_\circ$. Consequently

$u_n \to \bar{u}(\cdot - y_\circ)$ in $H^1(\mathbb{R}_N)$. Since $\|u_n^\pm\|_{1,2} \geq \beta > 0$ (for some constant β), we see that $\bar{u}^+ \not\equiv 0$ and $\bar{u}^- \not\equiv 0$ contradicting (8.25). If $\|v_n\|_{1,2} \geq \delta > 0$ for some constant δ and large n, then by (8.22) we can write

$$\int_{\mathbb{R}_N} (|\nabla \bar{u}^\pm|^2 + |u^\pm|^2)\, dx - \int_{\mathbb{R}_N} b(\mu x + \mu y_n)|u^\pm|^p\, dx = o(1)$$

and

$$\int_{\mathbb{R}_N} (|\nabla \tilde{u}_n|^2 + \tilde{u}_n^2)\, dx - \int_{\mathbb{R}_N} b(\mu x + \mu y_n)|\tilde{u}_n|^p\, dx = 0.$$

We deduce from this with the aid of Lemma 1.3 that

$$\int_{\mathbb{R}_N} (|\nabla v_n|^2 + v_n^2)\, dx - \int_{\mathbb{R})N} b(\mu x + \mu y_n)|v_n|^p\, dx = o(1).$$

Since $\|v_n\|_{1,2} \geq \delta$ we can find $s_n > 0$ such that $s_n \to 1$ and $s_n v_n \in \mathcal{N}_{b(\mu x + \mu y_n)}$. This allows us to show that

$$\frac{1}{2}\int_{\mathbb{R}_N} (|\nabla v_n + v_n^2)\, dx - \frac{1}{p}\int_{\mathbb{R}_N} b(\mu x + \mu y_n)|v_n|^p\, dx \geq M_{b_*} + o(1).$$

In a similar way we show that

$$\frac{1}{2}\int_{\mathbb{R}_N} (|\nabla \bar{u}^\pm|^2 + |\bar{u}^\pm|^2)\, dx - \frac{1}{p}\int_{\mathbb{R}_N} b(\mu x + \mu y_n)|u^\pm|^p\, dx \geq M_{b_*} + o(1).$$

Applying Lemma 1.3 we derive from the last two inequalities that

$$\begin{aligned}
G_{b_\mu}(u_n) &= \frac{1}{2}\int_{\mathbb{R}_N} (|\nabla \tilde{u}_n|^2 + \tilde{u}_n^2)\, dx - \frac{1}{p}\int_{\mathbb{R}_N} b(\mu x + \mu y_n)|\tilde{u}_n|^p\, dx \\
&= \frac{1}{2}\int_{\mathbb{R}_N} (|\nabla \bar{u}^+|^2 + |\bar{u}^+|^2)\, dx = \frac{1}{p}\int_{\mathbb{R}_N} b(\mu x + \mu y_n)|\bar{u}^+|^p\, dx \\
&+ \frac{1}{2}\int_{\mathbb{R}_N} (|\nabla \bar{u}^-|^2 + |\bar{u}^-|^2)\, dx - \frac{1}{p}\int_{\mathbb{R}_N} b(\mu x + \mu y_n)|\bar{u}^-|^2\, dx \\
&+ \frac{1}{2}\int_{\mathbb{R}_N} (|\nabla v_n|^2 + v_n^2)\, dx - \frac{1}{2}\int_{\mathbb{R}_N} b(\mu x + \mu y_n)|v_n|^p\, dx + o(1) \\
&\geq 3M_{b_*} + o(1).
\end{aligned}$$

Thus $\lim_{n\to\infty} G_{b_\mu}(u_n) \geq M_{b_*}$ which is impossible.

Case 2) We only consider the case $u^- \equiv 0$. First we observe that $\tilde{u}_n^+ \to \bar{u}^+$, otherwise $\|\tilde{u}_n^+ - \bar{u}^+\|_{1,2} \geq \gamma > 0$ for some constant γ and large n. This would lead to $\lim_{n\to\infty} G_{b_\mu}(u_n) \geq 3M_{b_*}$ which is impossible. As in Lemma 7.3 we show that there exists a sequence $\{\tilde{y}_n\} \subset \mathbb{R}_N$ such that

$$\int_{B(0,1)} |\tilde{u}_n^-(x+\tilde{y}_n)|^2\,dx \geq \delta > 0$$

for some constant δ and large n. We set $w_n(x) = \tilde{u}_n(x+\tilde{y}_n) - u_n(x+y_n+\tilde{y}_n)$. Since $\{w_n\}$ is bounded in $H^1(\mathbb{R}_N)$ we may assume that $w_n \rightharpoonup w_\circ$ in $H^1(\mathbb{R}_N)$ and $w_n^- \rightharpoonup w_\circ^- \not\equiv 0$ in $H^1(\mathbb{R}_N)$. If $w_\circ^+ \not\equiv 0$ we get a contradiction as in Case 1). Hence we must have $w_\circ^+ \equiv 0$. Therefore $w_n^- \to w_\circ^-$ in $H^1(\mathbb{R}_N)$ since otherwise we can show as in Case 1) that $\lim_{n\to\infty} G_{b_\mu}(u_n) \geq 3M_{b_*}$ which is impossible. Thus we have arrived at the following situation

$$\tilde{u}_n^+ \to \bar{u}^+ \text{ in } H^1(\mathbb{R}_N) \tag{8.28}$$

and

$$w_n^+ \to w_\circ^- \text{ in } H^1(\mathbb{R}_N). \tag{8.29}$$

Arguing as in Case 1) we show that $\{y_n\}$ and $\{y_n+\tilde{y}_n\}$ are bounded. Hence we may assume that $y_n \to y_\circ$ and $\tilde{y}_n \to \tilde{y}_\circ$ and $u_n^+ \to \bar{u}^+(\cdot + y_\circ)$ in $H^1(\mathbb{R}_N)$ and $u_n^- \to w_\circ^-(\cdot - y_\circ - \tilde{y}_\circ)$ in $H^1(\mathbb{R}_N)$. This means that $u_\circ^+ = \bar{u}^+(\cdot - y_\circ) \not\equiv 0$ and $u_\circ^- = w_\circ^-(\cdot - y_\circ - \tilde{y}_\circ) \not\equiv 0$ and this contradicts (8.23) and completes the proof of claim (8.21). Finally, to show (8.22) we argue by contradiction showing that $\lim_{n\to\infty} G_{b_\mu}(u_n) \geq 3M_{b_*}$ which is impossible. □

We now establish the following existence theorem.

Theorem 8.1 *There exists $\lambda_\circ > 0$ such that equation (8.1) admits at least k positive and k nodal solutions for each $\lambda \geq \lambda_\circ$.*

Proof. By virtue of Proposition 8.2 there exists $\mu_\circ > 0$ such that to each $\mu \leq \mu_\circ$ there corresponds a minimizing sequences $\{u_n^j\}$ and $\{v_n^j\}$ for J_μ^j and I_μ^j, respectively, which are relatively compact in $H^1(\mathbb{R}_N)$. Therefore we may may assume that $u_n^j \to u^j$ in $H^1(\mathbb{R}_N)$ and $v_n^j \to v^j$ in $H^1(\mathbb{R}_N)$. Obviously the limiting functions u^j and v^j satisfy the equation (8.2). It is clear that u^j are nodal solutions, that is, $(u^j)^+ \not\equiv 0$ and $(u^j)^- \not\equiv 0$. We now show that either $(v^j)^+ \not\equiv 0$ or $(v^j)^- \not\equiv 0$. Indeed, suppose that $(v^j)^+ \not\equiv 0$ and $(v^j)^- \not\equiv 0$. Since $(v^j)^\pm \in \mathcal{N}_{b_\mu}$, we see that

$$I_\mu^j = \lim_{n\to\infty} G_{b_\mu}(v_n^j) = G_{b_\mu}(v^j) = G_{b_\mu}((v^j)^+) + G_{b_\mu}((v^j)^-) \geq 2M_{b_*},$$

which is impossible. Thus we may assume that $v^j \geq 0$ on $\mathbb{R}_N$ and the positivity of v^j on $\mathbb{R}_N$ follows from the Harnack inequality. Since $g^j_\mu(u^j)$, $g^j_\mu(v^j) \in \bar{Q}^j_{\frac{1}{\mu}}$, solutions u^j and v^j are distinct. Letting $\lambda_\circ = \mu_\circ^{-2}$ and $U_j(x) = \lambda^{\frac{1}{2(p-2)}} u^j(\sqrt{\lambda}, x)$ and $V_j(x) = \lambda^{\frac{1}{2(p-2)}} v^j(\sqrt{\lambda} x)$ we obtain positive solutions V_j and nodal solutions U_j of equation (7.1) for $\lambda \geq \lambda_\circ$. □

8.4 Concentration of positive solutions

According to Theorem 8.1 for each $0 < \mu \leq \mu_\circ$ there exist positive solutions u^j_μ of equation (8.2) satisfying

$$G_{b_\mu}(u^j_\mu) = I^j_\mu = \inf\{G_{b_\mu}(u);\ u \in N^j_\mu\},\ j = 1, \ldots, k.$$

Motivated by results of Chapter 7 we study the behaviour of u^j_μ as $\mu \to 0$.

Theorem 8.2 *For each sequence $\mu_n \to 0$ there exists a subsequence μ_{n_k}, relabelled again by μ_n, such that each u^j_μ has a global maximum point x^j_n and $x^j_n \to a^j$.*

Proof. For the ease of notation we set $u_\mu = u^j_\mu$. We first observe that u_μ is bounded in $H^1(\mathbb{R}_N)$. We claim that there exists $\beta > 0$ such that to each $\mu \leq \mu_\circ$ there corresponds y_μ such that

$$\liminf_{\mu \to 0} \int_{B(y_\mu, 1)} u_\mu(x)^2\, dx \geq \beta. \tag{8.30}$$

For otherwise there exists a sequence $\mu_n \to 0$ such that

$$\lim_{n \to \infty} \sup_{y \in \mathbb{R}_N} \int_{B(y,1)} u_{\mu_n}(x)^2\, dx = 0.$$

By Lemma 1.1 $u_{\mu_n} \to 0$ in $L^q(\mathbb{R}_N)$ for $2 < q < 2^*$. Since $u_{\mu_n} \in \mathcal{N}_{b_{\mu_n}}$ we see that $u_{\mu_n} \to 0$ in $H^1(\mathbb{R}_N)$. On the other hand by the Sobolev inequality we have

$$\|u_{\mu_n}\|_{1,2} \geq C > 0$$

for some constant C independent of n, which is impossible. It therefore follows from (8.28) that there exist sequences $\{\mu_n\}$, with $\mu_n \to 0$, and $\{y_n\}$ such that $u_{\mu_n}(\cdot + y_n) \rightharpoonup u^j_\circ \not\equiv 0$ in $H^1(\mathbb{R}_N)$, $u^j_{\mu_n}(\cdot + y_n) \to u_\circ$ in $L^s_{\text{loc}}(\mathbb{R}_N)$ for $2 < s < 2^*$ and a.e. on $\mathbb{R}_N$. Let $Q^j_{l-\delta}$ and $Q^j_{l+\delta}$ be cubes satisfying (8.11).

As in Lemma 8.3 we rule out cases (iii) and (iv). Therefore, $\{\mu_n y_n\} \subset Q^j_{l+\delta}$ and we may assume that $\mu_n y_n \to x^\circ \in \bar{Q}^j_{l+\delta}$. It is easy to see that

$$\int_{\mathbb{R}_N} (|\nabla u_\circ|^2 + u_\circ^2)\, dx = b(x^\circ) \int_{\mathbb{R}_N} u_\circ^p\, dx \le b_* \int_{\mathbb{R}_N} u_\circ^p\, dx.$$

Therefore, there exists $t \in (0,1]$ such that $tu_\circ \in \mathcal{N}_{b_*}$. It then follows from the lower semicontinuity of a norm with respect to a weak convergence that

$$\begin{aligned} \lim_{n\to} I^j_{\mu_n} &= \lim_{n\to\infty} \left(\frac{1}{2} - \frac{1}{p}\right) \int_{\mathbb{R}_N} (|\nabla u_{\mu_n}|^2 + u_{\mu_n}^2)\, dx \qquad (8.31)\\ &\ge \lim_{n\to\infty} \left(\frac{1}{2} - \frac{1}{p}\right) t^2 \int_{\mathbb{R}_N} (|\nabla u_{\mu_n}(x+y_n)|^2 + u_{\mu_n}(x+y_n)^2)\, dx \\ &\ge \left(\frac{1}{2} - \frac{1}{p}\right) t^2 \int_{\mathbb{R}_N} (|\nabla u_\circ|^2 + u_\circ^2)\, dx \\ &= G_{b_*}(tu_\circ) \ge M_{b_*}. \end{aligned}$$

This combined with Lemma 8.2 implies that $\lim_{n\to\infty} I^j_{\mu_n} = M_{b_*}$. It therefore follows from (8.31) that $t = 1$. This means that $b_* = b(a^j)$ and necessarily $x^\circ = a^j$. Let $v_n(x) = u_{\mu_n}(x + y_n)$. Since $t = 1$ and $\lim_{n\to\infty} I^j_{\mu_n} = M_{b_*}$, we deduce from (8.31) that $v_n \to u_\circ$ in $H^1(\mathbb{R}_N)$. Therefore, $\alpha_\infty = 0$, where α_∞ is a quantity from Proposition 1.2 associated with $\{v_n\}$ in norm L^{2^*}. Applying the one sided Harnack inequality [60] we get

$$\max_{B(y,1)} v_n \le c \left(\int_{B(y,2)} |v_n|^{2^*}\, dx \right)^{\frac{1}{2^*}}$$

for every $y \in \mathbb{R}_N$, where $c > 0$ is a constant depending only on N and bound of $\|v_n\|_{L^{2^*}}$. From this we deduce that $\lim_{|x|\to\infty} v_n(x) = 0$ uniformly in n. Let x_n be a global maximum of u_{μ_n}. Since $u_{\mu_n}(x_n) \ge (b_*)^{\frac{1}{p-1}}$ we see that $\{x_n\}$ must be bounded and by the previous argument $x_n \to a^j$. □

With the aid of a comparison principle we establish an exponential decay of positive solutions of (8.2).

Proposition 8.3 *Suppose that b is locally Hölder continuous on $\mathbb{R}_N$. Then there exist constants $\delta_i > 0$, $i = 1, 2$, and $C > 0$ such that*

$$u_\mu(x) \le C \exp(-\delta_1 |x|) \quad \textit{on } \mathbb{R}_N \qquad (8.32)$$

and

$$\int_{\mathbb{R}_N} |\nabla u_\mu|^2 \exp(\delta_2|x|)\,dx < \infty. \tag{8.33}$$

Proof. By regularity theory $u_\mu \in C^2(\mathbb{R}_N)$. It is evident from the proof of Theorem 8.2 that $\lim_{|x|\to\infty} u_\mu(x) = 0$. Let $H(x,t) = \prod_{i=1}^N \cosh \delta x_i$, $\delta > 0$. By straightforward calculations we check that $\Delta(H^{-1}) = H^{-1}O(\delta^2)$. Therefore there exists $\delta_o > 0$ such that for all $\delta < \delta_o$ and all $x \in \mathbb{R}_N$ we have

$$-\Delta(H^{-1}) + \frac{1}{2}H^{-1} = H^{-1}\left(O(\delta^2) + \frac{1}{2}\right) > 0.$$

We now choose $R > 0$ such that $b_\mu u_\mu^{p-1} - \frac{1}{2}u_\mu < 0$ for $|x| \geq R$. Since $H^{-1} \geq \exp(-\delta_o N|x|)$ on $\mathbb{R}_N$, there exists $M = M(R, \|u_\mu\|_\infty, \delta_o)$ such that $MH^{-1} \geq u$ on $|x| = R$. We now observe that

$$\begin{aligned} -\Delta\big(u_\mu - MH^{-1}\big) \quad &+ \quad \frac{1}{2}\big(u_\mu - MH^{-1}\big) = b_\mu|u_\mu|^{p-2}u_\mu - \frac{1}{2}u_\mu \\ &+ \quad H^{-1}MO(\delta^2) - \frac{M}{2}H^{-1} < 0 \end{aligned}$$

on $\mathbb{R}_N$. Hence the maximum principle implies that $u_\mu \leq MH^{-1}$ for $|x| \geq R$ and the estimate (8.32) readily follows. To show (8.33) we take as a test function $v = u_\mu H^2\theta_R^2$, where θ_R is a function introduced in the proof of Proposition 4.1 in Chapter 4. Multiplying (8.2) by v and integrating by parts we obtain

$$\begin{aligned} &\int_{\mathbb{R}_N} |\nabla u_\mu|^2 H^2\theta_R^2\,dx + 2\int_{\mathbb{R}_N} \nabla u_\mu u_\mu H\nabla H\theta_R^2\,dx \\ &\quad +2\int_{\mathbb{R}_N} \nabla u_\mu u_\mu H^2\nabla\theta_R\theta_R\,dx + \int_{\mathbb{R}_N} u_\mu^2 H^2\theta_R^2\,dx \\ &\quad = \int_{\mathbb{R}_N} b_\mu|u_\mu|^p H^2\theta_R^2\,dx. \end{aligned} \tag{8.34}$$

Let us denote the second and third integral on the left side by J_1 and J_2, respectively. An application of the Hölder inequality gives the following estimates

$$|J_1| \leq \frac{1}{4}\int_{\mathbb{R}_N} |\nabla u_\mu|^2 H^2\theta_R^2\,dx + 4\int_{\mathbb{R}_N} u_\mu^2|\nabla H|^2\theta_R^2\,dx$$

and

$$|J_2| \le \frac{1}{4}\int_{\mathbb{R}_N} |\nabla u_\mu|^2 H^2 \theta_R^2\, dx + 4\int_{\mathbb{R}_N} u_\mu^2 H^2 |\nabla \theta_R|^2\, dx.$$

Inserting these estimates into (8.34) we get

$$\frac{1}{2}\int_{\mathbb{R}_N} |\nabla u_\mu|^2 H^2 \theta_R^2\, dx + \int_{\mathbb{R}_N} (H^2 - \frac{1}{4}|\nabla H|^2)\theta_R^2 u_\mu^2\, dx$$
$$\le 4\int_{\mathbb{R}_N} u_\mu^2 K^2 |\nabla \theta_R|^2\, dx + 4\int_{\mathbb{R}_N} u_\mu^2 |\nabla H|^2 \theta_R^2\, dx + \int_{\mathbb{R}_N} b_\mu u_\mu^p H^2 \theta_R^2\, dx.$$

Choosing $\delta = \delta_2 < \delta_1$ so that $H^2 - \frac{1}{4}|\nabla H|^2 > 0$, the result follows letting $R \to \infty$. □

8.5 Bibliographical notes

The effect of the shape of the graph of the coefficient $b(x)$ on the existence of positive and nodal solutions, presented in this chapter, has been established by Cao-Noussair [26]. For further related results we refer to the papers [3], [1], [9], [13], [12], [11], [7], [38], [79], [97], [104].

Bibliography

[1] Ambrosetti, A. Badiale, M. and Cingolani, S. (1996) "Semiclassical states of nonlinear Schrödinger equations with bounded potentials", *Rend. Mat. Acc. Lincei* **s.9, v.7**, 155–160.

[2] Ambrosetti, A. and Rabinowitz, P. (1976) "Dual variational methods in critical point theory and applications", *J. Funct. Anal.* **14**, 349–381.

[3] Alama, S. and Li, Y.Y. (1992) "On "multibump" bound states for certain semilinear elliptic equations", *Indiana Univ. Math. J.* **41**, 983–1026.

[4] Garcia Azorero, J.P. and Peral Alonso, I. (1998) "Hardy inequalities and some critical elliptic and parabolic problems", *J. Diff. Equ.* **144**, 441–476.

[5] Badiale, M. and Citti, G. (1991) "Concentration–compctness principle and quasilinear elliptic equations in $\mathbb{R}^n$", *Commun. in PDE* **16(1)**, 1795–1818.

[6] Bahri, A. and J.M. Coron, J.M. (1998) "On a nonlinear elliptic equation involving Sobolev exponent: the effect of the topology of the domain", *Comm. Pure Appl. Math.* **41**, 255–294

[7] Bahri, A. and Yan Yan Li (1990) "On a min-max procedure for the existence of a positive solution for certain scalar field equations in $\mathbb{R}^N$", *Rev. Mat. Ibero.* **6** *N^{os}* **1 y 2**, 1–15.

[8] Bahri, A. (1989) *Critical points at infinity in some variational problems*, Research Notes in Mathematics 182 Longman - Pitman.

[9] Bahri, A. and Lions, L.P. (1997) "On the existence of positive solutions of semilinear elliptic equations in unbounded domains", *Ann. I.H.P. Analyse non linéaire* **14(3)**, 365–413.

[10] Bartsch, T. (1993) "Infinitely many solutions of a symmetric Dirichlet problem", *Nonlinear Analysis TMA* **20**, 1205–1215.

[11] Bartsch, Th. and Wang, Z.Q. (1995) "Existence and multiplicity results for some superlinear elliptic problems on $\mathbb{R}^N$", *Commun. in P.D.E.* **20 (9,10)**, 1725–1241.

[12] Bartsch, T. and M. Willem, T. (1993) "Infinitely many nonradial solutions of a Euclidean scalar field equations", *J. Funct. Anal.* **117**, 447–460.

[13] Ben - Naouma, A.K., C. Troestler, C. and Willem, M. (1995) "Extrema problem with critical Sobolev exponents on unbounded domain", *Nonlinear Analysis TMA* **26(4)**, 823–833.

[14] Benci, V. and Cerami, G. (1990) "Existence of positive solutions of the equation $-\Delta u + a(x)u = u^{\frac{N+2}{N-2}}$ in $\mathbb{R}_N$", *J. Funct. Anal.* **88**, 90–117.

[15] Benci, V. and Cerami, G. (1987) "Positive solutions of some nonlinear elliptic problems in exterior domain", *Arch. Rat. Mech. Anal.* **99**, 283–300.

[16] Benci, V. and Cerami, G. (1991) "The effect of the domain topology on the number of positive solutions", *Arch. Rat. Mech. Anal.* **114**, 79–93.

[17] Benci, V., Cerami, G. and Passaseo, D. (1991) "On the number of positive solutions of some nonlinear elliptic problems", *Nonlinear Analysis (*a tribute in honour of G. Prodi, Scuola Norm. Sup. Pisa 93–107.

[18] Berestycki, H. and Lions, P.L. (1983) "Nonlinear scalar field equations, I Existence of a ground state, II Existence of infinitely many solutions", *Arch. Rat. Mech. Anal.* **82**, 313–346, 347–376.

[19] Bianchi, G., Chabrowski, J. and Szulkin, A. (1995) "On symmetric solutions of an elliptic equation involving critical Sobolev exponent", *Nonlinear Analysis TMA* **25(1)**, 41–59.

[20] Brézis, H. and Lieb, E.H. (1983) "A relation between pointwise convergence of functions and convergence of functionals", *Proc. Am. Math. Soc.* **88**, 486–490.

[21] Brézis, H. and Nirenberg, L. (1983) "Positive solutions of nonlinear elliptic equations involving critical Sobolev exponents", *Comm. Pure Appl. Math.* **36**, 437–477.

[22] Buffoni, B., Jeanjean, L. and Stuart, C.A. (1993) "Existence of a nontrivial solution to a strongly indefinite semilinear equation", *Proc. Am. Math. Soc.* **19(1)**, 179–186.

[23] Cao, D. (1990) "Positive solutions and bifurcation from essential spectrum of semilinear elliptic equation on $\mathbb{R}_N$", *Nonlinear Anal. TMA* **15(11)**, 1045–1052.

[24] Cao, D. (1993) "Multiple solutions of semilinear elliptic equation in $\mathbb{R}^n$", *Ann. Inst. H.Poincaré, Analyse non linéaire* **10(6)**, 593–604.

[25] Cao, D. (1995) "Positive solution of a semilinear elliptic equation on $\mathbb{R}^N$", *J. Partial Diff. Eqs.* **8**, 261–272.

[26] Cao, D. and Noussair, E.S. (1996) "Multiplicity of positive and nodal solutions for nonlinear elliptic problems in $\mathbb{R}_N$", *Ann. I.H.P Analyse non linéaire* **13(5)**, 567–588.

[27] Cao, D., Noussair, E.S. and Yan, S. (1999) "Solutions with multiple peaks for nonlinear elliptic equations", *Proc. Roy. Soc. Edinburgh* **129A**, 235–264.

[28] Caldiroli, P. and Musina, R. (1999) "On the existence of extremal functions for weighted Sobolev spaces with critical Sobolev exponent", *Calc. Var. in PDE* **8**, 365–387.

[29] Cerami, G. and Passaseo, G. (1992) "Existence and multiplicity of positive solutions for nonlinear elliptic problems in exterior domains with "rich" topology", *Nonlinear Analysis TMA* **18(2)**, 109–119.

[30] Cerami, G. and Passaseo, D. (1995) "Existence and multiplicity results for semilinear elliptic Dirichlet problems in exterior domains", *Nonlinear Analysis TMA* **24(11)**, 1533–1547.

[31] Chabrowski, J. (1995) "Concentration–compactness principle at infinity and semilinear elliptic equations involving critical and subcritical Sobolev exponents", *Calc. Var. PDE* **3(4)**, 493–512.

[32] Chabrowski, J. (1996) "Degenerate elliptic equation involving a subcritical Sobolev exponent", *Portugaliae Math.* **53(2)**, 167–177.

[33] Chabrowski, J. (1997) *Variational methods for potential operator equations* Walter de Gruyter Studies in Mathematics 24, Berlin - New York.

[34] Chabrowski, J. and Yang, J. (1997) "Existence theorems for elliptic equations involving supercritical Sobolev exponent", *Adv. Diff. Equations* **2(2)**, 231–256.

[35] Chabrowski, J. and Yang, J. (1998) "Existence theorems for Schrödinger equation involving a critical Sobolev exponent" *Z. Angew. Math. Phys.* **49**, 276–293.

[36] Chabrowski, J. and Yang, J. "Multiple semiclassical solutions of the Schrödinger equation involving a critical Sobolev exponent" to appear in *Portugaliae Math.*

[37] Cingolani, S. and Lazzo, M. (1997) "Multiple semiclassical standing waves for a class of nonlinear Schrödinger equations", *Top. Methods in Nonlinear Anal.* **10**, 1–13.

[38] Citti, G. (1991) "Existence of positive solutions of quasilinear degenerate elliptic equations on unbounded domains", *Ann. Math. Pura Appl.* **158**,315–330.

[39] Coffman, C.V. (19969) "A minimum - maximum principle for a class of nonlinear integral equations", *J. Anal. Math.* **22**,391–419.

[40] Coffman, C.V. (1970) "On a class of nonlinear elliptic boundary value problems", *J. Math. Mech.* **19**, 351–356.

[41] Coleman, S., Glaser, V. and Martin, A. (1978) "Action minima among solutions to a class of Euclidean scalar equations", *Commun. Math. Phys.* **58**, 211-221.

[42] Coron, J.M. (1984) "Topologie et cas limite des injections de Sobolev", *C. R. Acad. Sc. Paris* **299 (7)**, 209–212.

[43] Cotti–Zelati, V. and Rabinowitz, P.H. (1992) "Homoclinic type solutions for a semilinear elliptic PDE on $\mathbb{R}^n$", *Commun. Pure Appl. Math.* **45**, 1217–1269.

[44] Cotti–Zelati, V. and Rabinowitz, P.H. (1991) "Homoclinic orbits for second order Hamiltonian possessing superquadratic potentials", *J. Amer. Math. Soc.* **44**, 693–727.

[45] Dancer, E.N. and Yan S. "Sigularly perturbed elliptic problem in exterior

domain", to appear in *Diff. Int. Equations*

[46] Dancer, E.N. Lam, K.L. and Yan, S. "The effect of the graph topology on the existence of multipeak solutions for nonlinear Schrödinger equation", to appear in *Abstract and Applied Analysis*

[47] Ding, V.Y. and Ni, W. (1991) "On the existence of positive entire solution of a semilinear elliptic equation", *Arch. Rat. Mech. Anal.* **92**, 283–308.

[48] Dall'Aglio, A. and Grossi, M. (1997) "A perturbation theorem for the equation $-\Delta u + \lambda u = u^p$ in unbounded domains", *Nonlinear Analysis TMA* **28(11)**, 1867–1877.

[49] Deng, Y.B. Zhou, H.S. and Zhu, X.P. (1991) "On the existence and $L^p(\mathbb{R}_N)$-bifurcation for semilinear elliptic equations", *J. Math. Anal. Appl.* **154**, 116–133.

[50] Ding Yi (1990) "Non-trivial solutions of a semilinear elliptic equation", *Acta Math. Scientia* **10(2)**, 229–239.

[51] Ding Xiaxi and Ding Yi (1990) "The concentration–compactness principle and inverse power method", *Acta Math. Scientia,* **10(4)**, 383–395.

[52] Ekeland, I. (1974) "On variational principle", *J. Math. Anal. Appl.* **47**, 324–353.

[53] Eydeland, A. and Spruck, J. (1988) "The inverse power method for semilinear elliptic equations", *Nonlinear diffusion equations and their equilibrium states I, Proc. Microprogram, Berkeley, Calif., Publ. Sci. Res. Inst.* **12**, 273–286.

[54] Esteban, M.J. and Lions, P.L. (1987) "Γ-convergence and the concentration–compactness method for some variational problems with lack of compactness", *Ricerche di Matematica* **36**, 73–101.

[55] Esteban, M.J. Lions, P.L. (1983) "A compactness lemma", *Nonlinear Analysis TMA* **7(4)**, 381–385

[56] Evans, C.V. "Weak convergence methods for nonlinear partial differential equations", *CBMS Am. Math. Soc.* **74**.

[57] Floer, A. and Weinstein, A. (1986) "Nonspreading wave packets for the cubic Schrödinger equation with a bounded potential", *J. Funct. Anal.* **69**, 397–408.

[58] Flucher, M. and Müller, S. (1998) "Radial symmetry and decay rate of variational ground state in the zero mass case", *SIAM J. Maths. Anal.* **29(3)**, 712–719.

[59] Gidas, B., Ni, W.N. and Nirenberg (1981) "Symmetry of positive solutions of nonlinear elliptic equations in $\mathbb{R}^N$", *Advances in Mathematics, Supplementary Studies, Academic Press* **7A**.

[60] Gilbarg, D. and Trudinger, N.S. (1983) *Elliptic partial differential equations of second order* (second edition) Springer Verlag, New York.

[61] Gonçalves, J.V. and Miyagaki, O.H. (1998) "Multiple positive solutions for semilinear elliptic equations in $\mathbb{R}^N$ involving subcritical exponents", *Nonlinear Analysis, TMA* **32(1)**, 41–51.

[62] Grossi, M. (1993) "Multiplicity results for semilinear elliptic equations

with lack of compactness", *Diff. Int. Equations* **6(4)**, 807–823.

[63] Grossi, M. (1995) "An existence result for a semilinear equation in some domains with unbounded boundary", *Ann. Mat. Pura Appl.* **168**, 17–35.

[64] Grossi, M. and Passasseo, D. (1995) "Nonlinear elliptic Dirichlet problems in exterior domains: the role of geometry and topology", *Commun. Appl. Nonlin. Anal.* **2**, 1–31.

[65] Gui, C.F. (1996) "Existence of multi-bump solutions for nonlinear Schrödinger equations via variational method", *Commun. in PDE,* **21 (5&6)**, 787–820.

[66] Gui, C.F. (1996) "Multipeak solutions for semilinear Neumann problem", *Duke Math. J.* **84(3)**, 739–769.

[67] Jeanjean, L. (1997) "Existence of solutions with prescribed norm for semilinear elliptic equations", *Nonlinear Anal. TMA* **28(10)**, 1633–1659.

[68] Jeanjean, L. (1997) "Two positive solutions for a class of nonhomogeneous elliptic equations", *Diff. Int. Equ.* **10(4)**, 609–624.

[69] Keller, E.F. and Segel, L.A. (1970) "Initiation of slime mold aggregation viewed as an instability", *J. Theoret. Biol.* **26**, 399–415.

[70] Kinderlehrer, D. and Stampacchia, G. (1980) *An introduction to variational inequalities and their applications,* Academic - Press, New York.

[71] Kwong, M.K. (1989) "Uniqueness of positive solutions of $\Delta u - u + u^p = 0$", *Arch. Rat. Mech. Anal.* **105**, 243–266.

[72] Li, G. and Yan, S. (1989) "Eigenvalue problems for quasilinear elliptic equations on $\mathbb{R}^N$", *Commun. in PDE* **14(8 & 9)**, 1291–1341.

[73] Yi Li (1988) "Remarks on a semilinear elliptic equations", *J. Diff. Equations* **74**, 34–49.

[74] Lien, W.C., Tzeng, S.Y. and Wang, H.C. (1993) "Existence of solutions of semilinear elliptic problems in unbounded domains", *Diff. Int. Equations* **6(6)**, 1281–1298.

[75] Lin, C.-S., W.-M. Ni, W.-M. Tagaki, I. (1988) "Large amplitude stationary solutions to chemotaxis systems", *J. Diff. Equations* **72**, 1–27.

[76] Lions, P.L. (1984) "The concentration–compactness principle in the calculus of variations. The locally compact case, part 1,2", *Ann. I.H.P. Analyse nonlinéaire* **1(2 & 3)**, 109–145, 223–283.

[77] Lions, P.L. (1985) "The concentration–compactness principle in the calculus of variations, the limit case, Part 1, the limit case Part 2", *Rev. Mat. Ibero.* **1(1 & 2)**, 145–201, 45–121.

[78] Lions, P.L. (1982) "Symétrie et compacité dans les espaces de Sobolev", *J. Funct. Anal.* **49**, 315–334.

[79] Lions, P.L. (1988)"On positive solutions of semilinear equations in unbounded domains", *Nonlinear Diffusion Equations, Springer, New York*

[80] Maz'ja, V.G. (1985) *Sobolev spaces*, Springer-Verlag, Berlin.

[81] Musina, R. (1996) "Multiple positive solutions of scalar field equations in $\mathbb{R}^n$", *Topological Methods in Nonlinear Analysis* **7**, 171–185.

[82] Nehari, Z. (1960) "On a class of nonlinear second - order differential equa-

tions", *Trans Am. Math. Soc.* **95**, 101–123.

[83] Ni, W.N. and Takagi, I. (1993) "Locating the peaks of least-energy solutions to a semilinear Neumann problem", *Duke Math. J.* **70(2)**, 247–281.

[84] Ni, W.N. and Takagki, I. (1991) "On the shape of least-energy solutions to a semilinear Neumann problem", *Commun. Pure Appl. Math.* **44**, 819–851.

[85] Noussair, E.S. and Yan, S. "On positive multipeak solutions of nonlinear elliptic equations", to appear in *Proc. London Math. Soc.*

[86] Oh, Y.-G. (1988) "Existence of semi-classical bound states of nonlinear Schrödinger equations with potentials of the class $(V)_a$", *Commun. PDE* **13**, 1499–1519.

[87] Oh,Y.-G. (1989) "Corrections to "Existence of semi-classical bound states of nonlinear Schrödinger equations with potentials of the class $(V)_a$" *Commun. PDE* **13**, 833–834.

[88] Oh, Y.-G. (1989) "Stability of semi-classical bound states of nonlinear Schrödinger equations with potential", *Commun. Math. Phys.* **121**, 11–33.

[89] Oh, Y.-G. (1990) "On positive multi-bump bound states of nonlinear Schrödinger equations under multiple well potential", *Commun. Math. Phys.* **132**, 223–253.

[90] Palais, R.S. (1970) "Critical point theory and the minmax principle and the calculus of variations", *Proc. Sym. Pure Math., Global Analysis (edited by S.S. Chern)* **15**, 185–212.

[91] del Pino, M.A. and Felmer, P.L. (1996) "Local mountain passes for semilinear elliptic problems in unbounded domains", *Calc. Var. PDE* **4**, 121–137.

[92] del Pino, M.A. and Felmer, P.L. (1996) "Least energy solutions for elliptic equations in unbounded domains," *Proc. Roy. Soc. Edinburgh* **126A**, 195–208.

[93] del Pino, M.A. and Felmer, P.L. (1998) "Multi-peak bound states for nonlinear Schrödinger equations", *Ann. I.H.P, Analyse non linéaire* **15(2)**, 127–149.

[94] del Pino, M. Felmer, P.L. and Miyagaki, O.H. (1998) "Existence of positive bound states of nonlinear Schrödinger equations with saddle-like potential", *Nonlinear Analysis, TMA* **34**, 979–989.

[95] Protter, P.H. (960) "Unique continuation for elliptic equation", *Trans. Am. Math. Soc.* **95**, 81–91.

[96] Rey, O. (1990) "The role of the Green's function in a nonlinear elliptic equation involving the critical Sobolev exponent", *J. Funct. Anal.* **89**, 1–52.

[97] Rabinowitz, P. (1992) "On a class of nonlinear Schrödinger equations", *Z. Angew. Math. Phys.* **43**, 270–291.

[98] Rey, O. (1990) "Concentration of solutions to elliptic equations with critical nonlinearity", *Ann. I.H.P. Analyse non lináire* **9(2)**, 201–218.

[99] Rose, H.A. and Weinstein M.I. (1988) "On bound states of nonlinear Schrödinger equation with a linear potential". *Physica D* **30**, 207–218.

[100] Schindler, I. (1992) "Quasilinear elliptic boundary-value problems in unbounded cylinders and related mountain-pass lemma", *Arch. Rat. Mech. Anal.* **20**, 363–374.

[101] Schindler, I. (19950 "A critical value function and applications to semilinear elliptic equations on unbounded domains", *Nonlinear Analysis, TMA* **24(6)**, 947–959.

[102] Séré, E. (1992) "Existence of infinitely many homoclinic orbits in Hamiltonian systems", *Math. Z.* **209**, 27–42.

[103] Smets, D. (1998) "A concentration–compactness lemma with applications to singular eigenvalue problems", to appear in *J. Funct. Anal.*

[104] Stuart, C.A. (1988) "Bifurcation in $L^p(\mathbb{R}_N)$ for a semilinear elliptic equation", *Proc. London Math. Soc.* **57(3)**, 511–541.

[105] Strauss, W. (1977) "Existence of solitary waves in higher dimensions", *Commun. Math. Phys.* **55**, 149–162.

[106] Struwe, M. (1990) *Variational methods*, Springer - Verlag, Berlin - Heidelberg.

[107] Struwe, M. (1984) "A global compactness result for elliptic boundary value problems involving limiting nonlinearities", *Math. Zeit.* **187**, 511–517,

[108] Szulkin, A. and Willem, M. (1998) "Eigenvalue problems with indefinite weight", *Studia Math.* **135(2)**, 191–201.

[109] Talenti, G. (1976) "Best constant in Sobolev inequality", *Ann. Mat. Pura Appl.* **101**, 353–372.

[110] Tarantello, G. (1992) "On nonhomogeneous elliptic equations involving critical Sobolev exponent", *Ann. I.H.P. Analyse non linéaire* **9**, 281–304.

[111] Tertikas, A. (1998) "Critical phenomena in linear elliptic problems", *J. Func. Anal.* **154**, 42–66.

[112] Troestler, Ch. (1996) "Noncompact variational problems", *Ph.D. Thesis, Dep. Maths, Univ. Catholique de Louvain.*

[113] Trudinger, N. (1967) "On Harnack type inequalities and their application to quasilinear elliptic equations", *Commun. Pure Appl. Math.* **20**, 721–747.

[114] Yang, J. (1993) "Positive solutions of semilinear elliptic problems in exterior domains", *J. Diff. Equations* **106(1)**, 40–69.

[115] Yang, J. and Zhu, X. (1987) "On the existence on nontrivial solutions of a quasilinear elliptic boundary value problem for unbounded domains (1) Positive mass case", *Acta Math. Scientia.* **7(3)**, 342–359.

[116] Yang, J. and X. Zhu, X. (1987) "On the existence of nontrivial solutions of a quasilinear elliptic boundary value problem for unbounded domains, (2) Zero mass case", *Acta Math. Scientia.* **7(4)**, 447–459.

[117] Wang, X. (1993) "On concentration of positive bound states of nonlinear Schrödinger equations", *Commun. Math. Phys.* **153**, 229–244.

[118] Wang, X. (1997) "On concentration of positive bound states of nonlin-

ear Schrödinger equations with competing potential functions", *SIAM J. Math. Anal.* **38(3)**, 633–655.

[119] Wang, H.C. (1989) "The generalized compactness lemma and its applications to nonlinear elliptic problems", *Nonlinear Analysis TMA* **13(11)**, 1351–1358.

[120] Wei, J. (1996) "On the construction of single - peaked solutions to a singularly perturbed semilinear Dirichlet problem", *J. Diff. Equations* **129**, 315–333.

[121] Willem, M. (1996) *Minimax theorems,* Birkhäuser.

[122] Zhu, X. and Cao, D. (1989) "The concentration–compactness principle in nonlinear elliptic equations", *Acta Math. Scientia* **9**, 307–328.

Index